SAT® Math Prep

RELATED TITLES

8 Practice Tests for the SAT

SAT Reading & Writing Prep

PSAT Prep with 2 Practice Tests

SAT Total Prep

SAT Prep Plus with 5 Practice Tests

SAT Prep with 2 Practice Tests

SAT® Math Prep

Acknowledgments

Special thanks to those who made this book possible including Arthur Ahn, Laura Aitcheson, Becky Berthiaume, Michael Boothroyd, Matthew Callan, Potoula Chresomales, Dom Eggert, Kate Fisher, Katy Haynicz-Smith, Adam Hinz, Kate Hurley, Brandon Jones, Rebecca Knauer, Celina Lasota, Terrence McMullen, James Radkins, Justin Starr, Bob Verini, Lee Weiss, Devon Wible, Daniel Wittich, and many others who contributed materials and advice.

Published by Kaplan Publishing, a division of Kaplan, Inc.
750 Third Avenue
New York, NY 10017

Printed in the United States of America

10 9 8 7 6 5 4 3

ISBN-13: 978-1-5062-2873-0

Kaplan Publishing books are available at special quantity discounts to use for sales promotions, employee premiums, or educational purposes. For more information or to purchase books, please call the Simon & Schuster Special Sales Department at 866-506-1949.

Table of Contents

SECTION ONE: FOUNDATIONAL SKILLS AND KAPLAN METHODS & STRATEGIES

SECTION TWO: SAT MATH PRACTICE

SECTION THREE: PRACTICE TEST

How to Use This Book

BECOME FAMILIAR WITH THE SAT MATH TEST

Learn the structure of the SAT Math Test—what kinds of questions are on it, how it's scored, and how best to approach it. Knowing what to expect will set you up for success as you prepare. You'll find all the information you need in the next few pages of Kaplan's SAT Math Prep.

FOUNDATIONAL SKILLS AND KAPLAN METHODS & STRATEGIES

Section One of SAT Math Prep contains instruction related to prerequisite math skills that you need to master before taking the SAT. Studying or reviewing this material is a great way to warm up. You'll also want to master the methods and strategies we offer: the Kaplan Method for Math, the Kaplan Strategy for Translating English into Math, and the Kaplan Method for Multi-Part Math Questions. These methods and strategies will help you maximize your efficiency and strengths—both of which will go a long way on Test Day.

PRACTICE SETS

This book contains 16 practice sets, arranged by topic. Each practice set begins with a brief instructional review, which includes useful definitions, theorems, formulas, and other relevant information. Before you begin each practice set, read through the review to refresh your memory. As you work through the questions, you'll notice that some have a calculator icon—this means that similar questions have appeared in the Calculator Section of the Practice Tests released by the College Board. It does not, however, mean that you must use a calculator to answer the question.

ANSWERS & EXPLANATIONS

Following each practice set, you will find detailed answers and explanations. Use these to determine why you answered a question incorrectly and how to avoid the same mistakes on similar questions in the future. Each explanation includes both advice on how to approach similar questions and a detailed step-by-step solution. You should also read the explanations for questions you answered correctly to ensure that you took the most efficient route to those correct answers.

PRACTICE TEST

When you have completed the practice sets, take the Practice Test under test-like conditions, using the timing indicated for each section of the test, which is just a little over an hour long. Make sure to check your answers, calculate your score using the conversion chart provided, and read the answers and explanations to reinforce what you've learned.

Each section of the Math test begins with a reference page. Take some time before Test Day to get acquainted with the formulas and other information that is provided on this page.

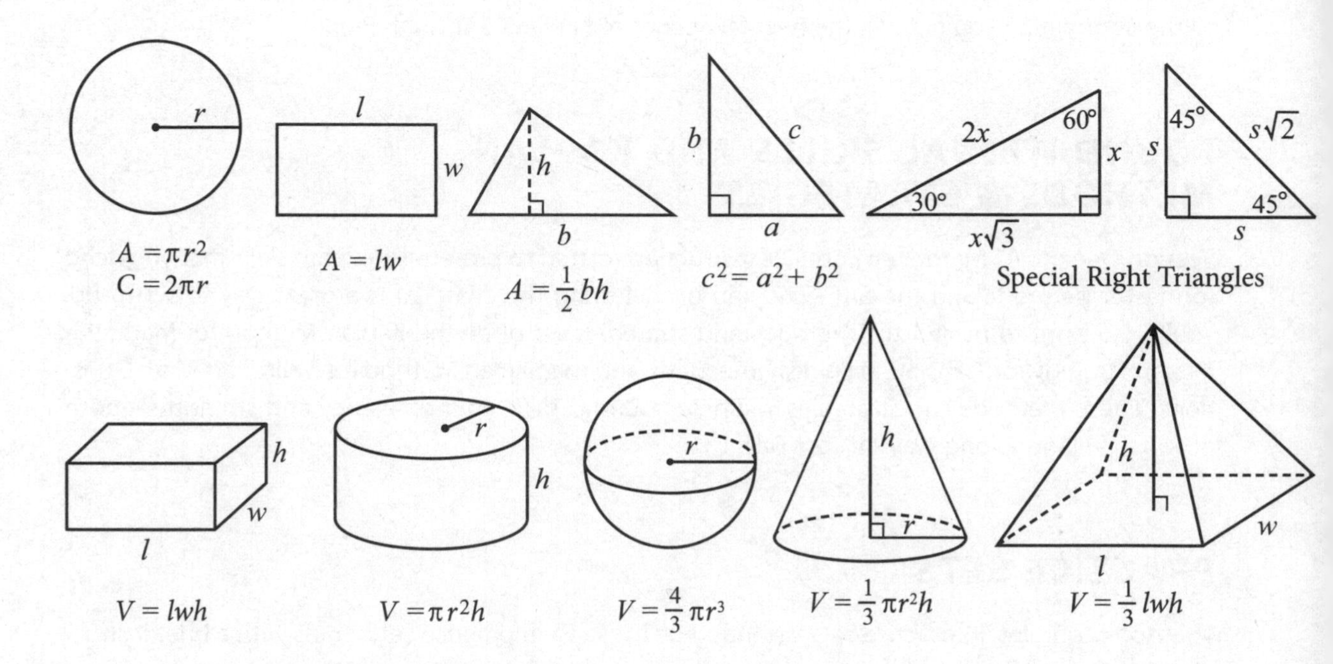

The sum of the degree measures of the angles in a triangle is 180.

The number of degrees of arc in a circle is 360.

The number of radians of arc in a circle is 2π.

Introduction to the SAT Math Test

The SAT Math Test is broken down into a no-calculator section and a calculator section. Questions across the sections consist of multiple-choice, student-produced response (Grid-in), and more comprehensive multi-part questions.

	No-Calculator Section	Calculator Section	Total
Duration (minutes)	25	55	80
Multiple-choice	15	30	45
Grid-in	5	8	13
Total Questions	20	38	58

The SAT Math Test is divided into four content areas: Heart of Algebra, Problem Solving and Data Analysis, Passport to Advanced Math, and Additional Topics in Math.

SAT Math Test Content Area Distribution	
Heart of Algebra (19 questions)	Analyzing and fluently solving equations and systems of equations; creating expressions, equations, and inequalities to represent relationships between quantities and to solve problems; rearranging and interpreting formulas
Problem Solving and Data Analysis (17 questions)	Creating and analyzing relationships using ratios, proportions, percentages, and units; describing relationships shown graphically; summarizing qualitative and quantitative data
Passport to Advanced Math (16 questions)	Rewriting expressions using their structure; creating, analyzing, and fluently solving quadratic and higher-order equations; purposefully manipulating polynomials to solve problems
Additional Topics in Math (6 questions)	Making area and volume calculations in context; investigating lines, angles, triangles, and circles using theorems; and working with trigonometric functions

ANALYSIS IN SCIENCE AND HISTORY/SOCIAL STUDIES QUESTIONS

A few math questions might look like something you'd expect to see on a science or history test. These "crossover" questions are designed to test your ability to use math in real-world scenarios. There are a total of 18 "crossover" questions that will contribute to subscores that span multiple tests. Nine of the questions will contribute to the Analysis in Science subscore, and nine will contribute to the Analysis in History/Social Studies subscore.

Test-Taking Strategies

The SAT is different from the tests you are used to taking in school. The good news is that you can use the SAT's particular structure to your advantage.

For example, on a test given in school, you probably go through the questions in order. You spend more time on the harder questions than on the easier ones because harder questions are usually worth more points. You probably often show your work because your teacher tells you that how you approach a question is as important as getting the correct answer.

This approach is not optimal for the SAT. On the SAT, you benefit from moving around within a section if you come across tough questions because the harder questions are worth the same number of points as the easier questions. It doesn't matter how you arrive at the correct answer—only that you bubble in the correct answer choice.

STRATEGY #1: TRIAGING THE TEST

You do not need to complete questions on the SAT in order. Every student has different strengths and should attack the test with those strengths in mind. Your main objective on the SAT should be to score as many points as you can. While approaching questions out of order may seem counterintuitive, it is a surefire way to achieve your best score.

Just remember, you can skip around within each section, but you cannot work on a section other than the one you've been instructed to work on.

To triage the test effectively, do the following:

- First, work through all the easy questions that you can do quickly. Skip questions that are hard or time-consuming
- Second, work through the questions that are doable but time-consuming
- Third, work through the hard questions
- If you run out of time, pick a Letter of the Day for remaining questions

A Letter of the Day is an answer choice letter (A, B, C, or D) that you choose before Test Day to select for questions you guess on.

STRATEGY #2: ELIMINATION

Even though there is no wrong-answer penalty on the SAT, Elimination is still a crucial strategy. If you can determine that one or more answer choices are definitely incorrect, you can increase your chances of getting the right answer by paring the selection down.

To eliminate answer choices, do the following:

- Read each answer choice
- Cross out the answer choices that are incorrect
- Remember: There is no wrong-answer penalty, so take your best guess

STRATEGY #3: GUESSING

Each multiple-choice question on the SAT has four answer choices and no wrong-answer penalty. That means if you have no idea how to approach a question, you have a 25 percent chance of randomly choosing the correct answer. Even though there's a 75 percent chance of selecting the incorrect answer, you won't lose any points for doing so. The worst that can happen on the SAT is that you'll earn zero points on a question, which means you should *always* at least take a guess, even when you have no idea what to do.

When guessing on a question, do the following:

- Always try to strategically eliminate answer choices before guessing
- If you run out of time, or have no idea what a question is asking, pick a Letter of the Day

Common Testing Myths

Since its inception, the SAT has gone through various revisions, but it has always been an integral part of the college admissions process. As a result of its significance and the changes it has undergone, a number of rumors and myths have circulated about the exam. In this section, we'll dispel some of the most common ones. As always, you can find the most up-to-date information about the SAT at the College Board website (https://www.collegeboard.org).

Myth: There is a wrong-answer penalty on the SAT to discourage guessing.

Fact: While this statement was true a few years ago, it is no longer true. Older versions of the SAT had a wrong-answer penalty so that students who guessed on questions would not have an advantage over students who left questions blank. This penalty has been removed; make sure you never leave an SAT question blank!

Myth: Answer choice C is most likely to be the correct answer.

Fact: This rumor has roots in human psychology. Apparently, when people such as high school teachers, for example, design an exam, they have a slight bias toward answer choice C when assigning correct answers. While humans do write SAT questions, a computer randomizes the distribution of correct choices; statistically, therefore, each answer choice is equally likely to be the correct answer.

Myth: The SAT is just like another test in school.

Fact: While the SAT covers some of the same content as your high school math, literature, and English classes, it also presents concepts in ways that are fundamentally different. While you might be able to solve a math problem in a number of different ways on an algebra test, the SAT places a heavy emphasis on working through questions as quickly and efficiently as possible.

Myth: You have to get all the questions right to get a perfect score.

Fact: Many students have reported missing several questions on the SAT and being pleasantly surprised to receive perfect scores. Their experience is not atypical: Usually, you can miss a few questions and still get a coveted perfect score. The makers of the SAT use a technique called scaling to ensure that a SAT score conveys the same information from year to year, so you might be able to miss a couple more questions on a slightly harder SAT exam and miss fewer questions on an easier SAT exam and get the same scores. Keep a positive attitude throughout the SAT, and in many cases, your scores will pleasantly surprise you.

Myth: You can't prepare for the SAT.

Fact: You've already proven this myth false by buying this book. While the SAT is designed to fairly test students regardless of preparation, you can gain a huge advantage by familiarizing yourself with the structure and content of the exam. By working through the questions and practice tests available to you, you'll ensure that nothing on the SAT catches you by surprise and that you do everything you can to maximize your score. Your Kaplan resources help you structure this practice in the most efficient way possible, and provide you with helpful strategies and tips as well.

SECTION ONE

Foundational Skills and Kaplan Methods & Strategies

Prerequisite Skills and Calculator Usage

This book focuses on the math skills that are tested on the SAT. It assumes a working knowledge of arithmetic, algebra, and geometry. Before you dive into the Practice Sets where you'll try test-like questions, there are a number of concepts—ranging from basic arithmetic to geometry—that you should master. The following pages contain a brief review of these concepts.

ALGEBRA AND ARITHMETIC

- Order of operations is one of the most fundamental of all arithmetic rules. A well-known mnemonic device for remembering this order is PEMDAS: Please Excuse My Dear Aunt Sally. This translates to Parentheses, Exponents, Multiplication/Division, Addition/Subtraction. Note: Multiplication and division have the same priority, as do addition and subtraction. Perform multiplication and division from left to right (even if it means division before multiplication), and treat addition and subtraction the same way.

$$(14 - 4 \div 2)^2 - 3 + (2 - 1)$$
$$(14 - 2)^2 - 3 + (1)$$
$$(12)^2 - 3 + 1$$
$$144 - 3 + 1$$
$$142$$

- Three basic properties of number (and variable) manipulation—commutative, associative, and distributive—will assist you with algebra on Test Day. These properties are outlined next.

  Commutative: Numbers can swap places and still provide the same mathematical result. This is valid only for addition and multiplication.

$$a + b = b + a \rightarrow 3 + 4 = 4 + 3;\text{ also, } 3 \times 4 = 4 \times 3$$

  Associative: Different number groupings will provide the same mathematical result. This is valid only for addition and multiplication.

$$(a + b) + c = a + (b + c) \rightarrow (4 + 5) + 6 = 4 + (5 + 6);\text{ also, } (4 \times 5) \times 6 = 4 \times (5 \times 6)$$

  Distributive: A number to be multiplied by a quantity inside a set of parentheses can be rewritten as the first number multiplied by each term inside the parentheses. This does *not* work with division.

Note: When subtracting an expression in parentheses, be particularly careful to distribute the negative sign.

$a(b+c) = ab+ac$; $a(b-c) = ab-ac \rightarrow 6(x - y + 3) = 6(x) + 6(-y) + 6(3) = 6x - 6y + 18$

- Make sure you are comfortable working with negative numbers and that you understand the additive inverse property: Subtracting a positive number is the same as adding its negative.

$$r-(-s) = r+s \rightarrow 22-(-15) = 22+15 = 37$$

- You should also practice manipulating both proper and improper fractions. To add and subtract fractions, first find a common denominator, then add the numerators together. Multiplication is straightforward: Multiply the numerators together, then repeat for the denominators. Cancel when possible to simplify your answer. Dividing by a fraction is the same as multiplying by its reciprocal. Once you've rewritten your division problem as multiplication, follow the rules for fraction multiplication to simplify.

addition/subtraction: $\frac{2}{3}+\frac{5}{4} \rightarrow \left(\frac{2}{3}\times\frac{4}{4}\right)+\left(\frac{5}{4}\times\frac{3}{3}\right)=\frac{8}{12}+\frac{15}{12}=\frac{23}{12}$

multiplication: $\frac{5}{8}\times\frac{8}{3}=\frac{5}{\cancel{8}_1}\times\frac{\cancel{8}^1}{3}=\frac{5\times 1}{1\times 3}=\frac{5}{3}$

division: $\frac{3}{4}\div\frac{5}{8}=\frac{3}{\cancel{4}_1}\times\frac{\cancel{8}^2}{5}=\frac{3\times 2}{1\times 5}=\frac{6}{5}$

- Know what absolute value is: the distance a number is from 0 on a number line. Because absolute value is a distance, it is always positive or 0. Absolute value can *never* be negative.

$$|-17| = 17,\ |21| = 21,\ |0| = 0$$

- Follow the properties of equality: Whatever you do to one side of an equation, you must also do to the other. For instance, if you multiply one side by 3, you must multiply the other side by 3 as well.

- The ability to solve straightforward one-variable equations is critical on the SAT. See the following example:

$\frac{4x}{5} - 2 = 6$ Add 2 to both sides.

$\frac{4x}{5} = 8$ Multiply both sides by 5.

$4x = 40$ Divide both sides by 4.

$x = 10$

- You'll need to be able to translate word problems, like the one that follows, into mathematical expressions or equations.

  Annabel bought eight pairs of shoes during a sale at her favorite boutique. If this purchase tripled the number of pairs of shoes she had before she went shopping, how many pairs of shoes did Annabel own before she visited the sale?

  Let p represent the number of pairs of shoes Annabel had before the sale. She adds eight pairs (+ 8) to this value, which is equal to triple her original number of pairs (= $3p$). Your equation should read $p + 8 = 3p$. Solving for p reveals $p = 4$, so Annabel had four pairs of shoes before she went shopping at the sale.

- You will encounter irrational numbers, such as radicals and π, on Test Day. You can carry an irrational number throughout your calculations as you would a variable (e.g., $4 \times \sqrt{2} = 4\sqrt{2}$). If a question asks for an approximate value, convert the irrational number to a decimal only after you have finished any intermediate steps.

FOR EXTRA EFFICIENCY

- Save time on Test Day by reviewing multiplication tables so that you don't have to use your calculator for simple computations. At a bare minimum, work up through the 10s. If you know them through 12 or 15, that's even better!
- You can save a few seconds of number crunching by memorizing perfect squares. Knowing perfect squares through 10 is a good start; go for 15 or even 20 if you can.
- The ability to recognize a few simple fractions masquerading in decimal or percent form will also save you time on Test Day. Memorize the content of the following table.

Fraction	Decimal	Percent
$\frac{1}{10}$	0.1	10%
$\frac{1}{5}$	0.2	20%
$\frac{1}{4}$	0.25	25%
$\frac{1}{3}$	$0.333\overline{3}$	$33.3\overline{3}\%$
$\frac{1}{2}$	0.5	50%

GRAPHING

- Basic two-dimensional graphing is performed on a coordinate plane. There are two axes, *x* and *y*, that meet at a central point called the origin. Each axis has both positive and negative values that extend outward from the origin at evenly spaced intervals. The axes divide the space into four sections called quadrants, which are labeled I, II, III, and IV. Quadrant I is always the upper-right section, and the rest follow counterclockwise.

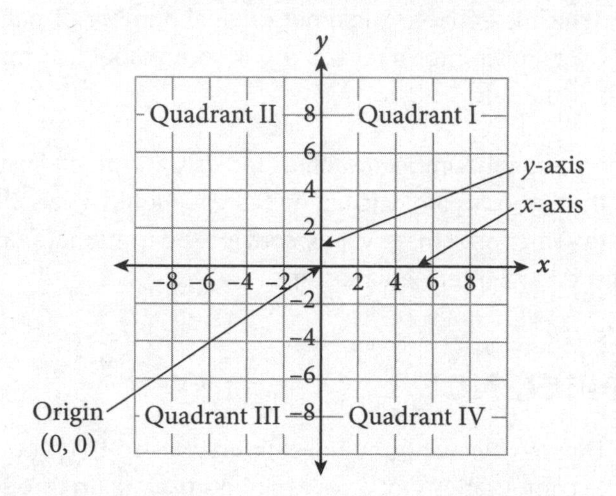

- To plot points on the coordinate plane, you use coordinates. The *x*-coordinate is where the point falls along the *x*-axis, and the *y*-coordinate is where the point falls along the *y*-axis. The two coordinates together make an ordered pair written as (x, y). When writing ordered pairs, the *x*-coordinate is always listed first. When plotting ordered pairs, you should always move horizontally first, and then vertically. Four points are plotted in the following figure as examples.

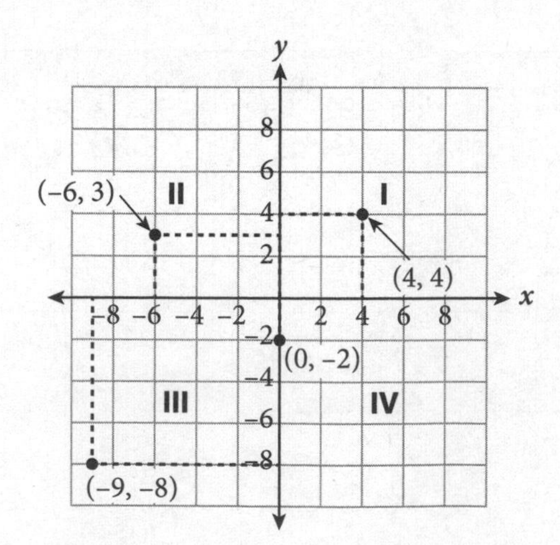

- When two points are vertically or horizontally aligned, calculating the distance between them is easy. For a horizontal distance, only the *x*-value changes; for a vertical distance, only the *y*-value changes. Take the positive difference of the *x*-coordinates (or *y*-coordinates) to determine the distance. Two examples are presented here.

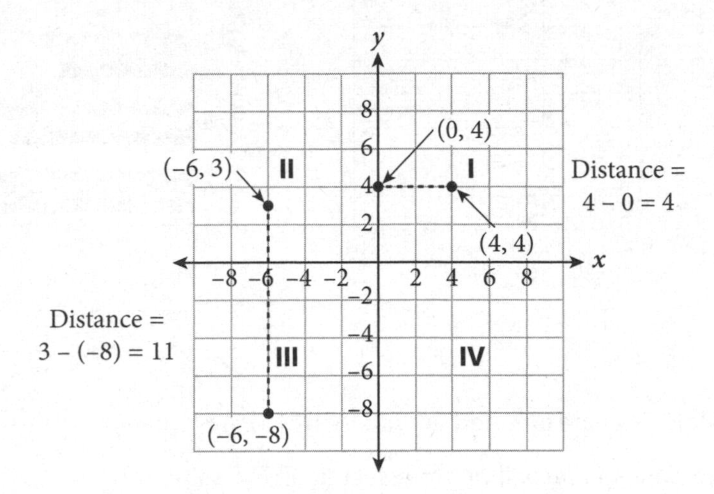

- Two-variable equations have an independent variable (input) and a dependent variable (output). The dependent variable (often *y*) depends on the independent variable (often *x*). For example, in the equation $y = 3x + 4$, *x* is the independent variable because each *y*-value depends on what you plug in for *x*. You can construct a table of values for the equation, which can then be plotted.

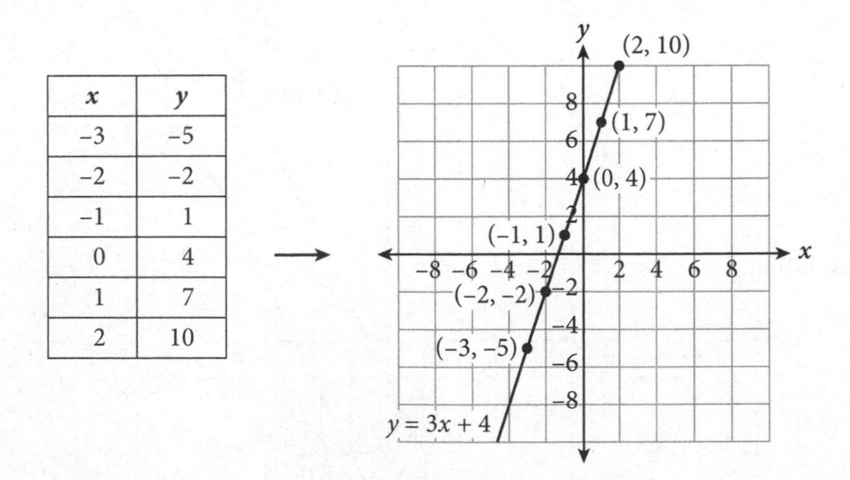

x	y
–3	–5
–2	–2
–1	1
0	4
1	7
2	10

- You can be asked to infer relationships from graphs. In the first of the following graphs, the two variables are time and population. Clearly, the year does not depend on how many people live in the town; rather, the population increases over time and thus depends on the year. In the second graph, you can infer that plant height depends on the amount of rain; thus, rainfall is the independent variable. Note that the independent variable for the second graph is plotted along the vertical axis; this can happen with certain nonstandard graphs.

However, on a standard coordinate plane, the independent variable is always plotted along the horizontal axis.

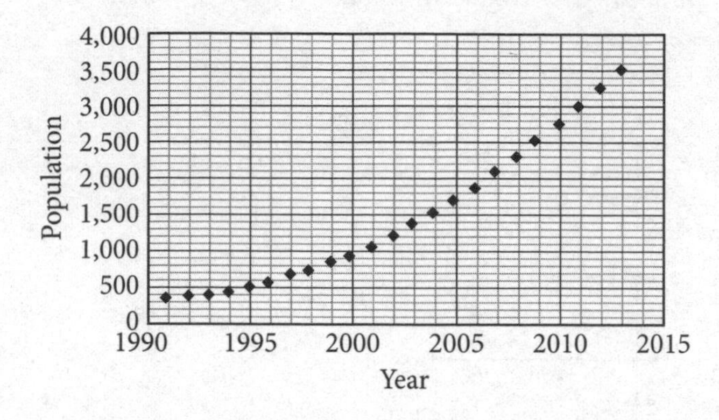

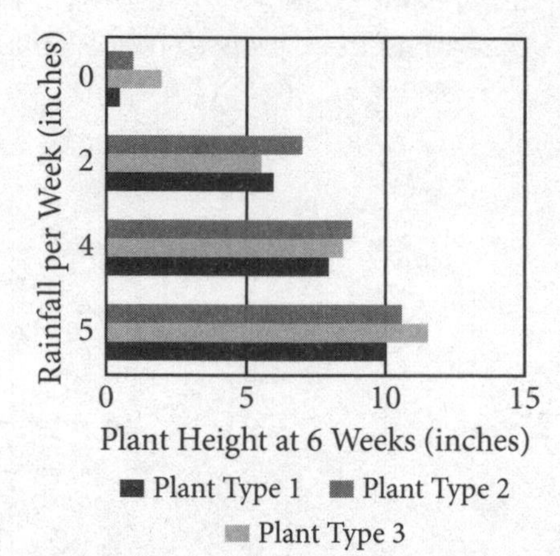

- When two straight lines are graphed simultaneously, one of three possible scenarios will occur:
  - The lines are parallel and will not intersect at all (no solution).
  - The lines will intersect at one point (one solution).
  - The lines will lie on top of each other (infinitely many solutions).

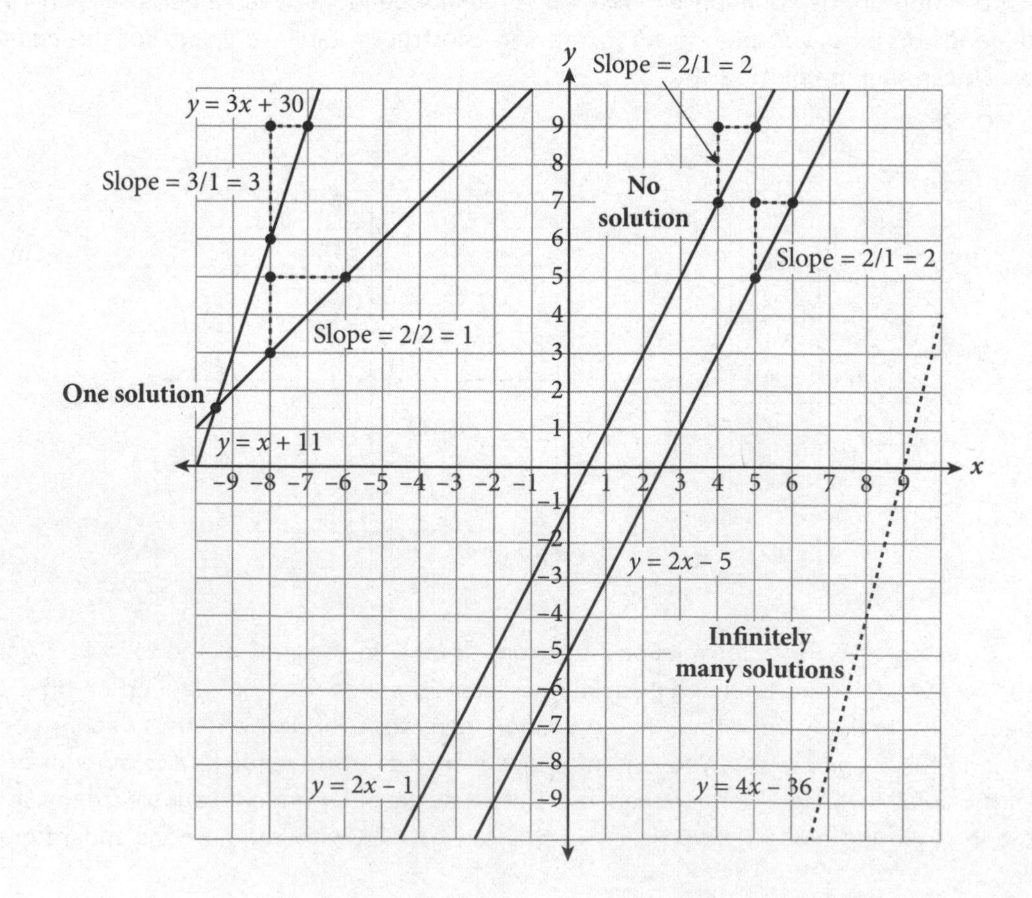

GEOMETRY

- Adjacent angles can be added to find the measure of a larger angle. The following diagram demonstrates this.

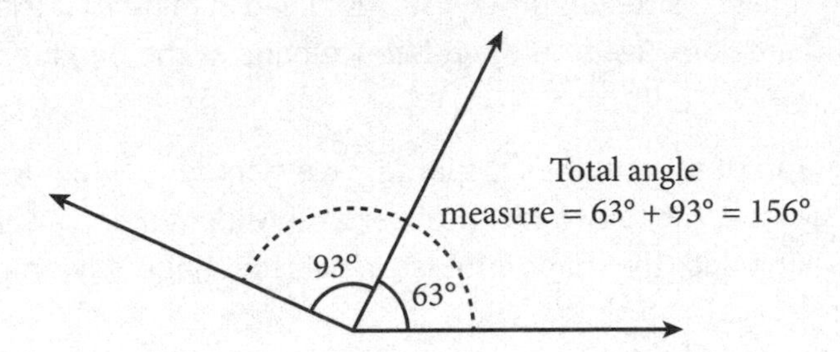

- Two distinct lines in a plane will either intersect at one point or extend indefinitely without intersecting. If two lines intersect at a right angle (90°), they are perpendicular and are denoted with ⊥. If the lines never intersect, they are parallel and are denoted with ||.

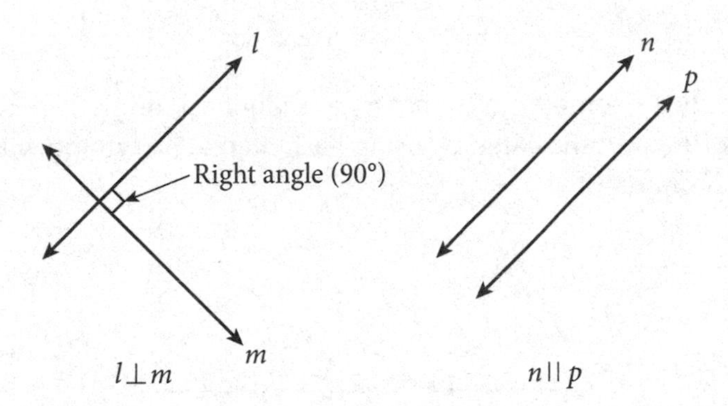

- Perimeter and area are basic properties that all two-dimensional shapes have. The perimeter of a polygon can easily be calculated by adding the lengths of all its sides. Area is the amount of two-dimensional space a shape occupies. The most common shapes for which you'll need these two properties on Test Day are triangles, parallelograms, and circles.
- The area (A) of a triangle is given by $A = \frac{1}{2}bh$, where b is the base of the triangle and h is its height. The base and height are always perpendicular. Any side of a triangle can be used as the base; just make sure you use its corresponding height (the longest perpendicular line you can draw within the triangle). You can use a right triangle's two legs as the base and height, but in non-right triangles, if the height is not given, you'll need to draw it in (from the vertex of the angle opposite the base down to the base itself at a right angle) and compute it.
- Parallelograms are quadrilaterals with two pairs of parallel sides. Rectangles and squares are subsets of parallelograms. You can find the area of a parallelogram using $A = bh$. As is the case with triangles, you can use any side of a parallelogram as the base; in addition, the height is still perpendicular to the base. Use the side perpendicular to the base as the height for a rectangle or square; for any other parallelogram, the height (or enough information to find it) will be given.

- A circle's perimeter is known as its circumference (C) and is found using $C = 2\pi r$, where r is the radius (distance from the center of the circle to its edge). Area is given by $A = \pi r^2$. The strange symbol is the lowercase Greek letter pi (π, pronounced "pie"), which is approximately 3.14. As mentioned in the algebra section, you should carry π throughout your calculations without rounding unless instructed otherwise. Also keep in mind that the formulas for finding circumference and area are given on the reference page at the beginning of each section of the math test.
- A shape (or the graph of an equation) is said to have symmetry when it can be split by a line (called an *axis of symmetry*) into two identical parts. Consider folding a shape along a line: If all sides and vertices align once the shape is folded in half, the shape is symmetrical about that line.

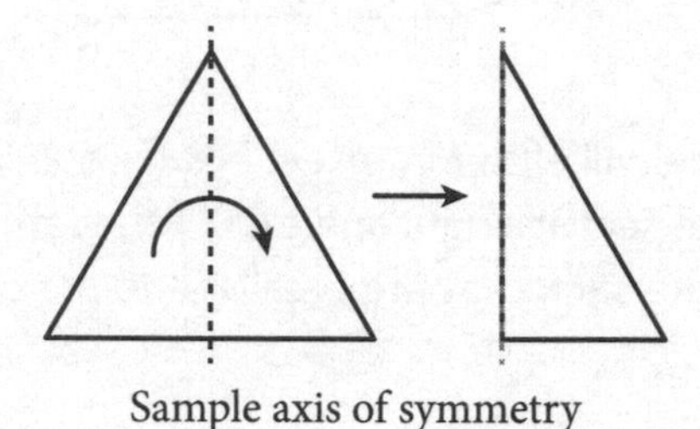

Sample axis of symmetry

- *Congruent* is simply a geometry term that means identical. Angles, lines, and shapes can be congruent. Congruence is indicated by using hash marks: Everything with the same number of hash marks is congruent.

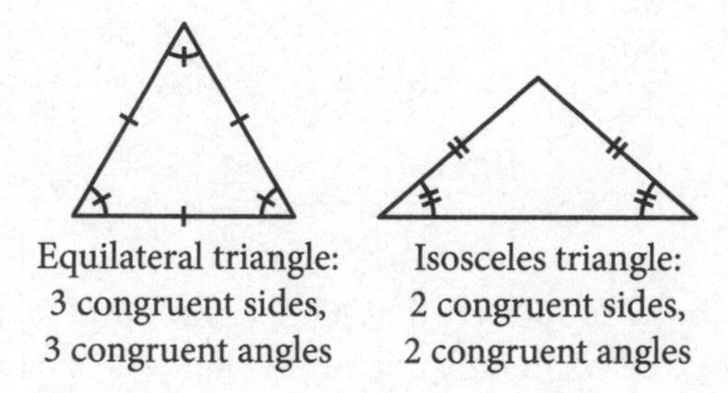

Equilateral triangle: 3 congruent sides, 3 congruent angles

Isosceles triangle: 2 congruent sides, 2 congruent angles

- *Similarity* between shapes indicates that they have identical angles and proportional sides. Think of taking a shape and stretching or shrinking each side by the same ratio. The resulting shape will have the same angles as the original. While the sides will not be identical, they will be proportional.

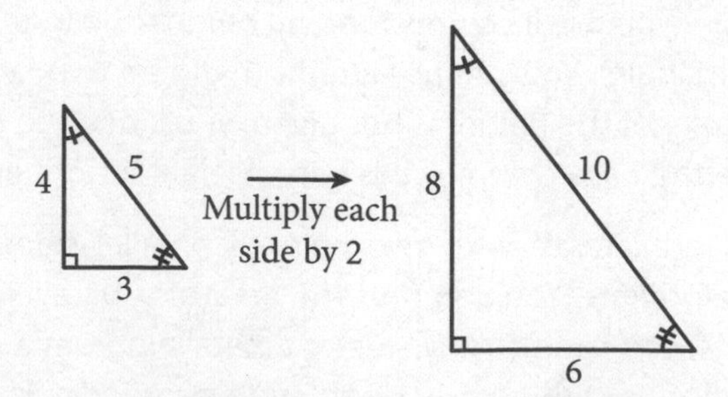

If you're comfortable with these concepts, read on for tips on calculator use.

Calculators and the SAT

The SAT has a no-calculator section and a calculator section, which means you'll need to be able to analyze and solve math problems both with and without technology. Even in the calculator section, you should ask yourself whether using a calculator is the most efficient way to solve a problem. The next couple of pages describe how the strongest test takers use their calculators strategically; that is, they carefully evaluate when to use the calculator and when to skip it in favor of a more streamlined approach. As you will see, even though you can use a calculator, sometimes it's more beneficial to save your energy by approaching a question more strategically. Work smarter, not harder.

WHICH CALCULATOR SHOULD YOU USE?

The SAT allows four-function, scientific, and graphing calculators. No matter which calculator you choose, start practicing with it now. You don't want to waste valuable time on Test Day looking for the exponent button or figuring out how to correctly graph equations. Due to the wide range of mathematics topics you'll encounter on Test Day, **we recommend using a graphing calculator**, such as the TI-83/84. If you don't already own one, see if you can borrow one from your school's math department or a local library.

A graphing calculator's capabilities extend well beyond what you'll need for the test, so don't worry about memorizing every function. If you're not already familiar with your graphing calculator, you'll want to get the user manual; you can find this on the Internet by searching for your calculator's model number. Identify the calculator functions necessary to answer various SAT math questions, then write down the directions for each to make a handy study sheet.

WHEN SHOULD YOU USE A CALCULATOR?

Some SAT question types are designed based on the idea that students will do some or all of the work using a calculator. As a master test taker, you want to know what to look for so you can identify when calculator use is advantageous. Questions involving statistics, determining approximate roots of complicated quadratic equations, and other topics are generally designed with calculator use in mind. Others are designed to be solved without a calculator. Solving some with a calculator can save you time and energy, but you'll waste both if you go for the calculator on others. You'll have to decide which method is best when you encounter topics such as the following:

- Long division and other extensive calculations
- Graphing equations
- Simplifying radicals and calculating roots
- Plane and coordinate geometry

Practicing long computations by hand and with the calculator will not only boost your focus and mental math prowess, but it will also help you determine whether it's faster to do the work by hand or reach for the calculator. If you tend to make quick work of long division and multiple-digit multiplication by hand, you can still use the calculator afterward to check your work.

Graphing equations (particularly quadratic equations) is probably a big reason most of you got that fancy calculator in the first place; it makes answering these questions a snap! This is definitely an area where you need to have an in-depth knowledge of your calculator's functions. The key to making these questions easy with the calculator is being meticulous when entering the equation. Errors in entry are likely to result in a trap answer.

Another stressful area for students is radicals, especially when the answer choices are written in decimal form. Those two elements are big red flags that trigger a reach for the calculator. Beware: Not all graphing calculators have a built-in radical simplification function, so consider familiarizing yourself with this process. For example, if you can't find a fourth root button on your calculator, you can always raise the quantity to the one-fourth power instead.

Geometry can be a gray area for students when it comes to calculator use. Consider working by hand when dealing with angles and lines, specifically when filling in information on complementary, supplementary, and congruent angles. You should be able to work fluidly through those questions without being completely dependent on your calculator.

✔ **Expert Tip**

If you choose to use trigonometric functions to get to the answer on triangle questions, make sure you have your calculator set to degrees or radians, depending on how the information is presented in the question.

Your calculator will come in handy when you need to work with formulas (volume, distance, arc, etc.) and when you want to check your work. Because the SAT uses π instead of 3.14..., there is no need to enter the decimal form into your calculator.

TO USE OR NOT TO USE?

A calculator may or may not be helpful on the SAT: It can be an asset for verifying work if you struggle when doing math by hand, but turning to it for the simplest computations will cost you time that you could devote to more complex questions. Practice solving questions with and without a calculator to get a sense of your personal style as well as strengths and weaknesses. Think critically about when a calculator saves you time and when mental math is faster. Use the exercises in this book to practice your calculations so that by the time Test Day arrives, you'll be in the habit of using your calculator as effectively as possible! Important note: If a question in a Practice Set has a calculator icon, this means that a similar question appeared in the calculator section of one of the practice tests released by the College Board. It does NOT mean that using a calculator is necessary or more efficient.

The Kaplan Method for Math

Because the SAT is a standardized test, students who approach each question in a consistent way will be rewarded on Test Day. Applying the same basic steps to every math question—whether it asks you about geometry, algebra, or even trigonometry—will help you avoid minor mistakes as well as tempting wrong answer choices.

Use the Kaplan Method for Math for every math question on the SAT. Its steps are applicable to every situation and reflect the best test-taking practices.

The Kaplan Method for Math has three steps:

Step 1: Read the question, identifying and organizing important information as you go

Step 2: Choose the best strategy to answer the question

Step 3: Check that you answered the *right* question

Let's examine each of these steps in more detail.

Step 1: Read the question, identifying and organizing important information as you go

This means:

- **What information am I given?** Take a few seconds to jot down the information you are given and try to group similar items together.
- **Separate the question from the context.** Word problems may include information that is unnecessary to solve the question. Feel free to discard any unnecessary information.
- **How are the answer choices different?** Reading answer choices carefully can help you spot the most efficient way to solve an SAT math question. If the answer choices are decimals, then painstakingly simplifying your final answer as a fraction is a waste of time; you can just use your calculator instead.

- **Should I label or draw a diagram?** If the question describes a shape or figure but doesn't provide one, sketch a diagram so you can see the shape or figure and add notes to it. If a figure is provided, take a few seconds to label it with information from the question.

> ✔ **Expert Tip**
>
> **Don't assume you understand a question as soon as you see it. Many students see an equation and immediately begin solving. Solving math questions without carefully reading can take you down the wrong path on Test Day.**

Step 2: Choose the best strategy to answer the question

- **Look for patterns.** Every SAT math question can be solved in a variety of ways, but not all strategies are created equally. To finish all of the questions, you'll need to solve questions as *efficiently* as possible. If you find yourself about to do time-consuming math, take a minute to look for time-saving shortcuts.
- **Pick numbers or use straightforward math.** While you can always solve an SAT math question with what you've learned in school, doing so won't always be the fastest way. On questions that describe relationships between numbers (such as percentages) but don't actually use numbers, you can often save time on Test Day by using techniques such as Picking Numbers instead of straightforward math.

> ✔ **Expert Tip**
>
> **The SAT won't give you any extra points for solving a question the hard way.**

Step 3: Check that you answered the *right* question

- When you get the final answer, **resist the urge to immediately bubble in the answer.** Take a moment to:
    - Review the question stem
    - Check units of measurement
    - Double-check your work
- The SAT will often ask you for quantities such as $x + 1$ or the product of x and y. **Be careful on these questions!** They often include tempting answer choices that correspond to the values of x or y individually. There's no partial credit on the SAT, so take a moment at the end of every question to make sure you're answering the right question.

The Kaplan Strategy for Translating English into Math

Expressions and equations are often disguised in "real-world" word problems, where it's up to you to extract the information you need. When you're solving these problems, you may run into trouble translating English into math. The following table shows some of the most common phrases and mathematical equivalents you're likely to see on the SAT.

Word Problems Translation Table	
English	**Math**
equals, is, equivalent to, was, will be, has, costs, adds up to, the same as, as much as	=
times, of, multiplied by, product of, twice, double, by	×
divided by, per, out of, each, ratio	÷
plus, added to, and, sum, combined, total, increased by	+
minus, subtracted from, smaller than, less than, fewer, decreased by, difference between	−
a number, how much, how many, what	*x*, *n*, etc.

Don't get frustrated—word problems can be broken down in predictable ways. To keep you organized on Test Day, use the **Kaplan Strategy for Translating English into Math:**

- **Define any variables, choosing letters that make sense.**
- **Break sentences into short phrases.**
- **Translate each phrase into a mathematical expression.**
- **Put the expressions together to form an equation.**

Let's apply this to a straightforward example: Colin's age is three less than twice Jim's age.

- **Define any variables, choosing letters that make sense:** We'll choose C for Colin's age and J for Jim's age.
- **Break sentences into short phrases:** The information about Colin and the information about Jim seem like separate phrases.
- **Translate each phrase into a mathematical expression:** Colin's age = C; 3 less than twice Jim's age = $2J - 3$.
- **Put the expressions together to form an equation:** Combine the results to get $C = 2J - 3$.

This strategy fits into the larger framework of the Kaplan Method for Math: When you get to **Step 2: Choose the best strategy to answer the question** and are trying to solve a word problem as efficiently as possible, switch over to this strategy to move forward quickly.

The Kaplan Method for Multi-Part Math Questions

The SAT contains multiple-choice and Grid-in questions, as well as multi-part math question sets. These question sets have multiple parts that are based on the same scenario and may require more analysis and planning than multiple-choice questions. To help you answer these questions effectively, use the Kaplan Method for Multi-Part Math Questions.

Step 1: Read the first question in the set, looking for clues

Step 2: Identify and organize the information you need

Step 3: Based on what you know, plan your steps to navigate the first question

Step 4: Solve, step-by-step, checking units as you go

Step 5: Did I answer the *right* question?

Step 6: Repeat for remaining questions, incorporating results from the previous question if possible

The next few pages will walk you through each step in more detail.

Step 1: Read the first question in the set, looking for clues

- **Focus all your energy here** instead of diluting it over the whole question; solving a Multi-Part Math question in pieces is far simpler. Further, you may need the results from earlier parts to solve subsequent ones. Don't even consider the later parts of the question until you've solved the first part.
- **Watch for hints** about what you'll need to use or how to solve. Underlining key quantities is often helpful to separate what you need from extraneous prose.

Step 2: Identify and organize the information you need

If you think this sounds like the Kaplan Method for Math, you're absolutely correct. You'll use some of those same skills. The difference: The Multi-Part Math question is just more involved with multiple pieces.

- **What information am I given?** Jot down key notes and group related quantities to develop your strategy.
- **What am I solving for?** This is your target. As you work your way through subsequent steps, keep your target at the front of your mind. This will help you avoid unnecessary work (and subsequent time loss). You'll sometimes need to tackle these problems from both ends, so always keep your goal in mind.

Step 3: Based on what you know, plan your steps to navigate the first question

- **What pieces am I missing?** Many students become frustrated when faced with a roadblock such as missing information, but it's an easy fix. Sometimes you'll need to do an intermediate calculation to elucidate the missing piece or pieces of the puzzle.

Step 4: Solve, step-by-step, checking units as you go

- **Work quickly but carefully**, just as you've done on other SAT math questions.

Step 5: Did I answer the *right* question?

- As is the case with the Kaplan Method for Math, **make sure your final answer is the requested answer**.
- Review the first part of the question.
- Double-check your units and your work.

Step 6: Repeat for remaining questions, incorporating results from the previous question if possible

- Now take your results from the first question and think critically about how they fit into the subsequent questions in the set. Previous results won't always be applicable, but when they are, they often lead to huge time savings.

When you've finished, congratulate yourself for persevering through such a challenging task. The Multi-Part Math questions are the toughest on the SAT. If you can ace these questions, you'll be poised for a great score on Test Day.

> ✔ **Expert Tip**
>
> **As these questions take substantially more time, consider saving Multi-Part Math questions for last.**

SECTION TWO

SAT Math Practice

PRACTICE SET 1: LINEAR EQUATIONS

Solving Linear Equations

An equation is a statement in algebra that says two expressions are equivalent. A *linear equation* is one in which the highest power on any variable is 1. Solving an equation is like solving a puzzle or playing a strategy game. The object of the game is to find the appropriate number value for the variable(s). Like any game, algebra has rules. The rules state that an algebraic equation is like a set of scales that you have to keep balanced at all times. If you want to balance a set of scales, whatever you do to one side you must do to the other. For instance, you have to add or take away the same amount from both sides. The same goes for an algebraic equation.

Many students inadvertently switch on "math autopilot" when solving linear equations, automatically running through the same set of steps on every equation without looking for the best way to solve the question. On the SAT, however, every second counts. You will want to use the most efficient strategy for solving questions. Looking carefully at how questions on the SAT use fractions and decimals can guide your strategy. Here are some examples:

- If the equation in the question has fractions, but the answers don't, try clearing the fractions in the equation by multiplying by a least common denominator.
- The presence of fractions in the answer choices likely means you'll need to rely on techniques for combining and simplifying fractions to get to the right answer.
- Seeing decimals in the answer choices likely indicates that using your calculator will save time on Test Day.

Modeling Real-World Scenarios

Linear equations and linear graphs can be used to model relationships and changes, such as those concerning time, temperature, or population. When a linear equation is presented in the context of a "real-world" word problem, it's up to you to extract and solve an equation. When you're answering these questions, you'll need to translate from English into math. The following table shows some of the most common phrases and mathematical equivalents you're likely to see on the SAT.

Word Problems Translation Table	
English	**Math**
equals, is, equivalent to, was, will be, has, costs, adds up to the same as, as much as	$=$
times, of, multiplied by, product of, twice, double, by	$\times$
divided by, per, out of, each, ratio	$\div$
plus, added to, and, sum, combined, total, increased by	$+$
minus, subtracted from, smaller than, less than, fewer, decreased by, difference between	$-$
a number, how much, how many, what	x, n, etc.

Working with Linear Graphs

Working with linear equations algebraically is only half the battle. The SAT will also expect you to work with graphs of linear equations. Here are some important facts to remember about graphs of linear equations:

- When a linear equation is written in slope-intercept form, $y = mx + b$, the variable m gives the slope of the line, and b represents the point at which the line intersects the y-axis.
- In a real-world scenario, slope represents a unit rate and the y-intercept represents a starting amount.
- The rate of change (slope) for a linear relationship is constant (does not vary).
- Slope is given by the formula $m = \frac{y_2 - y_1}{x_2 - x_1}$, where (x_1, y_1) and (x_2, y_2) are coordinates of points on the line. To help you remember the slope formula, think "rise over run."
- A line with a positive slope runs up and to the right ("uphill"), and a line with a negative slope runs down and to the right ("downhill").
- A horizontal line has a slope of 0 (because it does not rise or fall from left to right).
- A vertical line has an undefined slope.
- Parallel lines have the same slope.
- Perpendicular lines have negative reciprocal slopes (for example, 3 and $-\frac{1}{3}$).

To choose a graph that matches a given equation (or vice versa), find the slope of the line and its y-intercept. You can also use this strategy to write the equation of a line given a context, but you may need to translate two data points into ordered pairs before you find the slope.

PRACTICE SET

Easy

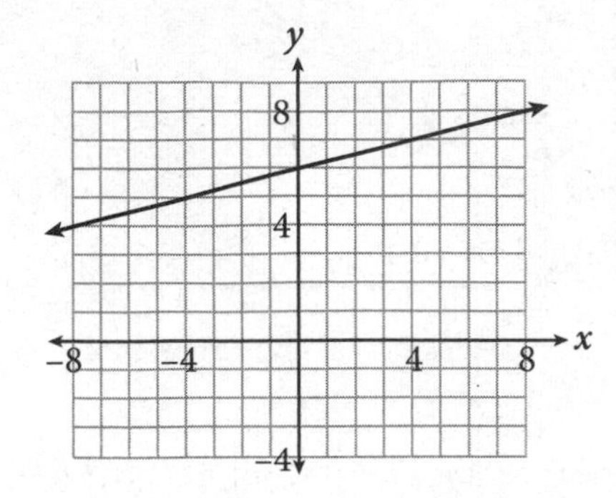

1. Which of the following equations represents the line shown in the graph?

   A) $y = 6x + \frac{1}{4}$

   B) $y = \frac{x}{4} + 6$

   C) $y = 4x + 6$

   D) $y = 6x + 4$

2. A lemonade stand's profit is given by the equation $p = 2c - 8.5$. Which of the following does the number 2 most likely represent?

   A) The price of one cup of lemonade

   B) The profit generated from the sale of one cup of lemonade

   C) The minimum number of cups of lemonade that must be sold to earn a profit

   D) The costs that must be recuperated before the lemonade stand earns any profits

3. Darien needs to buy several white dress shirts for his new job. He finds one he likes for \$35 that is on sale for 40% off. He also likes a blue tie that costs \$21. Which of the following represents the total cost, not including tax, if Darien buys x of the white shirts that are on sale and two of the blue ties?

   A) $C = 14x + 42$

   B) $C = 21x + 21$

   C) $C = 21x + 42$

   D) $C = 35x + 42$

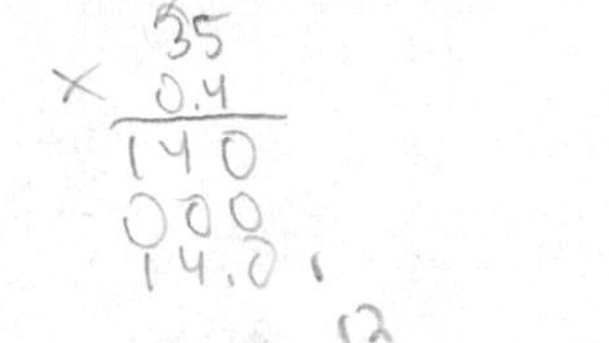

$$\frac{z}{6}\left(\frac{3}{2}\right) - 7 = -2(3z - 4)$$

4. What value of z satisfies the equation above?

   A) $-\frac{12}{5}$

   B) $-\frac{4}{25}$

   C) $\frac{4}{25}$

   D) $\frac{12}{5}$

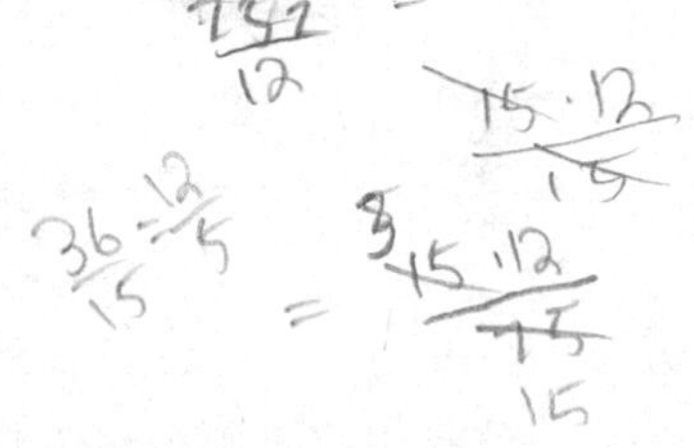

5. Line L passes through the coordinate points $\left(-\frac{7}{2}, 3\right)$ and $\left(-\frac{3}{2}, 5\right)$. What is the slope of line L?

   A) -1

   B) $-\frac{2}{5}$

   C) $\frac{2}{5}$

   D) 1

$$17(6x - 50) = 204\left(\frac{7}{24}x\right)$$

6. For what value of x is the equation above true?

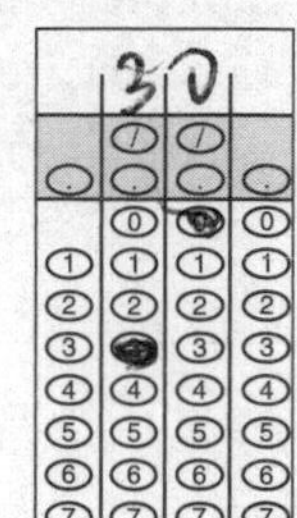

Medium

7. Line L has an undefined slope. Line M is perpendicular to line L. Which of the following could be the equation of line M?

A) $x = y$

B) $y = 7$

C) $x = -3$

D) $xy = 4$

8. A line in the xy-plane that passes through the coordinate points $(3, -6)$ and $(-7, -4)$ will never intersect a line that is represented by which of the following equations?

A) $x + 5y = 6$

B) $x + \frac{y}{2} = 7$

C) $y - 2x = -9$

D) $2y - x = -8$

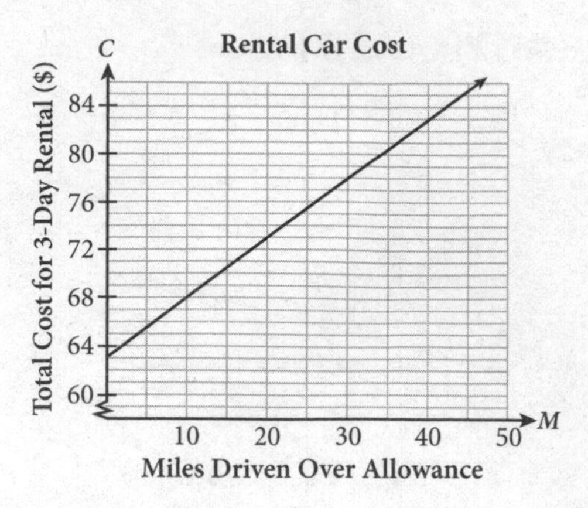

9. A car rental agency charges a per day rental fee which includes a daily mileage allowance plus a certain amount per mile driven over the allowance. The graph above compares the miles driven over the allowance and the total cost for a 3-day rental. What does the C-intercept most likely represent in this scenario?

A) The per day rental fee for renting the car

B) The number of miles a renter may drive the car per day

C) The penalty a renter must pay if the daily mileage allowance is exceeded

D) The total cost of a 3-day rental assuming the car is not driven over the allowance

$$\frac{2}{3}x + cy = 2$$

10. If the slope of the equation shown above is 6, what is the value of c?

A) -4

B) $-\frac{1}{9}$

C) $\frac{1}{3}$

D) 4

11. Anneke is competing in a 500-meter freestyle swim event, which consists of swimming the length of a pool 20 times. If Anneke averages 26.4 seconds per length of the pool, which of the following equations could be used to determine the number of meters (m) Anneke has left in the event after swimming for s seconds?

A) $m = 500 - \frac{25s}{26.4}$

B) $m = 500 - 25s$

C) $m = \frac{25s}{26.4}$

D) $m = 500 - 20s$

12. If the graph of the equation $y = 5x + 3$ is shifted down 4 units, what is the x-intercept of the new line?

A) -1

B) $\frac{1}{5}$

C) 1

D) $\frac{5}{4}$

13. A new color copier purchased for \$8,500 is expected to depreciate (lose value) according to the equation $y = -1{,}250x + 8{,}500$, where y is the value of the copier x years after it was purchased. The company that bought the copier plans to sell it when the value is \$1,000 and upgrade to a new one. How many years after the copier is purchased will the company sell it?

$$\frac{3(h+2)-4}{6} = \frac{h(7 \times 2 - 5)}{2}$$

14. In the equation above, what is the value of h?

15. If $\frac{2}{3}j - \frac{1}{4}k = \frac{5}{2}$, what is the value of $8j - 3k$?

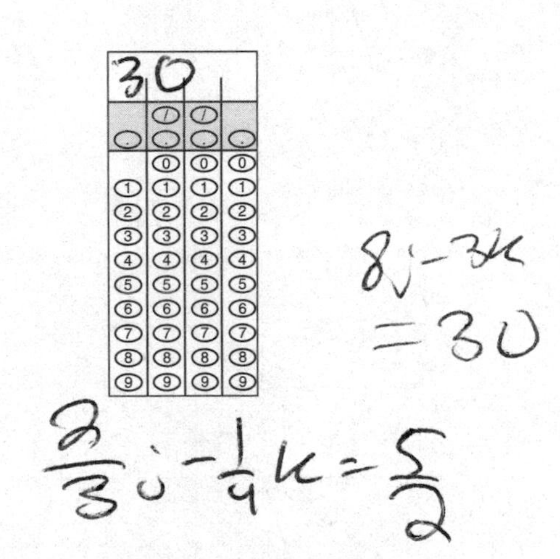

Hard

16. If w is an integer less than -1, which of the following could be the graph of $x + wy = wx - y - 3$?

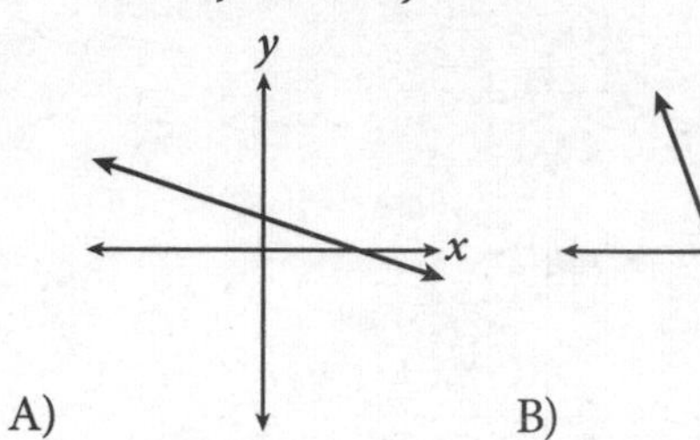

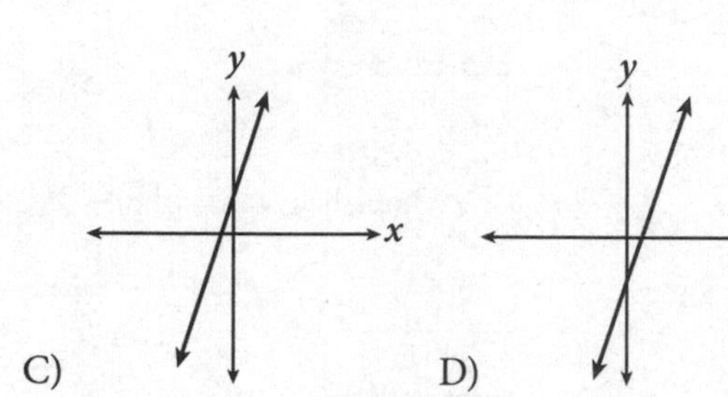

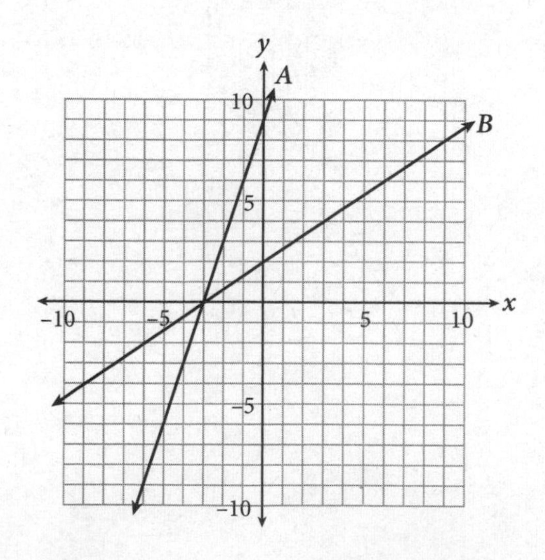

17. If the equation of line A shown on the graph is given by $y = mx + b$, and the equation of line B is given by $y = k(mx + b)$, what is the value of k?

    A) 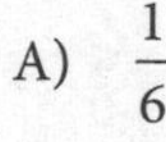 $\frac{1}{6}$

    B) $\frac{2}{9}$

    C) 6

    D) 9

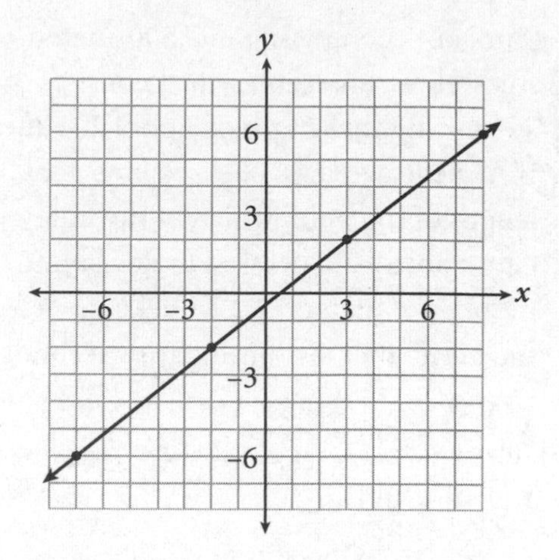

18. If the equation of the line shown on the graph is written in standard form, $Ax + By = C$, and $A = 3$, what is the value of B?

    A) $-\frac{15}{4}$

    B) $-\frac{5}{4}$

    C) $\frac{4}{5}$

    D) 15

19. Two garages each charge a fixed amount, plus an hourly rate, to service a car. The garage on Main Street charged one customer $153 for a 2-hour service appointment, and it charged a second customer $315 for a 5-hour service appointment. The garage on 2nd Street charges $5 less per hour than the garage on Main Street and $10 more for the fixed amount. How much would the garage on 2nd Street charge for a 3-hour service appointment?

A) $157

B) $174

C) $181

D) $202

$$\frac{1}{2}(6x-4)-(3-x)=ax+x+b$$

20. If the equation above has infinitely many solutions, what is the value of $a-b$?

ANSWER KEY

1. B	6. 20	11. A	16. C
2. A	7. B	12. B	17. B
3. C	8. A	13. 6	18. A
4. D	9. D	14. 1/12	19. D
5. D	10. B	15. 30	20. 8

ANSWERS & EXPLANATIONS

1. B

Difficulty: Easy

Category: Heart of Algebra / Linear Equations

Strategic Advice: When asked to match a graph to an equation, always start by examining the answer choices. Here, the equations are written in slope-intercept form ($y = mx + b$), so find the slope of the line and its y-intercept and match them to m and b in the correct equation.

Getting to the Answer: The line crosses the vertical axis at (0, 6), making the y-intercept of the line 6, which means you can immediately eliminate A and D. To find the slope of the line, you can count the rise and the run from the y-intercept to the next point that lands on the intersection of two grid-lines, (4, 7), or you can use the slope formula. Either way, you'll find that the slope is $\frac{1}{4}$, which makes (B) the correct answer.

2. A

Difficulty: Easy

Category: Heart of Algebra / Linear Equations

Strategic Advice: In a real-world scenario, the slope of a line represents a unit rate and the y-intercept represents a flat fee or a starting amount.

Getting to the Answer: The profit equation is linear and is given in slope-intercept form ($y = mx + b$). The number 2 is the slope (m), which represents a unit rate. This means you can eliminate C and D because neither describes a rate of change. To choose between (A) and B, compare the descriptions to the equation—the 2 in the equation is the coefficient of c (which represents cups), so 2 is the price per cup of lemonade. Choice (A) is correct.

3. C

Difficulty: Easy

Category: Heart of Algebra / Linear Equations

Strategic Advice: When matching an equation to a real-world scenario, take a quick look at the answer choices to see what variables are being used. Next, write an equation in words, and then use the Kaplan Strategy for Translating English into Math.

Getting to the Answer: Keep in mind that the shirts are on sale but the tie is not. The shirts are 40% off, which means that Darien pays only $100 - 40 = 60\%$ of the price, or 0.6(35). If he buys x shirts, the cost is $0.6(35)x$. He also buys two ties, which cost \$21 each. This means the total cost is $C = 0.6(35)x + 21(2)$. This is not one of the answer choices, so simplify to get $C = 21x + 42$, which is (C).

4. D
Difficulty: Easy

Category: Heart of Algebra / Linear Equations

Strategic Advice: Some questions, like this one, are intended to test your fluency with algebraic manipulations, so work carefully and you'll arrive at the correct answer.

Getting to the Answer: Simplify the left side of the equation using multiplication and the right side using the distributive property. Be sure to distribute the negative sign to both terms inside the parentheses on the right side. Next, collect the variable terms on one side of the equation and the constants on the other, and then solve for z using inverse operations:

$$\frac{z}{6}\left(\frac{3}{2}\right)-7=-2(3z-4)$$
$$\frac{z}{4}-7=-6z+8$$
$$\frac{z}{4}=-6z+15$$
$$4\cdot\left[\frac{z}{4}\right]=4\cdot\left[-6z+15\right]$$
$$z=-24z+60$$
$$25z=60$$
$$z=\frac{60}{25}=\frac{12}{5}$$

The correct answer is (D).

5. D
Difficulty: Easy

Category: Heart of Algebra / Linear Equations

Strategic Advice: When you're given two points on a line, find the slope using the slope formula, $m=\frac{y_2-y_1}{x_2-x_1}$. Remember, slope is rise over run, so the y's (which rhymes with "rise") go in the numerator and the x's go in the denominator.

Getting to the Answer: Be careful when you substitute the values into the formula and when you subtract negative numbers:

$$m=\frac{y_2-y_1}{x_2-x_1}=\frac{5-3}{-\frac{3}{2}-\left(-\frac{7}{2}\right)}=\frac{2}{\frac{-3+7}{2}}=\frac{2}{\frac{4}{2}}=\frac{2}{2}=1$$

The slope of line L is 1, which is (D).

6. 20
Difficulty: Easy

Category: Heart of Algebra / Linear Equations

Strategic Advice: When large numbers are being multiplied by quantities on both sides of an equation, check to see if the larger one is divisible by the smaller one. If so, you can divide both sides by the smaller one and end up working with much easier numbers.

Getting to the Answer: Because 204 is divisible by 17, start by dividing both sides of the equation by 17 instead of using the distributive property. Then, simplify both sides and solve for x using inverse operations:

$$17(6x-50)=204\left(\frac{7}{24}x\right)$$
$$6x-50=12\left(\frac{7}{24}x\right)$$
$$6x-50=\frac{7}{2}x$$
$$\frac{12}{2}x-\frac{7}{2}x=50$$
$$\frac{5}{2}x=50$$
$$5x=100$$
$$x=20$$

7. B
Difficulty: Medium

Category: Heart of Algebra / Linear Equations

Strategic Advice: A line with an undefined slope is a vertical line, or in other words, parallel to the y-axis.

Getting to the Answer: A line that is perpendicular to a vertical line must be a horizontal line, which has a slope of 0. Its equation, therefore, will be of the form $y = 0x + b$, or just $y = b$, where b is a constant. The only equation that meets this criterion is choice (B).

8. A

Difficulty: Medium

Category: Heart of Algebra / Linear Equations

Strategic Advice: Two lines that never intersect are parallel lines, which means they have the same slope.

Getting to the Answer: Start by finding the slope of the line between the two given points: $m = \frac{y_2 - y_1}{x_2 - x_1} = \frac{-6-(-4)}{3-(-7)} = \frac{-2}{10} = -\frac{1}{5}$. Now, find the equation that has a slope of $-\frac{1}{5}$ by rewriting the equations in slope-intercept form. Start with (A):

$$x + 5y = 6$$
$$5y = -x + 6$$
$$y = -\frac{1}{5}x + 6$$

The slope of the line is $-\frac{1}{5}$, and there can only be one correct answer, so there is no need to check the other choices—(A) is correct.

9. D

Difficulty: Medium

Category: Heart of Algebra / Linear Equations

Strategic Advice: In a real-world scenario, the slope of a line represents a unit rate and the y-intercept represents a flat fee or a starting amount. In some questions, you may need to use the axis titles to interpret the meaning of the intercept.

Getting to the Answer: With miles driven over the allowance plotted along the horizontal axis and the total cost for a 3-day rental plotted along the vertical axis (the C-axis), the point (0, 63) indicates that driving 0 miles over the allowance (or not driving over the allowance) results in a cost of \$63. This makes (D) correct.

10. B

Difficulty: Medium

Category: Heart of Algebra / Linear Equations

Strategic Advice: Don't overthink a question like this. It is asking about slope and you have an equation, so write the equation in slope-intercept form and see where that takes you.

Getting to the Answer: When an equation is written in slope-intercept form ($y = mx + b$), m is the slope of the line. Rewrite the given equation in this form by subtracting $\frac{2}{3}x$ from both sides and then dividing both sides by c:

$$\frac{2}{3}x + cy = 2$$
$$cy = -\frac{2}{3}x + 2$$
$$y = \frac{-\frac{2}{3}x}{c} + \frac{2}{c}$$

Now, set the coefficient of x equal to the given slope (6) and solve for c. The coefficient is a fraction, so write 6 as $\frac{6}{1}$ and cross-multiply:

$$\frac{6}{1} = \frac{-\frac{2}{3}}{c}$$
$$6c = -\frac{2}{3}$$
$$c = -\frac{2}{3} \div 6$$
$$c = -\frac{2}{3} \times \frac{1}{6} = -\frac{1}{9}$$

The value of c is $-\frac{1}{9}$, which is (B).

11. A

Difficulty: Medium

Category: Heart of Algebra / Linear Equations

Strategic Advice: Make use of the given units to make sure your equation is set up properly.

Getting to the Answer: You're told that the units of m are meters, so you must find a way to combine the various quantities so that, when simplified, the result will be in meters as well. Anneke will swim 20 lengths in the 500-meter event, which means each length of the pool is $\frac{500}{20} = 25$ m. Because each length is 25 m and Anneke can swim one length in 26.4 seconds on average, her average speed is $\frac{25}{26.4}$ m/s. Multiplying this quantity by s, the elapsed race time in seconds, will eliminate seconds from the units. The quantity $\frac{25}{26.4}s$ can therefore be subtracted from 500 (which is in meters) to yield the number of meters Anneke has remaining, making the final equation $m = 500 - \frac{25s}{26.4}$. (A) is the correct answer.

12. B

Difficulty: Medium

Category: Heart of Algebra / Linear Equations

Strategic Advice: You aren't given a graph to shift, so either draw a quick sketch of the given line or mentally picture what happens when the line is shifted. Don't forget—when a linear equation is written in the form $y = mx + b$, m is the slope of the line and b is the y-intercept (not the x-intercept).

Getting to the Answer: The graph of the given line has a slope of 5 and a y-intercept of 3. Shifting the line down 4 units doesn't change the slope, but the y-intercept moves down 4, so the new y-intercept is $3 - 4 = -1$. Wait, though—choice A is not the correct answer: The question asks for the x-intercept of the new line, not the y-intercept. The equation of the new line is $y = 5x - 1$. Set $y = 0$ and solve for x:

$$\begin{aligned} 0 &= 5x - 1 \\ 1 &= 5x \\ \frac{1}{5} &= x \end{aligned}$$

This means (B) is correct.

13. 6

Difficulty: Medium

Category: Heart of Algebra / Linear Equations

Strategic Advice: In a word problem, being able to match each variable to the quantity that it represents is often the key to getting started.

Getting to the Answer: The copier will be worth \$1,000 when its value (y) is equal to 1,000. Plug 1,000 in for y and solve for x using inverse operations:

$$\begin{aligned} y &= -1{,}250x + 8{,}500 \\ 1{,}000 &= -1{,}250x + 8{,}500 \\ -7{,}500 &= -1{,}250x \\ 6 &= x \end{aligned}$$

The company will sell the copier 6 years after its purchase.

14. 1/12

Difficulty: Medium

Category: Heart of Algebra / Linear Equations

Strategic Advice: One equation with one variable usually means straightforward algebra will get you to the answer.

Getting to the Answer: When both sides of an equation are written over a denominator, cross-multiplication is a great strategy for eliminating the fractions. However, simplifying the numerators first will save steps and reduce potential errors:

$$\frac{3(h+2)-4}{6}=\frac{h(7\times2-5)}{2}$$
$$\frac{3h+6-4}{6}=\frac{h(14-5)}{2}$$
$$\frac{3h+2}{6}=\frac{9h}{2}$$
$$2(3h+2)=6(9h)$$
$$6h+4=54h$$
$$4=48h$$
$$h=\frac{4}{48}=\frac{1}{12}$$

15. 30
Difficulty: Medium

Category: Heart of Algebra / Linear Equations

Strategic Advice: Don't panic when you see two variables and only one equation. Instead, look for a way to manipulate the given equation, or at least one side of it, so that it looks like the desired expression.

Getting to the Answer: The desired expression ($8j - 3k$) doesn't have fractions, so multiply both sides of the given equation by the least common denominator (12) to eliminate the fractions:

$$\frac{2}{3}j-\frac{1}{4}k=\frac{5}{2}$$
$$12\cdot\left[\frac{2}{3}j-\frac{1}{4}k\right]=12\cdot\left[\frac{5}{2}\right]$$
$$8j-3k=30$$

Notice that the coefficients of j and k are 8 and –3, respectively, which is exactly what you're looking for. The number on the other side of the equation, 30, is the correct answer.

16. C
Difficulty: Hard

Category: Heart of Algebra / Linear Equations

Strategic Advice: When matching a graph to a linear equation, focus on putting the equation in slope-intercept form ($y = mx + b$) so you can determine what the graph should look like.

Getting to the Answer: Start by collecting the y terms on the left side of the equation and the x terms on the right (don't worry about w yet):

$$x+wy=wx-y-3$$
$$wy+y=wx-x-3$$

Next, factor out x and y, and then divide both sides by the coefficient of y so that it's in slope-intercept form:

$$y(w+1)=x(w-1)-3$$
$$y=\frac{w-1}{w+1}x-\frac{3}{w+1}$$

The equation represents a line with a slope of $\frac{w-1}{w+1}$ and a y-intercept of $-\frac{3}{w+1}$. Because you're given that w is an integer less than –1, both the slope and the y-intercept of the line are positive (pick an easy number, such as –2, to plug in for w if you're not certain). This means the line is increasing (going up) from left to right and crosses the vertical axis above 0. The only choice that contains a graph with these characteristics is (C).

17. B
Difficulty: Hard

Category: Heart of Algebra / Linear Equations

Strategic Advice: The equation of line B is given as a multiple of line A, so start by finding the equation of each line using its slope and y-intercept. Then compare the equations.

Getting to the Answer: First, determine the equation of line *A*. Remember that slope is equal to rise over run, so count the change in *y* and the change in *x* from one point on the line to the next. The slope is $\frac{3}{1}$ or just 3. Next, find the *y*-intercept: 9. Now, find the equation of line *B*: The slope is $\frac{2}{3}$, and the *y*-intercept is 2. So, the equations are: Line *A*: $y = 3x + 9$ and Line *B*: $y = \frac{2}{3}x + 2$. Notice that the numbers in the equation of line *A* are larger, so *k* must be a fraction, which means you can eliminate C and D. To choose between A and (B), multiply the number by the equation of line *A* to see if it matches the equation of line *B*. Because $\frac{2}{9}(3x+9) = \frac{2}{3}x + 2$ (the equation of line *B*), $k = \frac{2}{9}$. Therefore, (B) is correct.

18. A

Difficulty: Hard

Category: Heart of Algebra / Linear Equations

Strategic Advice: Whenever there is more than one strategy that could get you to the answer, always consider which one will get you there faster. Here, you could use two points on the line to write its equation and then convert the equation to standard form, or you can convert standard form to slope-intercept form and compare the slope to that of the line in the graph. Use whichever method will get you to the answer quicker.

Getting to the Answer: Rewrite $Ax + By = C$ in slope-intercept form:

$$\begin{aligned} Ax + By &= C \\ By &= -Ax + C \\ y &= \frac{-A}{B}x + \frac{C}{B} \end{aligned}$$

The slope of the line is the coefficient of *x*, which is $\frac{-A}{B}$. Now, find the slope of the line shown in the graph by counting the rise and the run from one point to the next; the slope is $\frac{4}{5}$.

You're given that $A = 3$, so $\frac{-A}{B} = \frac{-3}{B}$. Set this quantity equal to $\frac{4}{5}$ and cross-multiply to solve for *B*:

$$\begin{aligned} \frac{-3}{B} &= \frac{4}{5} \\ -3(5) &= 4B \\ -\frac{15}{4} &= B \end{aligned}$$

(A) is the correct answer.

19. D

Difficulty: Hard

Category: Heart of Algebra / Linear Equations

Strategic Advice: In a question like this, take a few seconds to plan a route to the desired quantity. For example, before you can determine the amount that the garage on 2nd Street would charge, you'll need to find the hourly rate and the fixed amount charged by the garage on Main Street. To do this, think about what aspect of a linear relationship could tell you the hourly rate.

Getting to the Answer: In a real-world scenario, a unit rate (here, the hourly rate) is the same as slope. To find the slope, start by writing the amounts given as ordered pairs in the form (hours, amount); the ordered pairs are (2, 153) and (5, 315). Now, plug these values into the slope formula: $m = \frac{y_2 - y_1}{x_2 - x_1} = \frac{315 - 153}{5 - 2} = \frac{162}{3} = 54$. The garage on Main Street charges \$54 per hour. The garage on 2nd Street charges \$5 less per hour, or \$54 – \$5 = \$49 per hour. Now you need to find the fixed amount charged by the garage on Main Street. If a customer was charged \$153 for 2 hours, and the hourly rate is \$54 per hour, then the fixed amount is \$153 – 2(\$54) = \$45. The garage on 2nd Street charges \$10 more for the fixed amount, or \$45 + \$10 = \$55. Put the two pieces of information together to find that the garage on 2nd Street would charge \$55 + 3(\$49) = \$202 for a 3-hour service appointment, which is (D).

20. 8
Difficulty: Hard

Category: Heart of Algebra / Linear Equations

Strategic Advice: A linear equation has infinitely many solutions when, once simplified, the left side of the equation is equal to the right side (known as an identity). For example, if the simplified form of an equation is $5 = 5$, or even $2x + 5 = 2x + 5$, then the equation has infinitely many solutions.

Getting to the Answer: Simplify both sides of the equation as much as possible. Then examine the terms that remain:

$$\begin{aligned} \frac{1}{2}(6x - 4) - (3 - x) &= ax + x + b \\ 3x - 2 - 3 - (-x) &= ax + x + b \\ 3x - 2 - 3 + x &= ax + x + b \\ 4x - 5 &= x(a + 1) + b \end{aligned}$$

For the left side of the equation to equal the right side, the coefficient of x must be the same on each side, and the constant must be the same on each side:

$$\begin{aligned} a + 1 = 4, \text{ so } a &= 3 \\ b &= -5 \end{aligned}$$

The question asks for $a - b$, so the correct answer is $3 - (-5) = 3 + 5 = 8$.

PRACTICE SET 2: SYSTEMS OF LINEAR EQUATIONS

What Is a System of Equations?

A system of equations is a set of two or more equations that usually contains multiple variables. Systems of equations are extremely useful in modeling and simulation. Complex mathematical problems, such as weather forecasting or crowd control predictions, often require 10 or more equations to be simultaneously solved for multiple variables. Fortunately, virtually all systems of equations on the SAT that require solving will contain only two equations and two variables.

In general, when you are given a situation involving n variables, you need a system of n equations to arrive at fixed values for these variables. Thus, if you have two variables, you need two equations. Three variables would require three equations, and so on.

To solve a two-variable system of equations, the equations that make up the system must be independent. *Independent equations* are equations for which no algebraic manipulations can transform one of the equations into the other.

Consider the equation $4x + 2y = 8$. We could use properties of equality to rewrite this equation in a number of different ways. For example, we could multiply both sides by 2, resulting in the equation $8x + 4y = 16$. While it seems as though we've just created an additional equation, this is misleading, as our second equation has the same core variables and relationships as our first equation. These two equations are called *dependent equations*. Two dependent equations cannot be used to solve for two variables because they really represent the same equation, just written in different forms. If you attempt to solve a system of dependent equations, you will end up with the same thing on both sides of the equal sign (e.g., $16 = 16$), which is always true and indicates that the system has infinitely many solutions.

At other times, you'll encounter equations that are fundamentally incompatible with each other. For example, given the two equations $4x + 2y = 8$ and $4x + 2y = 9$, it should be obvious that there are no values for x and y that will satisfy both equations at the same time. Doing so would violate fundamental laws of math. In this case, the system of equations has no solution.

Solving Systems of Equations

The two main methods for solving a system of linear equations are substitution and combination (sometimes referred to as elimination or elimination by addition).

Substitution is the most straightforward method for solving systems and can be applied in every situation. However, the process can get messy if none of the variables in either of the equations has a coefficient of 1. To use substitution, solve the simpler of the two equations for one variable, and then substitute the result into the other equation. Unfortunately, substitution is often the longer and more time-consuming route for solving systems of equations.

Combination involves adding the two equations, or a multiple of one or both equations, together to eliminate one of the variables. You're left with one equation and one variable, which can be solved using inverse operations.

Caution: Although most students prefer substitution, questions on the SAT that involve systems of equations are often designed to be quickly solved using combination. For instance, combination is almost always the fastest technique to use when solving a system of equations question that only asks for the value of one of the variables. To really boost your score on Test Day, practice combination as much as you can on practice tests and in homework questions so that it becomes second nature.

Graphing Systems of Equations

Knowing how many solutions a system of equations has tells you how graphing the equations on the same coordinate plane should look. Conversely, knowing what the graph of a system of equations looks like tells you how many solutions the system has.

Recall that the solution(s) to a system of equations is the point or points where the graphs of the equations intersect. The table below summarizes three possible scenarios:

If your system has ...	**... then it will graph as:**	**Reasoning**
no solution	two parallel lines	Parallel lines never intersect.
one solution	two lines intersecting at a single point	Two straight lines have only one intersection.
infinitely many solutions	a single line (one line directly on top of the other)	One equation is a manipulation of the other—the graphs are the same line.

Translating Word Problems into Multiple Equations

Sometimes questions on the SAT will present information in a real-world context, and you'll need to translate it into a system of equations and then solve. This might sound scary, but solving systems of equations can be relatively straightforward once you get the hang of it. Just remember to use the Kaplan Strategy for Translating English into Math to set up your equations as you did in the previous practice set, and then solve using either substitution or combination.

PRACTICE SET

Easy

$$x + y = 29$$
$$x + 2y = 12$$

1. If the ordered pair (x, y) is the solution to the system of equations above, what is the value of y?

   A) -17

   B) 12

   C) 46

   D) 75

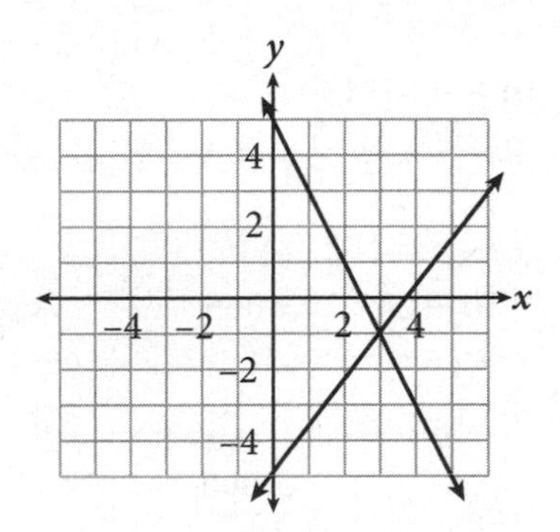

2. The graph of a system of linear equations is shown above. If the ordered pair (x, y) represents the solution to this system, what is the value of $x - y$?

   A) -4

   B) -2

   C) 2

   D) 4

3. If $5b = 6a + 16$ and $9a = 7b - 20$, then what is the value of $3a - 2b$?

   A) -8

   B) -4

   C) 4

   D) 8

$$4x - 2y + 3 = 8$$
$$3x + 6y = 8y - x + 5$$

4. How many solutions does the system of equations shown above have?

   A) 0

   B) 1

   C) 2

   D) Infinitely many

5. An office has 27 employees. If there are seven more women than men in the office, how many employees are men?

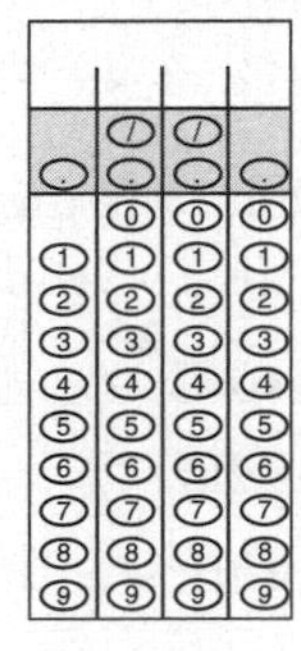

Medium

6. The total fare for two adults and three children on an excursion boat is $14. If each child's fare is one-half of each adult's fare, what is the total cost for one adult and one child?

   A) $4.00

   B) $5.25

   C) $6.00

   D) $6.50

$$y = \frac{1}{5}x + 4$$

$$y = \frac{3}{7}x - 4$$

7. If the ordered pair (x, y) satisfies the system of equations above, what is the value of y?

A) 0

B) 7

C) 10

D) 11

$$2x - 3y = -3$$

$$-12 = -4x + y$$

8. In what quadrant will the lines represented by the equations above intersect?

A) Quadrant I

B) Quadrant II

C) Quadrant III

D) Quadrant IV

9. If $10a = 6b + 7$ and $a - 6b = 34$, then what is the value of $-\frac{1}{3}a$?

A) -1

B) 1

C) $\frac{41}{27}$

D) $\frac{41}{9}$

10. In addition to the standard airfare, a particular airline charges passengers for two kinds of travel services: \$25 to check a bag and \$15 to upgrade to priority boarding. If the airline collected \$3,065 in baggage and priority boarding fees from 145 travel services on two flights combined, which of the following systems of equations could be used to determine the number of bags checked (b) and the number of priority boarding upgrades (p) purchased on the two flights?

A) $b + p = 145 \times 2$
$25b + 15p = 3{,}065 \times 2$

B) $b + p = 145$
$25b + 15p = 3{,}065$

C) $b + p = 145$
$15b + 25p = 3{,}065$

D) $b + p = \frac{145}{2}$
$15b + 25p = \frac{3{,}065}{2}$

11. The most popular items at a bakery are its raspberry scones and its lemon poppy seed muffins. The shop sells both items in boxes of 12 at a cost of $15 per box of raspberry scones and $9 per box of lemon poppy seed muffins. On Friday and Saturday, the shop earned $396 by selling a total of 46 boxes of these two items. If r and l represent the number of boxes of raspberry scones and lemon poppy seed muffins sold over the two-day period, respectively, which of the following systems of equations could be used to find the number of boxes of each type of item sold?

A) $r + l = 46$
$15r + 9l = 396$

B) $r + l = 12 \times 46$
$15r + 9l = 396$

C) $r + l = 46$
$15r + 9l = \frac{396}{2}$

D) $r + l = 46 \times 12$
$15r + 9l = \frac{396}{12}$

12. Tricia manages a health bar and wants to add a new fruit-and-protein smoothie to the menu. To decide on the first new flavor she plans to offer, Tricia sold trial-sized banana smoothies and kiwi smoothies. She charged $2 for a banana smoothie and $2.50 for a kiwi smoothie, and she sold 40 in all, totaling $87. How much more money did Tricia make on the banana smoothies than the kiwi smoothies?

A) $12

B) $17

C) $26

D) $52

13. A street vendor sells two types of newspapers, one for $0.25 and the other for $0.40. If she sold 100 newspapers for $28.00, how many newspapers did she sell at $0.25?

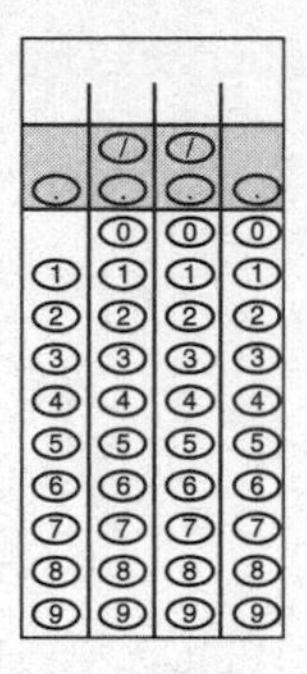

$$x + y = -6$$
$$y - 4x = 4$$

14. If the ordered pair (x, y) satisfies the system of equations shown above, what is the value of xy?

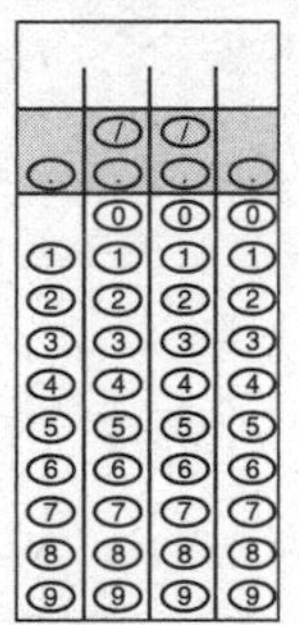

Hard

$$4x + 7y = 24$$
$$6x + \frac{21}{2}y = g$$

15. In the system of equations above, g is a constant. If the system has infinitely many solutions, what is the value of g?

A) 16

B) 32

C) 36

D) 72

16. A small office supply store sells paper clips in packs of 100 and packs of 250. If the store has 84 packs of paper clips in stock totaling 12,300 paper clips, how many paper clips would a customer buy if he buys half of the packs of 250 that the store has in stock?

A) 2,900

B) 3,250

C) 5,800

D) 6,500

17. In a college art class, 76 students are painting a mural on one wall of the campus amphitheater. The wall has been divided into 23 sections, and each section will be painted by a group of either 2 or 4 students. How many more sections will be painted by a group of 4 students than by a group of 2 students?

A) 6

B) 7

C) 8

D) 9

$$\frac{5}{8}x + \frac{7}{2}y = \frac{3}{2}$$

$$\frac{1}{6}x - \frac{2}{3}y = 1$$

18. If the ordered pair (x, y) satisfies the system of equations above, what is the sum of the values of x and y?

A) $\frac{5}{24}$

B) $\frac{5}{2}$

C) $\frac{29}{8}$

D) $\frac{33}{8}$

$$11x - 24y = 8$$

$$kx - 36y = 5$$

19. In the system of equations above, k is a constant. If the system has no solution, what is the value of k?

20. Two moonflower vines are growing on a trellis in Mallory's backyard. She bought the first vine when it was 11 inches long and found that it grows at a rate of 0.25 inches per day. Exactly 20 days later, Mallory bought the second vine, which started at 24 inches long and has a growth rate of 0.125 inches per day. How many days will Mallory have had the first vine when the lengths of the two vines are the same?

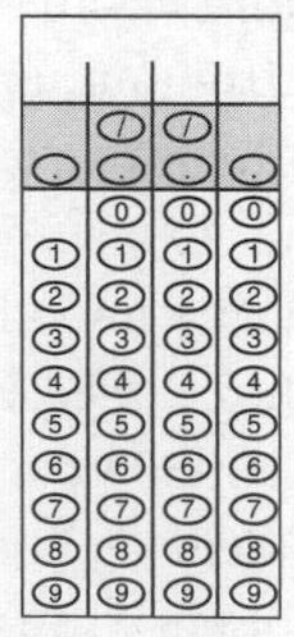

ANSWER KEY

1. A	**6. C**	**11. A**	**16. B**
2. D	**7. D**	**12. B**	**17. B**
3. B	**8. A**	**13. 80**	**18. D**
4. D	**9. B**	**14. 8**	**19. 16.5** or **33/2**
5. 10	**10. B**	**15. C**	**20. 84**

ANSWERS & EXPLANATIONS

1. A

Difficulty: Easy

Category: Heart of Algebra / Systems of Linear Equations

Strategic Advice: When solving a system of linear equations, always look at the coefficients of the variables to decide on a strategy. Here, because you're looking for the value of *y*, subtracting the first equation from the second equation will get you to the answer the quickest (because the *x*'s will be eliminated and you'll be left with *y*).

Getting to the Answer: Rewrite the equations with the second equation on top. Then, carefully subtract:

$$\begin{array}{r} x + 2y = 12 \\ -(x + y = 29) \\ \hline y = -17 \end{array}$$

The correct answer is (A).

2. D

Difficulty: Easy

Category: Heart of Algebra / Systems of Linear Equations

Strategic Advice: Don't overthink a question like this. You do not need to write the equations of the lines. The solution to a system of linear equations represented graphically is the point of intersection, which you can clearly see on the graph.

Getting to the Answer: According to the graph, the lines intersect, or cross each other, at (3, −1). The question asks for the value of $x - y$, so subtract to arrive at the answer $3 - (-1) = 3 + 1 = 4$, which is (D).

3. B

Difficulty: Easy

Category: Heart of Algebra / Systems of Linear Equations

Strategic Advice: Whenever you're asked for an unusual quantity, rather than just the value of one variable or the other, there's a good chance that a time-saving shortcut (like simply adding or subtracting the equations) exists.

Getting to the Answer: Start by rewriting the equations so that they are in the same form. After doing so, if you look closely at the coefficients, you'll see that adding the two equations will yield exactly what you're asked to find:

$$\begin{array}{rr} 5b = 6a + 16 \rightarrow & -6a + 5b = 16 \\ 9a = 7b - 20 \rightarrow & 9a - 7b = -20 \\ \hline & 3a - 2b = -4 \end{array}$$

The sum of the two equations tells you that $3a - 2b$ is −4, making (B) the correct answer.

4. D

Difficulty: Easy

Category: Heart of Algebra / Systems of Linear Equations

Strategic Advice: When a question about a system of equations asks for the *number* of solutions, rather than the actual solution, don't worry about finding the value of x and y right away—you might not even need them to answer the question. Just simplify the two equations as much as possible, and then compare them.

Getting to the Answer: For each equation, combine like terms and write the results in standard form ($Ax + By = C$):

$$4x - 2y + 3 = 8 \rightarrow 4x - 2y = 5$$

$$3x + 6y = 8y - x + 5 \rightarrow 4x - 2y = 5$$

The equations are identical (dependent) and therefore represent the same line, which means the system has infinitely many solutions, making (D) correct.

5. 10

Difficulty: Easy

Category: Heart of Algebra / Systems of Linear Equations

Strategic Advice: When a word problem presents you with two "types" of information (for example, the total number of employees versus how many are men and how many are women), creating a system of equations is usually necessary. Set up two equations each with two unknowns using variables that make sense.

Getting to the Answer: There are 27 total employees at the office, all either men or women, so m (the number of men) + w (the number of women) = 27. There are seven more women than men, so $w = m + 7$. Substitute $m + 7$ for w into the first equation:

$$\begin{aligned} m + w &= 27 \\ m + (m + 7) &= 27 \\ 2m + 7 &= 27 \\ 2m &= 20 \\ m &= 10 \end{aligned}$$

The question asks for the number of men (m), so you're done! There are 10 men in the office.

6. C

Difficulty: Medium

Category: Heart of Algebra / Systems of Linear Equations

Strategic Advice: Use the Kaplan Strategy for Translating English into Math to write two equations that represent the scenario. Use variables that make sense to help you stay organized (such as a for adult and c for child).

Getting to the Answer: Use the information given to write two equations in words first:

Two adult fares plus 3 child fares equals 14 $\rightarrow$ $2a + 3c = 14$

One child fare equals one-half an adult fare $\rightarrow$ $c = \frac{1}{2}a$

If you don't like working with fractions, multiply both sides of the second equation by 2 to get $2c = a$. Then, plug $2c$ in for a into the first equation and solve for c:

$$\begin{aligned} 2a + 3c &= 14 \\ 2(2c) + 3c &= 14 \\ 7c &= 14 \\ c &= 2 \end{aligned}$$

The cost of a child's fare is \$2, which is half the cost of an adult's fare. This means an adult's fare costs \$4. Don't fall for the trap answer, which is A. Add the two fares together to find that the total cost for one adult and one child is \$6, which is (C).

7. D

Difficulty: Medium

Category: Heart of Algebra / Systems of Linear Equations

Strategic Advice: When both equations in a system are written in the form $y = mx + b$, use substitution or graphing to find the solution to the system. Use whichever method gets you to the answer more quickly.

Getting to the Answer: Substitute the first expression for y into the second equation. Then clear the fractions by multiplying everything by 35 to make the numbers easier to work with, and solve for x:

$$\begin{aligned}\frac{1}{5}x + 4 &= \frac{3}{7}x - 4\\ 35\left(\frac{1}{5}x\right) + 35(4) &= 35\left(\frac{3}{7}x\right) - 35(4)\\ 7x + 140 &= 15x - 140\\ 280 &= 8x\\ 35 &= x\end{aligned}$$

The question asks for the value of y, so substitute 35 into either of the original equations and solve for y: $y = \frac{1}{5}(35) + 4 = 7 + 4 = 11$, which is (D).

8. A

Difficulty: Medium

Category: Heart of Algebra / Systems of Linear Equations

Strategic Advice: Whenever you're only asked to identify the quadrant in which the graphs of two lines intersect (not the exact values of x and y), drawing a quick sketch of the system of equations is likely to be the most time-efficient method to answer the question.

Getting to the Answer: Start by writing each equation in slope-intercept form ($y = mx + b$):

$$\begin{aligned}2x - 3y = -3 &\rightarrow -3y = -2x - 3\\ &\rightarrow y = \frac{2}{3}x + 1\\ -12 = -4x + y &\rightarrow y = 4x - 12\end{aligned}$$

Then plot each y-intercept and use the corresponding slope to rise (or fall) and run to the next point. You don't have to be exact—no need to waste time drawing tick marks or labeling the graph—a sketch is all you need. (You can also graph the lines in your calculator if you prefer.) Your graph might look something like this:

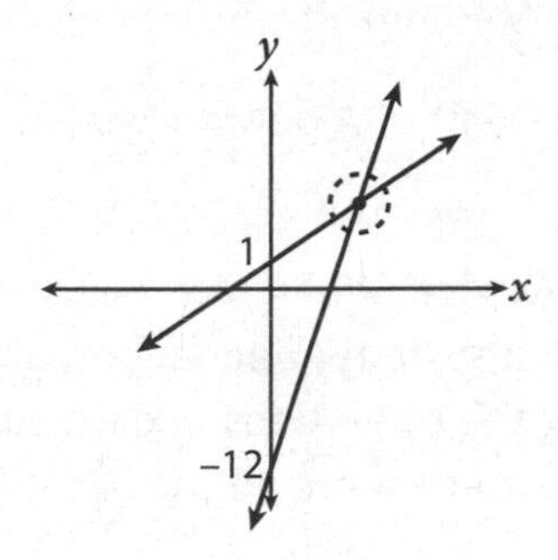

The lines intersect in quadrant I, which is (A).

9. B

Difficulty: Medium

Category: Heart of Algebra / Systems of Linear Equations

Strategic Advice: Sometimes, conventional substitution will get you to the answer, but there may be a faster (but cleverly hidden) alternate route. The clue in this question is the presence of $6b$ in both equations.

Getting to the Answer: With a little rearranging, you can isolate $a - 6b$ in the first equation, substitute 34 (from the second equation), and end up with only one equation and one variable. Start by moving $6b$ to the left side, and then rewrite $10a$ as $9a + a$. Once there, you can substitute and solve for a:

$$\begin{aligned} 10a &= 6b + 7 \\ 10a - 6b &= 7 \\ 9a + (a - 6b) &= 7 \\ 9a + 34 &= 7 \\ 9a &= -27 \\ a &= -3 \end{aligned}$$

You're asked to find $-\frac{1}{3}a$, so multiply -3 by $-\frac{1}{3}$ to get 1, making (B) the correct answer.

10. B

Difficulty: Medium

Category: Heart of Algebra / Systems of Linear Equations

Strategic Advice: When matching a real-world scenario to a system of equations, always look carefully at the operations in each equation as well as the coefficients on the variables.

Getting to the Answer: You're given two variables: b for bags checked and p for priority boarding upgrades. The question states that 145 travel services were charged for two flights combined (not on each flight), so the first equation should be $b + p = 145$. Eliminate A and D. Similarly, the airline collected $3,065 in fees on these two flights. Because a checked bag costs $25 and priority boarding costs $15, the second equation is $25b + 15p = 3{,}065$. Choice (B) is the only answer that contains both correct equations.

11. A

Difficulty: Medium

Category: Heart of Algebra / Systems of Linear Equations

Strategic Advice: In a lengthy word problem that contains lots of numbers, it may help to cross out information that does not impact the answer.

Getting to the Answer: Here, the number of items in each box does not matter because the entire question is about the boxes of items (not the number of items), so cross out the 12. This means you can eliminate B and D right away. The question states that r and l represent the number of boxes of raspberry scones and lemon poppy seed muffins sold and that the shop sold a total of 46 boxes of these items, so the first equation will be $r + l = 46$. The shop collects $15 for each box of raspberry scones, which translates to $15r$, and $9 for each box of lemon poppy seed muffins, which becomes $9l$. Put together, the second equation is $15r + 9l = 396$, making (A) the correct answer. Note that because the question doesn't ask about the number sold per day, you don't need to incorporate a 2 into either equation.

12. B

Difficulty: Medium

Category: Heart of Algebra / Systems of Linear Equations

Strategic Advice: When you're given two types of information, write and solve a system of equations to answer the question. Before selecting your answer, be sure to check that you answered the *right* question.

Getting to the Answer: Let b and k represent the number of banana and kiwi smoothies sold. Tricia sold 40 smoothies in all, so $b + k = 40$. She sold banana smoothies for $2 each and kiwi smoothies for $2.50 each for a total of $87, so $2b + 2.5k = 87$. Use either combination (elimination) or substitution to solve for b and k. To use substitution, solve the first equation for either variable ($b + k = 40 \rightarrow b = 40 - k$) and substitute the result into the second equation:

$$\begin{aligned} 2b + 2.5k &= 87 \\ 2(40 - k) + 2.5k &= 87 \\ 80 - 2k + 2.5k &= 87 \\ 0.5k &= 7 \\ k &= 14 \end{aligned}$$

From before, $b = 40 - k \rightarrow 40 - 14 = 26$, so Tricia sold 26 banana smoothies and 14 kiwi smoothies. Don't stop yet and don't fall for the trap answer, which is A. The question asks how much more money was made on the banana smoothies than the kiwi smoothies (not how many more banana smoothies were sold than kiwi smoothies). Tricia made $\$2 \times 26 = \$52$ on the banana smoothies and $\$2.50 \times 14 = \$35$ on the kiwi flavor, for a difference of $52 - 35 = \$17$. (B) is the correct answer.

13. 80
Difficulty: Medium

Category: Heart of Algebra / Systems of Linear Equations

Strategic Advice: Use the Kaplan Strategy for Translating English into Math to build a system of equations. Be sure to use consistent units throughout. Use cents (rather than dollars) to avoid having to work with decimals.

Getting to the Answer: Call the number of 25-cent newspapers x and the number of 40-cent newspapers y. Because the vendor sold 100 newspapers in total, $x + y = 100$. Also, the total cost of these newspapers was \$28.00, or 2,800 cents. So $25x + 40y = 2{,}800$. The question asks for x, the number of 25-cent newspapers sold. From the first equation, $y = 100 - x$. Substitute this value for y into the second equation, and solve for x:

$$\begin{aligned} 25x + 40y &= 2{,}800 \\ 25x + 40(100 - x) &= 2{,}800 \\ 25x + 4{,}000 - 40x &= 2{,}800 \\ -15x + 4{,}000 &= 2{,}800 \\ -15x &= -1{,}200 \\ x &= -1{,}200 \div (-15) \\ x &= 80 \end{aligned}$$

The question asks for the number of 25-cent newspapers sold (which is x), so you're all done. Grid in 80 and move on to the next question.

14. 8
Difficulty: Medium

Category: Heart of Algebra / Systems of Linear Equations

Strategic Advice: When answering a Grid-in question, it is particularly important to check that you answered the *right* question before you grid in your answer.

Getting to the Answer: Use substitution or combination (elimination) to find the solution to the system. If you choose combination, the second equation needs to be rearranged so that like terms are lined up vertically, and you'll need to multiply the first equation by 4 so that the x terms are eliminated when you add the equations:

$$\begin{aligned} 4[x + y = -6] &\rightarrow \quad 4x + 4y = -24 \\ y - 4x = 4 \quad &\rightarrow \quad \underline{-4x + y = 4} \\ &\qquad\qquad 5y = -20 \\ &\qquad\qquad\ \ y = -4 \end{aligned}$$

From the first equation, $x + y = -6$ gives $x + (-4) = -6$, so $x = -2$. The question asks for the value of xy, so multiply to get $(-2) \times (-4) = 8$.

15. C
Difficulty: Hard

Category: Heart of Algebra / Systems of Linear Equations

Strategic Advice: A system of equations that has infinitely many solutions indicates that the equations in the system actually represent the same line. This means you can manipulate one of the equations to look exactly like the other one.

Getting to the Answer: The easiest way to compare the equations is to rewrite them both in slope-intercept form ($y = mx + b$).

$$4x + 7y = 24 \rightarrow 7y = -4x + 24$$
$$\rightarrow y = -\frac{4}{7}x + \frac{24}{7}$$
$$6x + \frac{21}{2}y = g \rightarrow \frac{21}{2}y = -6x + g$$
$$\rightarrow y = \frac{2}{21} \cdot (-6x) + \frac{2}{21}g$$

Simplifying the x term in the second equation yields $-\frac{4}{7}x$ (which it should since the equations represent the same line), so all that's left to do is to set the y-intercepts (the constants) equal and solve for g:

$$\frac{2}{21}g = \frac{24}{7}$$
$$\frac{21}{2} \cdot \frac{2}{21}g = \frac{24}{7} \cdot \frac{21}{2}$$
$$g = \frac{12 \times 2}{7} \cdot \frac{7 \times 3}{2}$$
$$g = 12 \times 3$$
$$g = 36$$

The correct answer is (C).

16. B
Difficulty: Hard

Category: Heart of Algebra / Systems of Linear Equations

Strategic Advice: Create a system of linear equations where x represents the number of packs with 100 paper clips and y represents the number of packs with 250 paper clips. Before selecting your final answer, make sure you are answering the *right* question (how many paper clips a customer would buy if he buys half of the packs of 250).

Getting to the Answer: The first equation should represent the total number of *packs*, each with 100 or 250 paper clips, or $x + y = 84$. The second equation should represent the total number of *paper clips*. Because x represents packs with 100 paper clips and y represents packs with 250 paper clips, the second equation is $100x + 250y = 12{,}300$. Now solve the system using substitution. Solve the first equation for either variable, and substitute the result into the second equation:

$$x + y = 84$$
$$x = 84 - y$$
$$100(84 - y) + 250y = 12{,}300$$
$$8{,}400 - 100y + 250y = 12{,}300$$
$$8{,}400 + 150y = 12{,}300$$
$$150y = 3{,}900$$
$$y = 26$$

The store has 26 packs of 250 paper clips in stock. The question asks about packs of 250, so you don't need to find the value of x. You aren't done yet, however. The problem asks how many paper clips a customer would buy if he buys half of the packs of 250 the store has. The customer would buy $13 \times 250 = 3{,}250$ paper clips, which is (B).

17. B
Difficulty: Hard

Category: Heart of Algebra / Systems of Linear Equations

Strategic Advice: Assign variables to the unknowns. Then translate from English into math to create a system of equations—one equation for the number of *sections* and one equation for the number of *students*.

Getting to the Answer: Let t represent the number of sections of the wall that will be painted by a group of 2 students and f represent the number of sections that will be painted by a group of 4 students. Because there are 23 sections that will be painted by 76 students, it follows that $t + f = 23$ and $2t + 4f = 76$. Solve this system of equations using substitution: Solve the first equation for

t to get $t = 23 - f$. Then substitute this expression into the second equation for t and solve for f:

$$2(23 - f) + 4f = 76$$
$$46 - 2f + 4f = 76$$
$$46 + 2f = 76$$
$$2f = 30$$
$$f = 15$$

The number of groups of 4 students is 15, which means the number of groups of 2 students is $t = 23 - 15 = 8$. Therefore, $15 - 8 = 7$ more of the sections will be painted by a group of 4 students than by a group of 2 students, making (B) correct.

18. D
Difficulty: Hard

Category: Heart of Algebra / Systems of Linear Equations

Strategic Advice: When a system of equations involves lots of fractions, multiply each equation by its least common denominator to clear the fractions.

Getting to the Answer: Multiply the first equation by 8 and the second equation by 6 to clear the fractions:

$$8\left(\frac{5}{8}x + \frac{7}{2}y = \frac{3}{2}\right) \rightarrow 5x + 28y = 12$$
$$6\left(\frac{1}{6}x - \frac{2}{3}y = 1\right) \rightarrow x - 4y = 6$$

Now, examine the new coefficients. Notice that multiplying the second equation by 7 (which yields $7x - 28y = 42$) and adding the two equations will eliminate the y terms, leaving a single equation with one variable:

$$\begin{array}{r} 5x + 28y = 12 \\ 7x - 28y = 42 \\ \hline 12x = 54 \end{array}$$
$$x = \frac{54}{12} = \frac{9}{2}$$

Substitute $\frac{9}{2}$ for x into the easier of the modified equations (the equations without the fractions), and solve for y:

$$\frac{9}{2} - 4y = 6$$
$$-4y = \frac{3}{2}$$
$$y = \frac{3}{2} \cdot \left(-\frac{1}{4}\right) = -\frac{3}{8}$$

The question asks for the sum of x and y, so add to get $\frac{9}{2} + (-\frac{3}{8}) = \frac{36}{8} - \frac{3}{8} = \frac{33}{8}$, which is (D).

19. 16.5 or **33/2**
Difficulty: Hard

Category: Heart of Algebra / Systems of Linear Equations

Strategic Advice: A system of linear equations that has no solution indicates two parallel lines (because parallel lines never intersect).

Getting to the Answer: Parallel lines have equal slopes, so rewrite each equation in slope-intercept form ($y = mx + b$) and set their slopes equal to each other:

$$11x - 24y = 8 \rightarrow -24y = -11x + 8$$
$$\rightarrow y = \frac{11}{24}x - \frac{8}{24}$$
$$kx - 36y = 5 \rightarrow -36y = -kx + 5$$
$$\rightarrow y = \frac{k}{36}x - \frac{5}{36}$$

The slope of the first line is $\frac{11}{24}$ and the slope of the second line is $\frac{k}{36}$. Set the slopes equal and cross-multiply to solve for k:

$$\frac{11}{24} = \frac{k}{36}$$
$$24k = 11(36)$$
$$24k = 396$$
$$k = 16.5$$

Grid in 16.5 (or 33/2).

20. 84
Difficulty: Hard

Category: Heart of Algebra / Systems of Linear Equations

Strategic Advice: Use what you know about real-world scenarios and linear equations to write a system of equations. In a real-world scenario, a rate is the same as slope and a starting amount is the *y*-intercept.

Getting to the Answer: Let d represent the number of days of vine growth for the *second* vine, and let l represent the length of each vine. Mallory's first vine grows at a rate of 0.25 inches per day and was 11 inches long when she bought it (20 days before d), so the length of the first vine can be represented by the equation $l = 0.25(d + 20) + 11$, which simplifies to $l = 0.25d + 16$. The second vine started at 24 inches and grows 0.125 inches per day, so the equation that represents its length is $l = 0.125d + 24$. To determine when the lengths of the vines will be the same, set the two equations equal to each other and solve for d:

$$0.25d + 16 = 0.125d + 24$$
$$0.125d = 8$$
$$d = 64$$

Don't stop yet—the question asked how many days Mallory had the *first* vine when the lengths of the two vines are the same, so add 20 to d to account for the time that she had the first vine before buying the second vine. This means the correct answer is 84.

PRACTICE SET 3: INEQUALITIES

Solving Inequalities

Working with inequalities is similar to working with equations, but with a few key differences.

- The language used to describe inequalities tends to be more complex than the language used to describe equations. You "solve" an equation for x, but with an inequality, you might be asked to "describe all possible values of x" or provide an answer that "includes the entire set of solutions for x." This difference in wording exists because an equation describes a specific value of a variable, whereas an inequality describes a range of values.
- Instead of an equal sign, you'll see a sign denoting either "greater/less than" ($>$ and $<$) or "greater/less than or equal to" ($\geq$ and $\leq$). The open end of the inequality symbol should always point toward the greater quantity.
- When solving an inequality that involves multiplying or dividing by a negative number, the inequality symbol must be reversed. For example, if given $-4x < 12$, you must reverse the symbol when dividing by -4, which will yield $x > -3$ (NOT $x < -3$).

Compound Inequalities

Sometimes you'll see a variable expression wedged between two quantities. This is called a compound inequality. For example, $-5 < 2x + 1 < 11$ is a compound inequality. You solve it the same way (using inverse operations), keeping in mind that whatever you do to one piece, you must do to all three pieces. And, of course, if you multiply or divide by a negative number, you must reverse both inequality symbols.

Graphing Inequalities

Inequalities can be presented graphically in one or two dimensions. In one dimension, inequalities are graphed on a number line with a shaded region. For example, $x > 1$ is graphed like this:

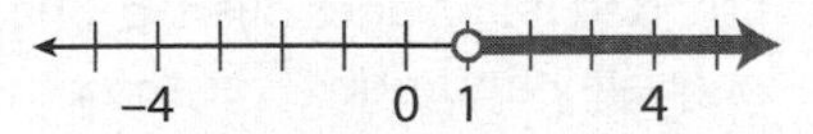

Notice the open dot at 1, which indicates that 1 is not a solution to the inequality. This is called a strict inequality. By contrast, the graph of $x \leq 0$ looks like this:

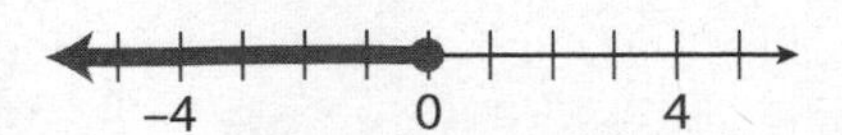

Notice the solid dot at 0, which indicates that 0 should be included in the solution set for the inequality.

In two dimensions, things get a bit more complicated. While linear equations graph as simple lines, *inequalities* graph as lines called *boundary lines* with shaded regions known as *half planes*. Solid lines indicate inequalities that have $\leq$ or $\geq$ because the values on the line itself are included in the solution set. Dashed lines involve strict inequalities that have > or < because, in these cases, the values on the line itself are not included in the solution set. The shaded region (and the line if it is solid) represents all points that make up the solution set for the inequality. For example, the graph below represents the solution set to the inequality $y < \frac{1}{4}x - 3$.

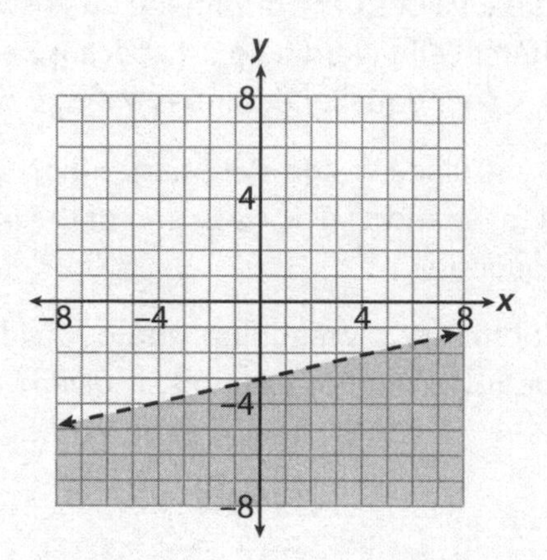

If you're not sure whether to shade above or below a boundary line, plug a pair of coordinates into the inequality. If the coordinates satisfy the inequality, then the region in which that point lies should be shaded. If the coordinates violate the inequality, then that region should not be shaded. An easy point to use when testing inequalities is the origin, (0, 0). In the example above $0 \nless -3$, so the half plane that contains (0, 0) should not be shaded (which it isn't).

Solving Systems of Inequalities

Multiple inequalities can be combined to create a system of inequalities. Systems of inequalities can also be represented graphically with multiple boundary lines and multiple shaded regions. Follow the same rules for graphing single inequalities, but keep in mind that the solution set to the system is the region where the shaded half planes overlap. Shading in different directions (e.g., parallel lines slanted up for one inequality and down for the other) makes the overlap easier to see. Just as with single inequalities, you can pick coordinates to plug into a system of inequalities to determine which side of the boundary lines should be shaded.

PRACTICE SET

Easy

1. Which of the following gives all values of j that satisfy the inequality $3j - 4 \leq 6j + 11$?

A) $j \leq -5$

B) $j \geq -5$

C) $j \leq 5$

D) $j \geq 5$

2. Which of the following numbers is not a solution to the inequality $6x - 9 \geq 7x - 5$?

A) −8

B) −5

C) −4

D) −2

3. To take the neighbor's children to the movies, Mellie charges $5 for gas and $8 per hour spent with the children. Ron charges $3 for gas and $8.50 per hour spent with the children. If h represents the number of hours spent with the children, what are all the values of h for which Ron's total charge is greater than Mellie's total charge?

A) $h < 3$

B) $3 < h \leq 4$

C) $4 \leq h < 5$

D) $h > 4$

4. Yasmine is a pharmaceutical sales representative. Her firm gives her a weekly allowance of $300 to spend on lunches with physicians and their office staffs. A restaurant from which Yasmine often buys the lunches charges $7 for a cold dish and $11 for a hot dish, including drinks. If each meal is subject to a 5.75% sales tax, which of the following inequalities represents the number of cold dishes (c) and hot dishes (h) that Yasmine can purchase for sales-related lunches in one week, assuming she purchases all the lunches from this restaurant?

A) $7c + 11h \leq 1.0575(300)$

B) $7c + 11h \geq 1.0575(300)$

C) $1.0575(7c + 11h) \leq 300$

D) $1.0575(7c + 11h) \geq 300$

5. If $\frac{5}{6} < \frac{1}{2}x - \frac{1}{2}y < \frac{3}{2}$, then what is one possible value of $x - y$?

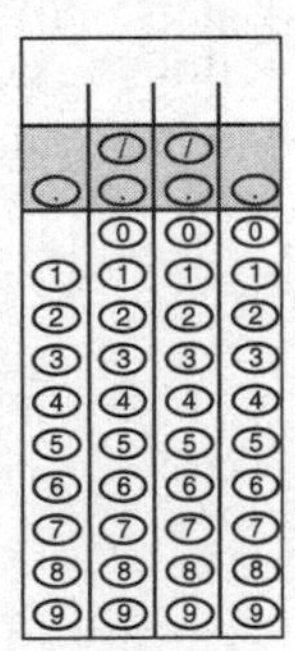

Medium

6. What is the least possible integer value for which 40% of that integer is greater than 9.6?

A) 4

B) 12

C) 20

D) 25

7. Body mass index (BMI) is a comparison of a person's body mass to his or her height. A high BMI can be an indicator of high body fat, which can lead to health problems. According to the American Heart Association, an adult is underweight if his or her BMI is less than 18.5, or overweight if it is greater than or equal to 25.0. Which of the following number lines could be used to model a healthy BMI range for an adult?

A) 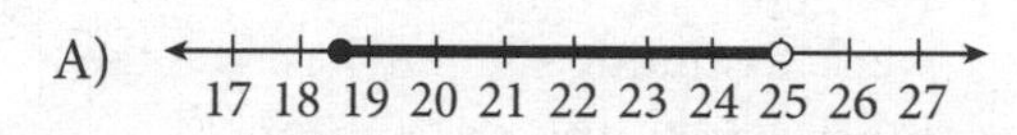

B) 17 18 19 20 21 22 23 24 25 26 27

C) 17 18 19 20 21 22 23 24 25 26 27

D) 17 18 19 20 21 22 23 24 25 26 27

$$\frac{2(4k+1)}{3} \geq \frac{k(6+5)-3}{2}$$

8. Which of the following correctly describes the possible values of k in the inequality above?

A) $k \leq -1$

B) $k \geq -1$

C) $k \leq \frac{13}{17}$

D) $k \geq \frac{13}{17}$

9. A math teacher decides to create several practice tests for her students before they take the SAT. She wants to make some non-calculator tests and some calculator tests so that her students will be able to practice both. She figures that each non-calculator test will take her 3 hours to create, and each calculator test will take 4 hours to create. If she is willing to devote at most 6 hours per week of her time for the next 5 weeks to create the practice tests, and she wants to provide at least 8 practice tests, which of the following systems of inequalities can help her determine how many of each type of test she can create?

A) $n + c \geq 8$
$3n + 4c \leq 6$

B) $n + c \geq 8$
$3n + 4c \leq 30$

C) $n + c \leq 8$
$3n + 4c \geq 30$

D) $n + c \geq 6$
$3n + 4c \leq 8$

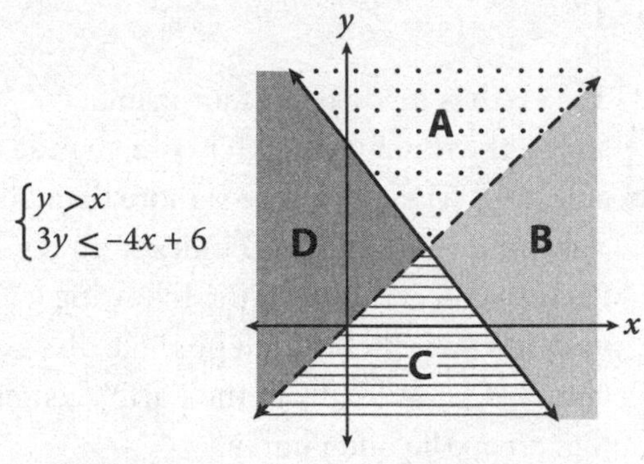

10. A system of inequalities and the corresponding graph are shown above. Which part of the graph could represent all of the solutions to the system?

    A) A

    B) B

    C) C

    D) D

11. A housing down payment is money that a prospective buyer provides up front when purchasing a home and is usually a percent of the purchase price of the home. A lender typically requires private mortgage insurance (PMI) when the buyer's down payment is less than 20% of the purchase price. To secure a mortgage, buyers also need to have additional cash on hand for closing costs and prepaid property tax. Suppose a buyer wants to purchase a \$375,000 house and must have \$7,200 on hand for closing costs and property tax. Which of the following inequalities represents the total funds (f) the buyer must have on hand to secure the mortgage without having to pay PMI?

    A) $f \le 0.2(375{,}000) + 7{,}200$

    B) $f \ge 0.2(375{,}000) + 7{,}200$

    C) $f \le 0.2(375{,}000 + 7{,}200)$

    D) $f \ge 0.2(375{,}000 + 7{,}200)$

12. If $-\frac{5}{2} < -2m + 1 < -\frac{7}{5}$, what is the greatest possible integer value of the expression $10m - 5$?

    A) 6

    B) 7

    C) 10

    D) 12

$$18{,}000 + x \le 72{,}000$$

13. The federal interstate weight limit for a particular four-axle transfer truck is 18,000 pounds per axle. The cab (front) of the truck weighs 11,000 pounds, and the trailer of the truck, when empty, weighs 7,000 pounds. The inequality above represents the legally permissible weight range for this truck when travelling on an interstate. What does the value 18,000 represent in the inequality?

    A) The weight of the truck when fully loaded

    B) The weight of the truck when the trailer is empty

    C) The maximum weight allowed per axle

    D) The maximum weight of the cargo being transported

14. A construction company prepares an estimate to install a new pool for a homeowner. The estimate includes h hours of labor, where $h > 80$. The company's goal is for the estimate to be within 8 hours of the actual number of hours of labor. If the company meets the goal and it takes a hours of actual labor, which inequality represents the relationship between the estimated number of hours of labor and the actual number of hours of labor?

A) $a + h \leq 8$

B) $a \geq h + 8$

C) $a \leq h - 8$

D) $-8 \leq a - h \leq 8$

15. Margo can peel and slice at least 10 dozen apples per hour and at most 15 dozen apples per hour. Based on this information, what is a possible amount of time, in hours, that it could take Margo to peel and slice 60 dozen apples?

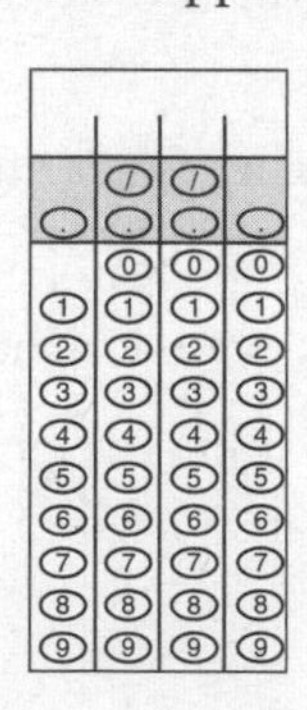

Hard

16. The earth is made up of four primary layers: the inner core, outer core, mantle, and crust. The outer core is more than 800 miles and less than 2,200 miles from the Earth's center. Which of the following inequalities represents all possible distances, d, in miles, from the Earth's center that are in the outer core?

A) $|d + 800| < 2,200$

B) $|d - 800| < 2,200$

C) $|d + 1,500| < 700$

D) $|d - 1,500| < 700$

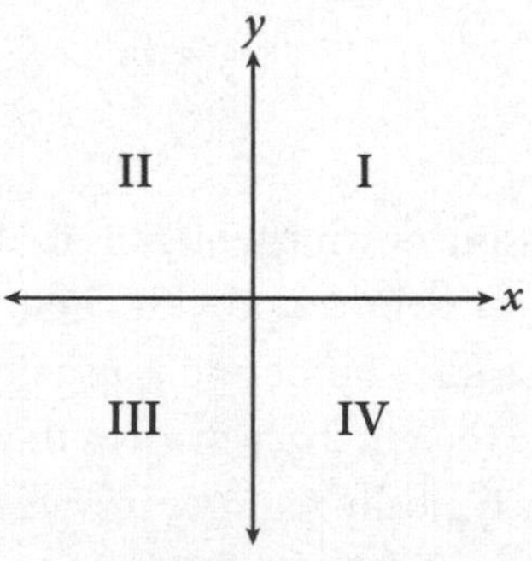

17. If the system of inequalities given by $x + 3y \leq 12$ and $2x - 3y \leq -3$ is graphed on the coordinate plane above, which quadrant of the plane contains no solutions to the system?

A) Quadrant I

B) Quadrant IV

C) Quadrants I and IV

D) Quadrants III and IV

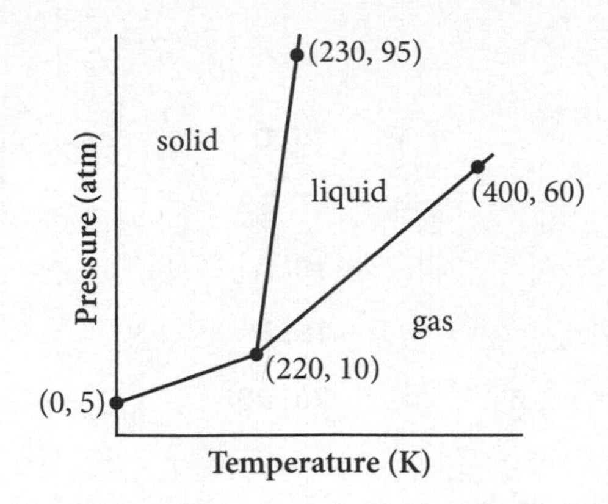

18. A phase diagram shows the temperatures and pressures at which a chemical substance exists in a certain phase (solid, liquid, or gas). A sample phase diagram for a fictional substance is shown in the figure, where T is temperature in Kelvin and P is pressure in atmospheres. Where on the diagram a certain temperature and pressure combination falls determines the state(s) in which the substance exists. For example, if a certain temperature-pressure pairing falls in the gas area (but not on the line segment between gas and liquid), the substance exists only as a gas. However, if the temperature-pressure pairing is on the line segment between gas and liquid, the substance exists as both a gas and a liquid. Assuming $T > 0$ and $P > 0$, which of the following systems of inequalities could be used to describe the temperature and pressure ranges in which this substance exists only as a liquid?

A) $P < \frac{17}{2}T - 1{,}860;\ P > \frac{5}{18}T - \frac{460}{9}$

B) $P > \frac{17}{2}T - 1{,}860;\ P < \frac{5}{18}T - \frac{460}{9}$

C) $P < \frac{17}{2}T - 1{,}860;\ P > \frac{11}{200}T - \frac{460}{9}$

D) $P \geq \frac{17}{2}T - 1{,}860;\ P \leq \frac{11}{200}T - \frac{460}{9}$

$$y < x + k_1$$
$$y > 2x + k_2$$

19. Suppose that on a coordinate plane, (0, 0) is a solution to the system of inequalities given above. Which of the following conclusions about k_1 and k_2 must be true?

A) $k_1 < k_2$

B) $k_2 < k_1$

C) $|k_1| < |k_2|$

D) $k_1 = -k_2$

20. The variables x and y represent numbers for which the statements $x - y > 300$ and $\frac{y}{x} = 0.625$ are true. What is the smallest integer that x can equal?

ANSWER KEY

1. B	**6. D**	**11. B**	**16. D**
2. D	**7. A**	**12. D**	**17. B**
3. D	**8. C**	**13. B**	**18. A**
4. C	**9. B**	**14. D**	**19. B**
5. $\frac{5}{3} < x - y < 3$	**10. D**	**15.** $4 \le r \le 6$	**20. 801**

ANSWERS & EXPLANATIONS

1. B
Difficulty: Easy

Category: Heart of Algebra / Inequalities

Strategic Advice: When an inequality has only one variable, straightforward algebra is the best route to the answer.

Getting to the Answer: Solve for j by moving the variable terms to one side of the inequality and the constants to the other, remembering to flip the inequality symbol if you multiply or divide by a negative number:

$$3j - 4 \le 6j + 11$$
$$-3j \le 15$$
$$\frac{-3j}{-3} \ge \frac{15}{-3}$$
$$j \ge -5$$

(B) is the correct answer.

2. D
Difficulty: Easy

Category: Heart of Algebra / Inequalities

Strategic Advice: Substituting each answer choice into the inequality to see if it produces a true statement is not the most time-efficient strategy here, especially due to all the negative numbers in the answer choices. Instead, use inverse operations to solve the inequality and then check the answer choices.

Getting to the Answer: Subtracting $6x$ and adding 5 to both sides of $6x - 9 \ge 7x - 5$ yields $-4 \ge x$. Turn the solution around so it's easier to read (keep the small end of the inequality pointed toward the x): $x \le -4$. This tells you that x is a solution if and only if x is less than or equal to -4. The question asks which value is *not* a solution, and the only answer choice that is not less than or equal to -4 is -2, making (D) the correct answer.

3. D
Difficulty: Easy

Category: Heart of Algebra / Inequalities

Strategic Advice: Use the Kaplan Strategy for Translating English into Math to write an expression (in words first) that represents each person's total charge.

Getting to the Answer: If h represents the number of hours spent with the children, and Mellie charges \$8 per hour, then she charges a total of 8 times the numbers of hours ($8h$) plus the flat fee for gas (\$5). Use the expression $8h + 5$ to represent Mellie's total charge. Similarly, Ron charges a total of 8.5 times the numbers of hours ($8.5h$) plus the flat fee for gas (\$3), so use the expression $8.5h + 3$ to represent Ron's total charge. Now, set Ron's expression greater (>) than Mellie's (per the question) and solve for h.

To make the numbers easier to work with, you can multiply each term by 10 (which moves the decimal point one place to the right):

$$8.5h + 3 > 8h + 5$$
$$85h + 30 > 80h + 50$$
$$5h > 20$$
$$h > 4$$

The correct inequality is given in (D).

4. C
Difficulty: Easy

Category: Heart of Algebra / Inequalities

Strategic Advice: Numbers sprinkled throughout a wordy question tell you to use the Kaplan Strategy for Translating English into Math. Note that in this question, you're only asked for an inequality and not a simplified solution set.

Getting to the Answer: The variables c and h represent the number of cold and hot dishes purchased, which cost $7 and $11 each, respectively. Therefore, the amount spent on cold dishes is $7c$, and the amount spent on hot dishes is $11h$. Sales tax is 5.75%, so the total spent on sales-related lunches is $1.0575(7c + 11h)$. Yasmine has a weekly allowance of $300 for lunches, so she can spend that much or less, making the inequality $1.0575(7c + 11h) \leq 300$, which is (C).

5. Any value between $\frac{5}{3}$ and 3
Difficulty: Easy

Category: Heart of Algebra / Inequalities

Strategic Advice: When there is only one equation (or inequality), but two variables, you can't solve for either variable. Instead, try to manipulate what is given to make it look like what you want (usually through multiplication or division).

Getting to the Answer: Compare the variable terms between the two inequality symbols to the expression you're looking for: $\frac{1}{2}x - \frac{1}{2}y$ is exactly half of $x - y$, so multiply the entire inequality by 2. The inequality becomes $\frac{5}{3} < x - y < 3$. Grid in any value between $1\frac{2}{3}$ and 3 (but not $1\frac{2}{3}$ or 3), such as 2 or 2.5.

6. D
Difficulty: Medium

Category: Heart of Algebra / Inequalities

Strategic Advice: Sometimes, questions are much easier than they sound. Translate from English into math and you'll see that this question actually involves solving a fairly simple inequality.

Getting to the Answer: Call the unknown integer x. Translate "40% of that integer" as $0.4x$. The rest is very straightforward: $0.4x > 9.6$. Divide both sides of the inequality by 0.4 to find that $x > 24$. The question specifically says *greater* (not greater than or equal to), so the correct answer is 25, which is (D).

7. A
Difficulty: Medium

Category: Heart of Algebra / Inequalities

Strategic Advice: When you see phrases like "less than" or "greater than," write an inequality to describe the situation. Then graph your inequality on a number line by shading to the left for $<$ or $\leq$ and shading to the right for $>$ or $\geq$.

Getting to the Answer: A person with a BMI that is less than 18.5 is underweight. If b represents BMI, the BMI of an underweight person can be expressed as $b < 18.5$, so draw an open dot at 18.5 and shade to the left. Likewise, a person with a BMI that is greater than or equal to 25.0 is overweight, which can be expressed as $b \geq 25.0$. Draw a solid dot at 25 and shade to the right. The number line in choice B represents the combination of these inequalities, but be careful—that's not what the question asks! The question asks for the healthy BMI range, so you're looking for the opposite of choice B—a solid dot at 18.5, an open dot at 25.0, and shaded between them, which is (A).

8. C

Difficulty: Medium

Category: Heart of Algebra / Inequalities

Strategic Advice: Although cross-multiplication is an appropriate strategy when you have a fractional expression on each side of the equal sign, simplifying the numerators first will save a few steps and avoid some potential errors.

Getting to the Answer: Start by distributing the 2 on the left side and by adding 6 + 5 on the right side. Then cross-multiply to eliminate the denominators. Finally, solve for *k* using inverse operations. Don't forget—if you multiply or divide by a negative number at any point during the process, you must reverse the inequality symbol:

$$\frac{2(4k+1)}{3} \geq \frac{k(6+5)-3}{2}$$

$$\frac{8k+2}{3} \geq \frac{11k-3}{2}$$

$$2(8k+2) \geq 3(11k-3)$$

$$16k+4 \geq 33k-9$$

$$-17k \geq -13$$

$$\frac{-17k}{-17} \leq \frac{-13}{-17}$$

$$k \leq \frac{13}{17}$$

(C) is correct.

9. B

Difficulty: Medium

Category: Heart of Algebra / Inequalities

Strategic Advice: Use the Kaplan Strategy for Translating English into Math. The clue "at least" means that many or more, so use the symbol ≥ when describing the number of tests. The clue "at most" means that many or less, so use the symbol ≤ when describing the number of hours.

Getting to the Answer: First assign variables: Let *n* represent the number of non-calculator tests and *c* represent the number of calculator tests. Write an inequality to describe the number of tests:

non-calculator plus calculator is at least 8: $n + c \geq 8$

You can already eliminate C and D. Next, write an inequality to describe the number of hours needed to make the tests. Be sure to incorporate all 5 weeks of time the teacher is willing to devote:

hours per non-calc. test times # of tests: $3n$
plus hours per calc. test times # of tests: $+ 4c$
is at most 5 hours per week times 6 weeks: ≤ 30

$$3n + 4c \leq 30$$

Put the two inequalities together to form the system given in (B).

10. D

Difficulty: Medium

Category: Heart of Algebra / Inequalities

Strategic Advice: The solution to a system of inequalities is the region of the graph that satisfies both inequalities. Pay careful attention to which line is dashed and which is solid as you analyze the regions of the graph.

Getting to the Answer: The solution set for the inequality $y > x$ is the union of the sections of the graph that are above the dashed line, which are A and D. The solution set for the inequality $3y \leq -4x + 6$ is the union of the sections that are below the solid line, which are C and D. The solution set for the system of inequalities consists of all the points that satisfy both inequalities, which are contained within region (D).

11. B
Difficulty: Medium

Category: Heart of Algebra / Inequalities

Strategic Advice: When you have a wordy question stem, turn to the Kaplan Strategy for Translating English into Math.

Getting to the Answer: If a buyer puts down 20% *or more* of the purchase price, he or she can avoid paying PMI, which means the down payment must be greater than or equal to (≥) 20% of 375,000, or $0.2(375{,}000)$ (remember, "of" means multiply). Add this amount to the \$7,200 for closing costs and property tax to find that the total funds needed are $f \geq 0.2(375{,}000) + 7{,}200$, which is (B).

12. D
Difficulty: Medium

Category: Heart of Algebra / Inequalities

Strategic Advice: You could solve the compound inequality for m and substitute the result into the expression $10m - 5$, but there is a quicker way to answer a question like this. Look for a relationship between what you're given, the possible values of $-2m + 1$, and what you're looking for, the possible values of $10m - 5$.

Getting to the Answer: Notice that $10m - 5$ is -5 times the expression $-2m + 1$. This means you can answer the question by multiplying all three pieces of the inequality by -5. Don't forget to flip the inequality symbols because you are multiplying by a negative number. Then write the inequality with increasing values from left to right:

$$-\frac{5}{2} < -2m + 1 < -\frac{7}{5}$$

$$-5\left(-\frac{5}{2}\right) < -5(-2m + 1) < -5\left(-\frac{7}{5}\right)$$

$$\frac{25}{2} > 10m - 5 > 7$$

$$7 < 10m - 5 < \frac{25}{2}$$

$$7 < 10m - 5 < 12.5$$

The question asks for the greatest possible *integer* value, so the correct answer is 12, which is (D).

13. B
Difficulty: Medium

Category: Heart of Algebra / Inequalities

Strategic Advice: When determining the meaning of a particular number, variable, or expression that represents a real-world scenario, read the question very carefully. Here, the maximum weight per axle is 18,000 pounds, but that doesn't necessarily mean that the 18,000 in the inequality represents the same thing.

Getting to the Answer: Try writing an inequality in words to represent the scenario: The weight of the truck (cab and trailer together) plus the weight of the cargo (which varies) must be less than or equal to the maximum weight allowed per axle times the number of axles. Now, translate from English into math: $(11{,}000 + 7{,}000) + x \leq 18{,}000 \times 4$, or $18{,}000 + x \leq 72{,}000$. This means the 18,000 represents the weight of the truck before any cargo is loaded, or in other words, when the trailer is empty, which is (B).

14. D
Difficulty: Medium

Category: Heart of Algebra / Inequalities

Strategic Advice: You can use an absolute value to represent an unknown difference that is described as "within" a certain amount—you don't know whether the end result is more or less, but you do know it is within a certain range.

Getting to the Answer: The difference between the actual number of hours of labor, a, and the number of hours the company estimated, h, is $|a - h|$. If the company meets their goal, the difference is less than 8, which can be represented as $|a - h| \leq 8$ or $-8 \leq a - h \leq 8$. This means (D) is correct.

15. Any value between 4 and 6, inclusive
Difficulty: Medium

Category: Heart of Algebra / Inequalities

Strategic Advice: When you see phrases like "at least" and "at most," consider writing an inequality in words first, and then translating from English into math. "At least" translates as $\geq$ because it means that much or more, and "at most" translates as $\leq$ because it means that much or less.

Getting to the Answer: Because Margo can peel and slice at least 10 dozen apples per hour and at most 15 dozen apples per hour, her rate is somewhere between 10 and 15 dozen apples per hour, or $10 \leq r \leq 15$. This means her rate for 60 dozen apples is somewhere between $60 \div 10 = 6$ hours and $60 \div 15 = 4$ hours. Grid in any number between 4 and 6, inclusive.

16. D
Difficulty: Hard

Category: Heart of Algebra / Inequalities

Strategic Advice: Given the interval in which a variable lies, you can find an equivalent absolute value inequality by finding the midpoint of the original interval and the distance that each endpoint is from the midpoint. For example, $1 < x < 5$ is equivalent to $|x - 3| < 2$ because 3 is the midpoint of the interval and each end is within 2 units of the midpoint.

Getting to the Answer: The question states that the outer core is more than 800 miles and less than 2,200 miles from the Earth's center, so the possible distances from the Earth's center that are in the outer core are given by the interval $800 < d < 2{,}200$. The midpoint of this interval is $\frac{800 + 2{,}200}{2} = \frac{3{,}000}{2} = 1{,}500$. The interval $800 < d < 2{,}200$ consists of all the points that are within 700 units ($1{,}500 - 800 = 700$ and $2{,}200 - 1{,}500 = 700$) of the midpoint. Thus, the equivalent absolute value inequality is $|d - 1{,}500| < 700$, which is (D).

17. B
Difficulty: Hard

Category: Heart of Algebra / Inequalities

Strategic Advice: The only way to solve a system of inequalities is to graph it, so draw a sketch on the coordinate plane provided to find the answer. Remember, you are looking for the quadrant that contains *no solutions* to the system.

Getting to the Answer: Start by rewriting each inequality in slope-intercept form, $y = mx + b$, where m is the slope and b is the y-intercept. Then, use m and b to graph each inequality.

$$x + 3y \leq 12 \rightarrow 3y \leq -x + 12 \rightarrow y \leq -\frac{1}{3} + 4$$

$$2x - 3y \leq -3 \rightarrow -3y \leq -2x - 3 \rightarrow y \geq \frac{2}{3}x + 1$$

To represent the first inequality, shade below its line ($\leq$); to represent the second inequality, shade above its line (≥). The solution to the system of inequalities is the region where the two shaded areas *overlap.* A sample graph follows:

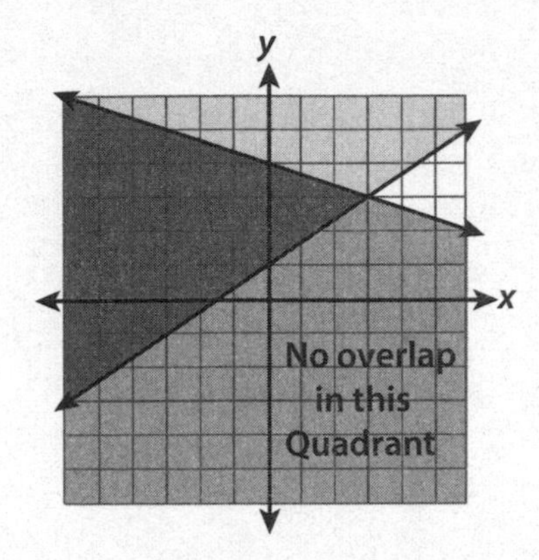

Notice that there is some darker shading (the overlap) in quadrants I, II, and III, making quadrant IV the only one that does not contain solutions to the system. Choice (B) is correct.

18. A

Difficulty: Hard

Category: Heart of Algebra / Inequalities

Strategic Advice: Sometimes, it is not necessary to apply brute force to a question. Here, you can get away with doing minimal math because the slopes of the lines are enough to answer the question—you do not need to find the *y*-intercepts.

Getting to the Answer: The question asks for a system of inequalities that describes the region where the substance exists as a liquid only, so you need two restricted (greater/less than but not equal to) inequalities. This means you can eliminate D. Next, examine the lines separating "solid" and "liquid," and "liquid" and "gas." The "liquid" region is below the line that passes through (220, 10) and (230, 95), and it is above the line that passes through (220, 10) and (400, 60). Find the slope of each line:

Below line (<) that has a slope of:

$$m = \frac{y_2 - y_1}{x_2 - x_1} = \frac{95 - 10}{230 - 220} = \frac{85}{10} = \frac{17}{2}$$

Above line (>) that has a slope of:

$$m = \frac{y_2 - y_1}{x_2 - x_1} = \frac{60 - 10}{400 - 220} = \frac{50}{180} = \frac{5}{18}$$

Now examine the inequality symbols and the slopes in the remaining answer choices: B is incorrect because the symbols are reversed, and C is incorrect because the slope of the second line is not right. You already eliminated D, so (A) must be the correct answer.

19. B

Difficulty: Hard

Category: Heart of Algebra / Inequalities

Strategic Advice: Quickly examine the answer choices—there are no *x*'s and no *y*'s. This is a hint to plug in the solution (0, 0), which will eliminate both *x* and *y*.

Getting to the Answer: You are given that (0, 0) is a solution to the system of inequalities, which means that substituting 0 for *x* and 0 for *y* into each inequality must yield a true statement. You can then compare the values of k_1 and k_2.

$0 < 0 + k_1 \rightarrow 0 < k_1$, or $k_1 > 0$, which means k_1 is a positive number.

$0 > 2(0) + k_2 \rightarrow 0 > k_2$, or $k_2 < 0$, which means k_2 is a negative number.

Every negative number is less than every positive number, which means $k_2 < k_1$, which is (B). (Note: Choice A is always false because it contradicts (B); C and D may or may not be true, depending on the values of k_1 and k_2, but the question asks which statement *must* be true.)

20. 801
Difficulty: Hard

Category: Heart of Algebra / Inequalities

Strategic Advice: When you have two unknowns, try to write one in terms of the other to simplify the situation.

Getting to the Answer: Solve the equation for y in terms of x:

$$\frac{y}{x} = 0.625$$
$$y = 0.625x$$

Substitute this value for y into the inequality and simplify:

$$x - y > 300$$
$$x - 0.625x > 300$$
$$0.375x > 300$$
$$x > \frac{300}{0.375}$$
$$x > 800$$

Be careful—the inequality specifies that x must be greater than 800, so the least possible integer value of x is 801.

PRACTICE SET 4: RATES, RATIOS, PROPORTIONS, AND PERCENTAGES

Rates, Measurement, and Unit Conversions

By now, you've become adept at using algebra to answer many SAT math questions, which is great because you'll need those algebraic skills to answer questions involving rates and other quantitative reasoning questions. You're likely already familiar with all kinds of rates—kilometers per hour, meters per second, and even miles per gallon are all considered rates.

Most rate questions can be solved using some form of the *DIRT equation*—Distance = rate × time. If you have two of the three components of the equation, you can easily find the third.

Units of measurement are also important when answering rate questions (and others that require a unit conversion). The *factor-label method* is a simple yet powerful way to keep calculations organized and to ensure that you arrive at an answer that has the requested units.

For example, suppose you need to find the number of cups in 2 gallons. To use the factor-label method, start by identifying the initial unit (gallons), and then identify the desired unit (cups). The next step is to piece together a path of relationships (measurement conversions) that will convert gallons into cups, cancelling units as you go. Keep in mind that you will often have multiple stepping-stones along the way:

$$2\ \cancel{\text{gallons}} \times \frac{4\ \cancel{\text{quarts}}}{1\ \cancel{\text{gallon}}} \times \frac{2\ \cancel{\text{pints}}}{1\ \cancel{\text{quart}}} \times \frac{2\ \text{cups}}{1\ \cancel{\text{pint}}} = (2 \times 4 \times 2 \times 2)\ \text{cups} = 32\ \text{cups}$$

Don't worry if you don't know all the measurement conversions—these will be provided on Test Day within the context of the question.

Ratios and Proportions

You most likely first encountered ratios and proportions before you even got to high school, and chances are that you will see them again in college math courses. You will certainly see them on the SAT, so be sure to spend adequate time making sure you have a good grasp of the material in this practice set.

A *ratio* is a relationship that compares the relative size of two amounts. You might see ratios written with a colon, 2:5, as a fraction, $\frac{2}{5}$, or using words, 2 to 5. Ratios can compare parts to parts or parts to wholes. For example, suppose you make a fruit salad using 6 oranges, 3 apples, and 2 pears. The ratio of apples to pears is 3:2. The ratio of apples to all the fruit is 3:11.

It is also possible to combine ratios. If you have two ratios, *a:b* and *b:c*, you can derive *a:c* by finding a common multiple of the *b* terms. For example, suppose the ratio of *a* to *b* is 3:4 and the ratio of *b* to *c* is 5:2. Take a look at the following chart to see how to find the ratio of *a* to *c*.

a	:	*b*	:	*c*
3	:	4		
		5	:	2
15	:	20		
		20	:	8
15		:		8

The number 20 is the least common multiple of 4 and 5, so write 20 in the middle column under the *b* terms. Next, multiply each ratio by the factor (use 5 for *a:b* and 4 for *b:c*) that gives $b = 20$. Finally, write the resulting relationship between *a* and *c*: The ratio of *a* to *c* is 15:8.

Proportions are simply two ratios set equal to each other. They are an efficient way to solve certain problems, but you must exercise caution when setting them up. Watching the units of each piece of the proportion is critical. Writing the proportion in words first is a good way to avoid careless errors. To solve a proportion, cross-multiply and then use inverse operations to isolate the unknown quantity. You can also use cross-multiplication to verify that two ratios are proportional. For example $\frac{a}{b} = \frac{c}{d}$ if and only if $ad = bc$.

Percentages

A percentage is a type of proportion that means "per 100." Not only are percentages a common occurrence on the SAT, they're also common in daily life. These two reasons alone are enough for you to know that percentages are a concept you're going to need to understand.

Here are some useful formulas to learn before Test Day:

- Percent $\times$ whole = part
- $\text{Percent} = \frac{\text{part}}{\text{whole}} \times 100\%$
- $\text{Percent change} = \frac{\text{amount of change}}{\text{initial amount}} \times 100\%$

Keep in mind that when using percentages in calculations, you must write the percent as a decimal number.

PRACTICE SET

You may use your calculator for all questions in this practice set.

Easy

1. An architect is working on a design for a solar power plant. While double-checking the blueprint, he realizes that he has made an error: The solar panels will not fit on the roof unless the dimensions of the roof are increased by 0.5%. Which of the following expresses this change in the dimensions of the roof as a fraction of the original dimensions?

   A) $\frac{1}{2{,}000}$

   B) $\frac{1}{200}$

   C) $\frac{1}{20}$

   D) $\frac{1}{2}$

2. A graphic designer created an image for a school banner. He uses this image for small and medium banners. To produce a large banner, the designer increases the size of the image by 12.5% to 18 inches tall by 22.5 inches wide. What was the size of the original image?

   A) 15 inches by 17.5 inches

   B) 15 inches by 18 inches

   C) 16 inches by 18.5 inches

   D) 16 inches by 20 inches

3. A typical song downloaded from the Internet is 4 megabytes in size. Lindy has satellite Internet, and her computer downloads music at a rate of 256 kilobytes per second. If 1 megabyte equals 1,024 kilobytes, about how many songs can Lindy download in 2 hours?

   A) 128

   B) 450

   C) 1,800

   D) 1,920

4. A subway car passes 3 stations every 10 minutes. At this rate, how many stations will it pass in 1 hour?

Medium

5. Cecilia leaves home for school on her bike at 9:05 AM, riding at an average speed of 10 miles per hour. Fifteen minutes later, her mother realizes that Cecilia forgot to take her lunch. Cecilia's mother immediately gets into her car and drives after Cecilia at an average speed of 25 miles per hour. At what time will Cecilia's mom catch up to her?

    A) 9:15 AM

    B) 9:20 AM

    C) 9:26 AM

    D) 9:30 AM

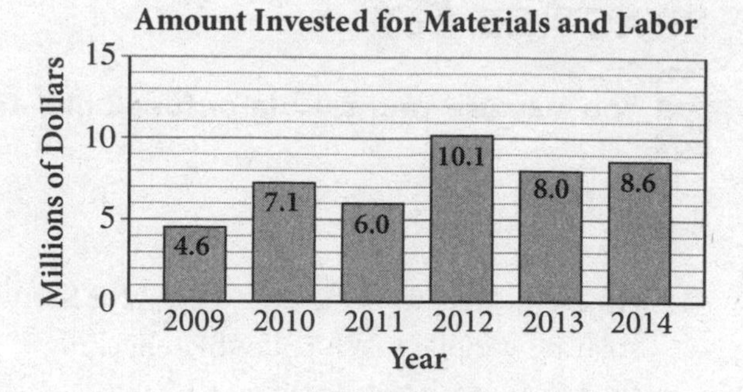

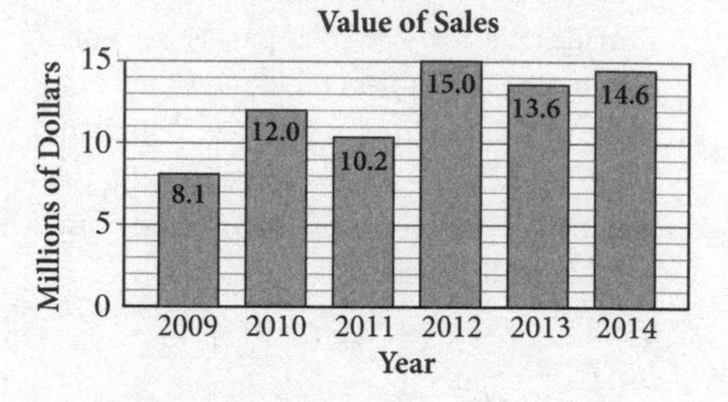

6. The bar graphs show the amounts, in dollars, invested by a company for materials and labor to produce a certain product and the value of the sales for that product. The company claims that the amount invested for materials and labor is never more than 60% of the value of the sales. Which year or years disprove this claim?

    A) 2010

    B) 2012

    C) 2014

    D) 2012 and 2014

7. A sporting goods store ordered an equal number of white and yellow tennis balls. The tennis ball company delivered 30 extra white balls, making the ratio of white balls to yellow balls 6:5. How many tennis balls did the store originally order?

A) 120

B) 150

C) 240

D) 300

Metal	Density (g/cm^3)
Copper	8.96
Iron	7.87
Nickel	8.90
Tin	7.26

8. The *density* of a substance can be found by dividing the mass of the substance by the volume of the substance. The table gives the density of several pure metals in grams per cubic centimeter. Suppose a rectangular sheet of a pure metal weighs about 515.9 grams and measures $\frac{1}{4}$ inch by 2 inches by 8 inches. Assuming the sheet is one of the metals in the table, which metal is it? (There are approximately 2.54 centimeters in 1 inch.)

A) Copper

B) Iron

C) Nickel

D) Tin

9. A bike messenger delivers a package from the courthouse to a law firm downtown, traveling at an average speed of 24 miles per hour. On the way back, the bike messenger gets stuck in traffic, which reduces his average speed for the return trip by 12.5 percent. What was the bike messenger's average speed, in miles per hour, for the round trip?

A) 22

B) 22.1

C) 22.4

D) 22.5

10. The mayor of Dunderville proposes a new ordinance that would give a tax credit to parents who homeschool their children. Ten percent of Dunderville's residents approve of the ordinance. Of these, 7 out of 10 are women. All of Dunderville's other residents are against the ordinance. If there are equal numbers of men and women in Dunderville, then what fraction of Dunderville's male residents approve of the ordinance?

A) $\frac{3}{10}$

B) $\frac{3}{47}$

C) $\frac{3}{50}$

D) $\frac{3}{100}$

11. Each crop duster who works for Gary's Crop Dustin' 4 Cheap can dust half an acre of crops in 25 minutes. The company must dust eight 0.75-acre lots and twelve 1.5-acre lots to complete a certain job. What is the minimum number of crop dusters needed to complete the job in 5 hours?

A) 4

B) 5

C) 6

D) 7

12. Comix Fanatix has 250 customers who subscribe to its newsletter, 68% of whom are male. After some female customers cancel their subscriptions, the total number of subscribers becomes 1.3 times the number of male subscribers. How many female customers canceled their subscriptions?

13. In preparation for buying a house, Sergey and Juanita draft a budget to see how much they can afford. Their monthly spending and income are shown in the table. Numbers in parentheses indicate expenses (money being spent); numbers without parentheses indicate income (money being earned). Sergey and Juanita will not buy a house unless they can save at least $400 a month after all expenses, including the new mortgage payment and property taxes, are paid.

Category	Monthly Amount
Sergey's income (after taxes)	$1,300
Juanita's income (after taxes)	$1,600
Car insurance and gas	($220)
Food	($400)
Entertainment	($100)
Student loans	($550)
Phone and Internet	($120)

The couple would like to buy a house that costs $230,400, for which the monthly mortgage payment would be $730, and the annual property taxes would be 2.5% of the purchase price. By what percent could Sergey and Juanita cut their monthly food spending in order to buy this particular house and meet their criteria for purchasing a house? (Ignore the percent sign and grid in your answer as a whole number.)

Hard

14. Amal, Geoff, and Julia each have some fireworks. If Amal gives half of his fireworks to Geoff, then Amal and Geoff will have fireworks in a 2:3 ratio. If Julia has one-fifth as many fireworks as Geoff (before Amal gives half his fireworks to Geoff), then what is the ratio of the number of fireworks Amal has to the number of fireworks Julia has (before Amal gives half his fireworks to Geoff)?

    A) 20:1
    B) 15:4
    C) 3:2
    D) There is insufficient information to determine the answer.

15. An oil tank has two pipes connected to it. If the tank is empty, Pipe A can fill it in 2 hours. If the tank is full, Pipe B can empty it in 3 hours. If both pipes are activated at the same time when the tank is empty, how many hours will it take for the tank to be filled to 60% of its capacity?

    A) 2.4
    B) 3.6
    C) 6
    D) 60

16. An all-natural health clinic stocks teas and herbs in a 5:11 ratio at one of its warehouses. In an effort to cut overhead costs, the owners close this warehouse and transfer its stock to a larger warehouse, increasing that warehouse's inventory of items by 20%. Which of the following could be the number of items at the larger warehouse after the transfer?

    A) 198
    B) 386
    C) 480
    D) 573

17. John buys c pounds of cheese to feed m people at a party. If $m + n$ people come to the party, how many more pounds of cheese must John buy in order to feed everyone at the original rate?

    A) $\frac{mn}{c}$
    B) $\frac{m}{cn}$
    C) $\frac{m+n}{c}$
    D) $\frac{nc}{m}$

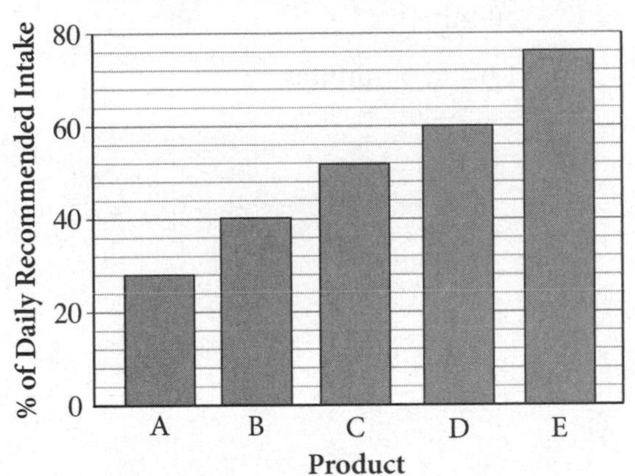

18. The bar graph above shows the percent of the daily recommended intake of calcium supplied by five different multivitamins, A, B, C, D, and E, per pill. Products A and D are sold in bottles of 75 pills and cost \$12 and \$18, respectively. Products B, C, and E are sold in bottles of 50 pills and cost \$9.50, \$10.25, and \$15.50, respectively. Which of the five products supplies the greatest percentage of the daily-recommended intake of calcium per dollar?

    A) A
    B) B
    C) C
    D) E

19. A typical 3-D printer creates objects by laying down layers of ink one on top of another. The average thickness of a layer depends on the caliber of the printer. For one particular 3-D printer, the average thickness of a layer is 102 μm (micrometers). When creating a 1-inch cube, the printer takes 6 seconds to lay down each layer of ink. Approximately how long should it take this printer to create a 1-inch cube? (There are 1,000,000 micrometers in a meter, and 1 inch is approximately equal to 0.0254 meters.)

A) 12 minutes

B) 25 minutes

C) 40 minutes

D) 1 hour, 6 minutes

20. Eli left his home in New York and traveled to Brazil on business. Before he left, he used his credit card to purchase these pewter vases:

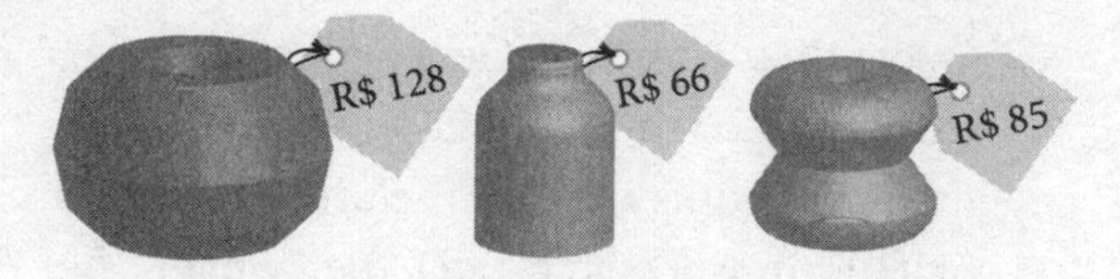

For daily purchases totaling less than 200 U.S. dollars, Eli's credit card company charges a 2% fee. If the total charge on his credit card for the vases was $126.48, what was the foreign exchange rate in Brazilian reais (R$) per U.S. dollar on the day that Eli bought the vases? If necessary, round your answer to the nearest hundredth.

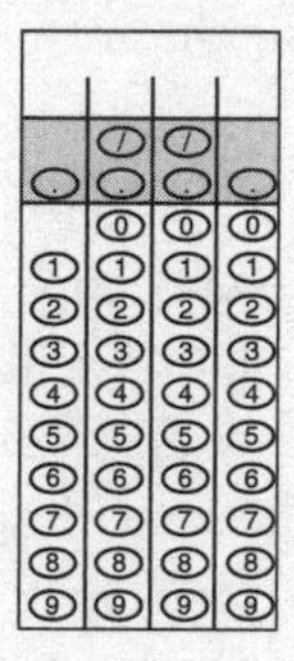

ANSWER KEY

1. B	**6. B**	**11. A**	**16. C**
2. D	**7. D**	**12. 29**	**17. D**
3. B	**8. B**	**13. 25**	**18. C**
4. 18	**9. C**	**14. A**	**19. B**
5. D	**10. C**	**15. B**	**20. 2.25**

ANSWERS & EXPLANATIONS

1. B
Difficulty: Easy

Category: Problem Solving and Data Analysis / Rates, Ratios, Proportions, and Percentages

Strategic Advice: Don't let yourself get overwhelmed by long problems. Focus on the question that is being posed.

Getting to the Answer: Reword the question in your head: 0.5% is what fraction of the original dimensions? In other words, this question is simply asking you to convert 0.5% to a fraction. (Don't answer too quickly, though: $\frac{1}{2}$ is 0.5, not 0.5%, so D is not correct.) Recall that "percent" means *per hundred*, which translates to *divide by 100*. Therefore, $0.5\% = 0.5 \div 100 = \frac{1}{2} \times \frac{1}{100} = \frac{1}{200}$. The answer is (B).

2. D
Difficulty: Easy

Category: Problem Solving and Data Analysis / Rates, Ratios, Proportions, and Percentages

Strategic Advice: You could set up and solve an equation to answer this question. However, sometimes trying the answer choices is a quicker route.

Getting to the Answer: You're looking for dimensions that, when multiplied by 112.5% (or 1.125), the result is 18 inches by 22 inches. Start with A: $1.125 \times 15 = 16.875$ (not 18), so you can eliminate both A and B. Move on to C: $1.125 \times 16 = 18$, but $1.125 \times 18.5 = 20.8125$ (not 22.5), so (D) must be correct. ($1.125 \times 20 = 22.5$✔)

3. B
Difficulty: Easy

Category: Problem Solving and Data Analysis / Rates, Ratios, Proportions, and Percentages

Strategic Advice: When converting lots of units, use the factor-label method and let the unit conversions guide you to the correct answer.

Getting to the Answer: The initial rate is given in kilobytes per second, the size per song is given in megabytes, and the question asks for the number of songs that can be downloaded in 2 hours. Arrange the given rates so that the units in the end result are what you're looking for:

$$\frac{256 \text{ kb}}{1 \cancel{\text{sec}}} \times \frac{60 \cancel{\text{sec}}}{1 \cancel{\text{min}}} \times \frac{60 \cancel{\text{min}}}{1 \cancel{\text{hr}}} \times 2 \cancel{\text{hr}} = 1{,}843{,}200 \text{ kb}$$

$$1{,}843{,}200 \cancel{\text{kb}} \times \frac{1 \cancel{\text{mb}}}{1{,}024 \cancel{\text{kb}}} \times \frac{1 \text{ song}}{4 \cancel{\text{mb}}} = 450 \text{ songs}$$

Lindy's computer can download about 450 songs in 2 hours, which is (B).

4. 18
Difficulty: Easy

Category: Problem Solving and Data Analysis / Rates, Ratios, Proportions, and Percentages

Strategic Advice: Think logically to answer a question like this—there are no complicated conversions to make.

Getting to the Answer: Because there are 60 minutes in an hour, the subway will pass $60 \div 10 = 6$ times as many stations in one hour as it passes in 10 minutes. In 10 minutes, it passes 3 stations, so in 60 minutes it must pass $6 \times 3 = 18$ stations.

5. D
Difficulty: Medium

Category: Problem Solving and Data Analysis / Rates, Ratios, Proportions, and Percentages

Strategic Advice: Almost all rate questions that involve distance and time can be solved using the formula Distance = rate × time. When you have two rates, consider organizing the information in a table.

Getting to the Answer: Think about each variable in the formula. You're given both rates (10 mph and 25 mph), so fill the rates in the table. When Cecilia's mother catches up to her, their distances will be the same (call the distance *d*). Because her mother leaves 15 minutes *after* Cecilia, call Cecilia's time *t* and her mother's time $t - 15$.

	Distance	Rate	Time
Cecilia	d	10	t
Cecilia's mother	d	25	$t - 15$

Reading across each row of the table, the two equations are $d = 10t$ and $d = 25(t - 15)$. Set the two equations equal and solve for t:

$$\begin{aligned} 10t &= 25(t-15) \\ 10t &= 25t - 375 \\ 375 &= 15t \\ t &= 25 \end{aligned}$$

Cecilia's mother will catch up at 9:05 AM + 25 minutes = 9:30 AM, choice (D).

6. B
Difficulty: Medium

Category: Problem Solving and Data Analysis / Rates, Ratios, Proportions, and Percentages

Strategic Advice: When lots of information is presented in a graphic, make sure you look at the answer choices before you start calculating. Chances are that you'll only need to use a small part of the data.

Getting to the Answer: Manipulate the three-part percent formula, Percent × whole = part, to read Percent = part ÷ whole (or here, Percent = amount invested ÷ sales), and check each year given in the answer choices. To *disprove* the claim, you're looking for the year or years in which the percent is *greater* than 60% (greater than 0.6).

$2010 \rightarrow \frac{7.1}{12} \approx 0.59 < 0.6$, so eliminate A.

$2012 \rightarrow \frac{10.1}{15} \approx 0.67 > 0.6$. Don't choose (B) yet because 2012 is also in D.

$2014 \rightarrow \frac{8.6}{14.6} \approx 0.59 < 0.6$, so eliminate C and D.

This means (B) is correct.

7. D
Difficulty: Medium

Category: Problem Solving and Data Analysis / Rates, Ratios, Proportions, and Percentages

Strategic Advice: When you're given a ratio and one amount of something, and asked to find another amount, setting up and solving a proportion usually gets you to the correct answer.

Getting to the Answer: Let the number of yellow balls received be y. Then the number of white balls received is 30 more than this, or $y + 30$. Set up a proportion (white to yellow) and cross-multiply to solve for y:

$$\frac{6}{5} = \frac{y+30}{y}$$
$$6y = 5(y + 30)$$
$$6y = 5y + 150$$
$$y = 150$$

Don't forget to check that you answered the *right* question (the number of balls ordered). Because the number of white balls ordered equals the number of yellow balls ordered, the total number of balls ordered is $2y$, which is 2×150, or 300, making (D) correct.

8. B

Difficulty: Medium

Category: Problem Solving and Data Analysis / Rates, Ratios, Proportions, and Percentages

Strategic Advice: Always pay careful attention to units. When converting from a linear measure, such as inches, to a cubic measure, such as cm^3 (or vice versa), you will need to write three unit conversions.

Getting to the Answer: Start by finding the volume of the metal sheet using the formula $V = lwh$. The volume is $\frac{1}{4} \times 2 \times 8 = 4$ cubic inches. Now, carefully convert this to cubic centimeters:

$$4 \cancel{in.^3} \times \frac{2.54 \text{ cm}}{1 \cancel{in.}} \times \frac{2.54 \text{ cm}}{1 \cancel{in.}} \times \frac{2.54 \text{ cm}}{1 \cancel{in.}} \approx 65.55 \text{ cm}^3$$

Skim the question again: To find the density of the metal, divide the mass (given in the question) by the volume: $515.9 \div 65.55 \approx 7.87$, which matches the density of iron, (B).

9. C

Difficulty: Medium

Category: Problem Solving and Data Analysis / Rates, Ratios, Proportions, and Percentages

Strategic Advice: In average speed problems, never take the average of the two average speeds. Instead, use the formula:

$$\text{Average speed} = \frac{\text{total distance}}{\text{total time}}$$

Getting to the Answer: The bike messenger's average speed in one direction is 24 mph. On the way back, it's reduced by 12.5%, or $0.125 \times 24 = 3$ mph, for an average speed of 21 mph. The answer is NOT the average of these two speeds, 22.5, because that fails to account for the fact that the bike messenger spent more time traveling at the slower speed. Instead, use the average speed formula to find the answer.

You're not given the distance to the law firm, so pick a convenient number to use in your calculations. A good number to pick is 168 because it's divisible by both 21 and 24. Given that $\text{Time} = \frac{\text{distance}}{\text{speed}}$, the time spent traveling from the courthouse is $\frac{168}{24} = 7$ hours, and the time spent traveling back is $\frac{168}{21} = 8$ hours, for a total time of $7 + 8 = 15$ hours. The total distance there and back is $168 + 168 = 336$ miles. You now have enough information to use the average speed formula:

$$\text{Average speed} = \frac{\text{total distance}}{\text{total time}} = \frac{336}{15} = 22.4,$$

choice (C).

10. C

Difficulty: Medium

Category: Problem Solving and Data Analysis / Rates, Ratios, Proportions, and Percentages

Strategic Advice: Don't panic when you see a lot of information. Take a deep breath and calmly assess the information one piece at a time.

Getting to the Answer: You're not given an amount for the population of Dunderville, so start by picking one. If Dunderville has 100 people (a nice number to pick whenever a problem involves percents), then according to the question, 50 are men and 50 are women. Furthermore, 10%, or $0.10 \times 100 = 10$ people, approve of the ordinance. The 10 people who approve of the ordinance are not evenly split by gender: 7 are women and 3 are men. Thus, overall, 3 men of the 50 in Dunderville approve the ordinance. The answer is (C).

11. A

Difficulty: Medium

Category: Problem Solving and Data Analysis / Rates, Ratios, Proportions, and Percentages

Strategic Advice: Questions that involve rates, especially rates related to time, often require logical thinking rather than the use of an algebraic equation.

Getting to the Answer: Start by determining how many acres need dusting: $(8)(0.75) + (12)(1.5) = 6 + 18 = 24$ acres. One crop duster can dust half an acre in 25 minutes, or a whole acre in 50 minutes. Thus, if one duster were to do the whole job, it would take 24 acres × 50 minutes/acre = 1,200 minutes which is $\frac{1{,}200}{60} = 20$ hours. However, the job needs to be completed in 5 hours, which is $\frac{1}{4}$ of 20. To get the job done in only one-quarter the time, Gary's needs to use $4 \times 1 = 4$ crop dusters, choice (A).

12. 29

Difficulty: Medium

Category: Problem Solving and Data Analysis / Rates, Ratios, Proportions, and Percentages

Strategic Advice: Step 3 of the Kaplan Method for Math (making sure you answered the right question) is especially important in Grid-in questions, where you don't have any choices to gently steer you back on track if you solved for the wrong thing.

Getting to the Answer: There are 250 subscribers, 68% of whom are male. That makes $250 \times 0.68 = 170$ male subscribers. After some females unsubscribe, the total subscriber count is 1.3 times the number of male subscribers, for a total of $170 \times 1.3 = 221$ subscribers. Here you have to pause and make sure you solve for the right thing: The question asks for the number of females who *un*subscribed, which is $250 - 221 = 29$ females.

13. 25

Difficulty: Medium

Category: Problem Solving and Data Analysis / Rates, Ratios, Proportions, and Percentages

Strategic Advice: When answering a Grid-in question, be sure to follow any directions that are given about how to enter your answer.

Getting to the Answer: First, compute the monthly property taxes of the home: $\$230{,}400 \times 2.5\% \div 12 = \$480$ per month. Next, tabulate Sergey and Juanita's monthly gains and losses if they buy the house: $\$1{,}300 + \$1{,}600 - \$220 - \$400 - \$100 - \$550 - \$120 - \$730 - \$480 = \$300$. This means they can save \$300 each month. However, they won't buy the house unless they can save \$400 per month. To make up the deficit, they could shave \$100 off their monthly \$400 food bill, which is $\frac{\$100}{\$400} = 0.25$.

Be sure to follow directions (enter the answer as a whole number): 0.25 = 25%, so grid in 25.

14. A
Difficulty: Hard

Category: Problem Solving and Data Analysis / Rates, Ratios, Proportions, and Percentages

Strategic Advice: When you see an answer choice indicating that there is insufficient information to answer the question, don't automatically assume that that choice is correct.

Getting to the Answer: Let A and G equal the number of fireworks that Amal and Geoff originally had. If Amal gives half of his fireworks to Geoff, then Amal will have $\frac{A}{2}$ fireworks left while Geoff will have $G+\frac{A}{2}$. You're told that these two values will be in a 2:3 ratio, so set up a proportion and cross-multiply to simplify:

$$\frac{\frac{A}{2}}{G+\frac{A}{2}}=\frac{2}{3}$$
$$3\left(\frac{A}{2}\right)=2\left(G+\frac{A}{2}\right)$$
$$\frac{3A}{2}=2G+A$$
$$3A=2(2G+A)$$
$$3A=4G+2A$$
$$A=4G$$

While this doesn't give you the exact numbers for either Geoff or Amal, it does tell you that originally Amal had 4 times as many fireworks as Geoff. If Geoff, in turn, originally had 5 times as many fireworks as Julia (because the question states that Julia had $\frac{1}{5}$ as many as Geoff), then Amal must originally have had $4 \times 5 = 20$ times as many fireworks as Julia, which makes the ratio 20:1. Choice (A) is correct.

15. B
Difficulty: Hard

Category: Problem Solving and Data Analysis / Rates, Ratios, Proportions, and Percentages

Strategic Advice: When you need to find an answer that involves a percent of an amount, but you're not given the amount, pick a convenient number and work through the problem using that number.

Getting to the Answer: Pick a number for the capacity of the oil tank, choosing one that divides nicely by 2 and 3. Let the capacity of the tank equal 600 gallons. Given that Pipe A can fill the tank in 2 hours, its rate is $600 \div 2 = 300$ gallons per hour. Likewise, Pipe B can empty the tank in 3 hours, so its rate is $600 \div 3 = 200$ gallons per hour. If Pipe A adds 300 gallons per hour while Pipe B removes 200 gallons per hour, then a net of 100 gallons of oil will be added to the tank each hour. At this rate, it will take $\frac{600}{100}=6$ hours to fill the tank. But wait—this isn't the answer. The question asks how long it will take to fill 60% of the tank, not the whole tank. Multiply the time by 0.6 to get the final answer: $(6)(0.6) = 3.6$ hours, choice (B).

16. C
Difficulty: Hard

Category: Problem Solving and Data Analysis / Rates, Ratios, Proportions, and Percentages

Strategic Advice: When given a ratio, you can always add an x to each term. For example, if the ratio of boys to girls is 2:3, then you know that there are $2x$ boys, $3x$ girls, and $5x$ boys and girls in total.

Getting to the Answer: The ratio of teas to herbs is 5:11, which means that the total number of teas and herbs is $5x + 11x = 16x$. Because this represents a 20% increase in the larger warehouse's stock, it follows that $16x$ is 20%, or $\frac{1}{5}$, of the large warehouse's initial stock: $16x \times 5 = 80x$. The final stock in the warehouse is therefore $80x + 16x = 96x$. This means that the answer must be a multiple of 96, or in other words, evenly divisible by 96. Use your calculator to quickly test each answer choice: Of the numbers given, 480 is the only one that is evenly divisible by 96, making (C) the correct answer.

17. D

Difficulty: Hard

Category: Problem Solving and Data Analysis / Rates, Ratios, Proportions, and Percentages

Strategic Advice: This is a great question in which you can pick convenient numbers to represent the variables given and work through the answer choices using those numbers.

Getting to the Answer: Say John buys 10 pounds of cheese for 5 people (that is, $c = 10$ and $m = 5$). Then everyone gets 2 pounds of cheese. Also, say 8 people come, 3 more than expected (that is, $n = 3$). Then John needs 16 pounds to have enough for everybody to consume 2 pounds of cheese. Because he already bought 10 pounds, he must buy an additional 6 pounds. Therefore, an answer choice that equals 6 when you substitute 10 for c, 5 for m, and 3 for n is the correct choice:

Choice A: $\frac{(5)(3)}{10} \neq 6$. Eliminate.

Choice B: $\frac{5}{(10)(3)} \neq 6$. Eliminate.

Choice C: $\frac{5+3}{10} \neq 6$. Eliminate.

Choice (D): $\frac{(3)(10)}{5} = 6$. Choice (D) is correct.

18. C

Difficulty: Hard

Category: Problem Solving and Data Analysis / Rates, Ratios, Proportions, and Percentages

Strategic Advice: Whenever you're comparing multiple rates (percent calcium per pill and percent calcium per dollar), especially when reading data from a graph, consider organizing the information in a table.

Getting to the Answer: The following table organizes the data provided by the graph and additional data needed to answer the question. Because there is more than one calculation to make per product, don't round any values until you reach the final column. (Note: Product D is not one of the answer choices, so save yourself a bit of time by skipping D in the calculations.)

Product	Cost per bottle	Number of pills	Cost per pill (not rounded)	Percent of calcium supplied per pill (from the graph)	Percent of calcium supplied per dollar (rounded to the nearest percent)
A	\$12.00	75	$12 \div 75 = 0.16$	28%	$28 \div 0.16 = 175$
B	\$9.50	50	$9.5 \div 50 = 0.19$	40%	$40 \div 0.19 = 211$
C	\$10.25	50	$10.25 \div 50 = 0.205$	52%	$52 \div 0.205 = 254$
E	\$15.5	50	$15.5 \div 50 = 0.31$	76%	$76 \div 0.31 = 245$

According to the calculations in the table, product C provides the greatest percentage of the daily-recommended intake of calcium per dollar, making (C) correct.

19. B

Difficulty: Hard

Category: Problem Solving and Data Analysis / Rates, Ratios, Proportions, and Percentages

Strategic Advice: When a question involves multiple unit conversions, make a plan to keep things organized. Using the factor-label method (canceling units) also helps.

Getting to the Answer: Think backward: To determine how long it will take to print the cube, you need to know how many layers are required. To determine the number of layers needed, convert micrometers to meters and then meters to inches. Don't forget to line up the conversions so that the units cancel nicely. Start with 1 inch and work your way toward layers:

$$1\ \cancel{\text{in.}} \times \frac{0.0254\ \cancel{\text{m}}}{1\ \cancel{\text{in.}}} \times \frac{1{,}000{,}000\ \cancel{\mu\text{m}}}{1\ \cancel{\text{m}}} \times \frac{1\ \text{layer}}{102\ \cancel{\mu\text{m}}}$$

≈ 249 layers

The printer takes 6 seconds to print one layer, so it takes $6 \times 249 = 1{,}494$ seconds to print the cube. There are 60 seconds in 1 minute, so it takes $1{,}494 \div 60 = 24.9$, or about 25 minutes to print the cube. Choice (B) is correct.

20. 2.25

Difficulty: Hard

Category: Problem Solving and Data Analysis / Rates, Ratios, Proportions, and Percentages

Strategic Advice: Some questions, especially ones that are based on real-world scenarios, require a step-by-step approach. Make a plan and move through the plan one step at a time.

Getting to the Answer:

Step 1: Find the total cost of the vases in Brazilian reais: $128 + 66 + 85 = 279$.

Step 2: Find the total cost of the vases in U.S. dollars. The charge amount of \$126.48 represents the conversion of 279 Brazilian reais plus the 2% fee that Eli's credit card company charged him. To find the original cost, c, of the vases in U.S. dollars (before the 2% fee), write and solve the equation, $1.02c = 126.48$. Dividing both sides of the equation by 1.02 results in a cost of $c = \$124$.

Step 3: Find the exchange rate. To find the rate, r, in Brazilian reais per U.S. dollar, let the units guide you:

$$124\ \cancel{\text{dollars}} \times \frac{r\ \text{reais}}{1\ \cancel{\text{dollar}}} = 279\ \text{reais}$$

$$124r = 279$$

$$r = \frac{279}{124} = 2.25$$

The exchange rate that day was 2.25 Brazilian reais per U.S. dollar.

PRACTICE SET 5: SCATTERPLOTS

Scatterplot Basics

Some students tend to associate scatterplots with complicated statistical analyses and consequently become nervous when they hear they'll likely encounter them on Test Day. However, these seemingly difficult plots are usually straightforward—if you know what to look for. Below are the fundamental parts of a scatterplot.

- You're already familiar with x- and y-axes, but something that might be new is their units. Most scatterplots based on real data have units on the axes; these are important when drawing conclusions and inferences based on the data.
- The *domain* of a set of data points is the set of inputs, which corresponds to the x-values of the data points when plotted on a graph. The set of values that make up the *range* corresponds to the y-values.
- The *line of best fit* is drawn through the approximate center of the data points to describe the relationship between the two variables. This line does not need to go through all, or even most, of the data points, but it should accurately reflect the trend of the data with about half the points above the line and half below.
- The *equation of the line of best fit* (also called the *regression equation*) describes the best-fit line algebraically. On Test Day, you'll most likely encounter this equation as linear, quadratic, or exponential, though it can also be other types of equations.
- As with linear equations, lines of best fit associated with real-world data contain valuable information. For example, the slope of a linear line of best fit gives the average rate of change, and the y-intercept represents an initial amount.

Growth and Decay

The real world is full of examples of growth and decay, and you're bound to see some examples on Test Day. The two most common types are linear and exponential.

Linear growth and decay are represented by the equation $y = kx + x_0$, where k is the rate of change and x_0 (pronounced "x-naught") is the initial amount. The equation is the same as the slope-intercept form you've seen in the past, just with different variables standing in for m and b. When $k > 0$, growth occurs, and when $k < 0$, decay occurs. Note that, as with linear equations, linear growth and decay have a constant rate of change.

Exponential growth and decay are represented by the equation $y = x_0(1 + r)^x$, where x_0 is the initial amount and r is the rate of change. Unlike linear growth and decay, exponential growth and decay have a variable rate of change. When $r > 0$, growth occurs, and when $r < 0$, decay occurs.

Modeling Data

The relationship between two variables presented in the form of data may be modeled by functions or equations, which can be used for drawing conclusions and making predictions. Following are the three types of models you are most likely to see:

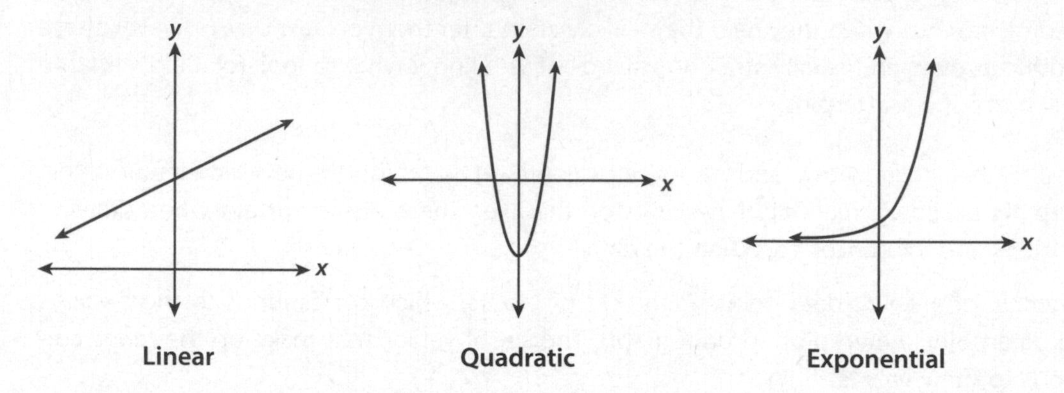

- A *linear model* will always increase (when its slope is positive) or decrease (when its slope is negative) at a constant rate, making it easy to spot.
- A *quadratic model* is U-shaped and the trend of the data changes from decreasing to increasing, or vice versa. The graph of a quadratic equation takes the shape of a parabola, which has either a minimum or a maximum called the vertex (although it is sometimes not shown on the graph). A parabola opens upward when the coefficient of the x^2 term is positive, and it opens downward when the coefficient of the x^2 term is negative.
- An *exponential model* typically starts with a gradual rate of change, which increases significantly over time. Unlike a quadratic model, the trend of the data does not change direction, and the graph does not have a vertex.

Using a Graphing Calculator to Model Data

There are times when using a graphing calculator on Test Day is prudent; deriving an equation that fits a data set is one of them. These best-fit equations are called regression equations and can take several shapes depending on the data's behavior. The correlation coefficient, r, indicates how well a regression equation fits the data; the closer r is to 1 for an increasing equation (or -1 for a decreasing equation), the better the fit. Make sure you are familiar with your calculator's data modeling functionalities *before* Test Day!

PRACTICE SET

You may use your calculator for all questions in this practice set.

Easy

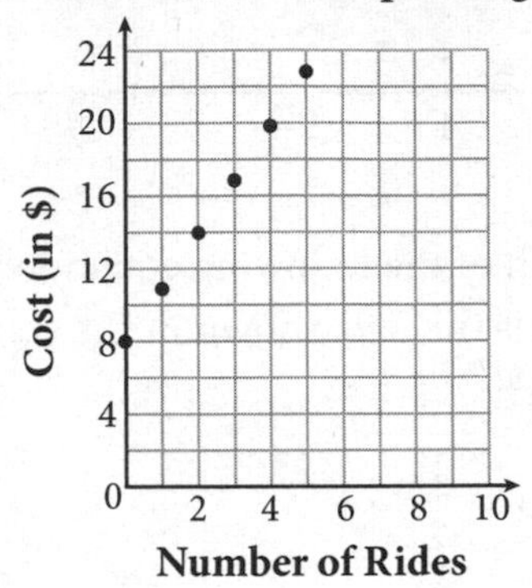

1. Wesley went to an amusement park with his family. He paid for his own admission ticket and all of the rides he rode. The scatterplot shows possible amounts that he could have paid. If a line of best fit (not shown) is used to model the data, the equation of the line would be $y = 3x + 8$. In this scenario, what does 3 represent?

   A) The cost of one ride

   B) The number of rides Wesley rode

   C) The cost of the admission ticket

   D) The number of hours Wesley stayed at the park

2. The data in which of the following scatterplots would be best modeled by a quadratic function in which the x^2 term has a negative coefficient?

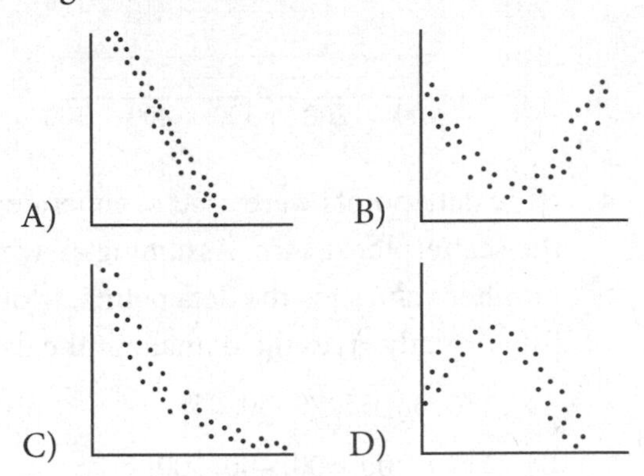

Round	Total Number of Participants Eliminated
1	256
2	384
3	448
4	480
5	496

3. The table above shows the cumulative number of participants eliminated in a national miniature golf tournament by the end of each round. Which of the following best describes the relationship between the round and the number of participants remaining in the tournament?

   A) Exponential growth

   B) Exponential decay

   C) Linear growth

   D) Linear decay

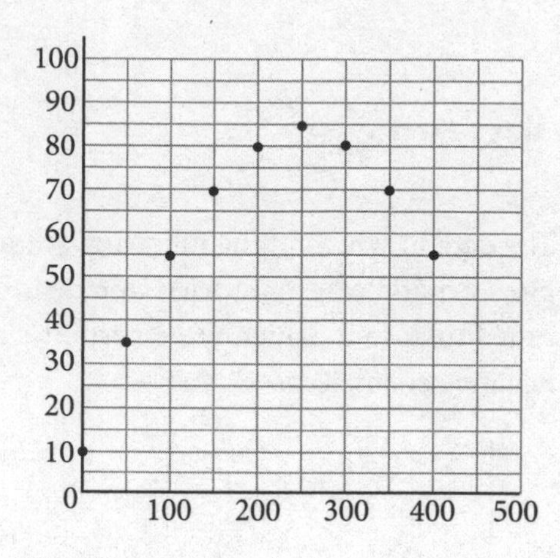

4. Nine data points were used to generate the scatterplot shown. Assuming all whole number values for the data points, which list correctly gives the domain of the data?

   A) {10, 35, 55, 70, 80, 85}

   B) {100, 200, 300, 400, 500}

   C) {0, 50, 100, 150, 200, 250, 300, 350, 400}

   D) {0, 10, 20, 30, 40, 50, 60, 70, 80, 90, 100}

Medium

5. When water is heated in a closed system, the vapor pressure increases slowly at first and then more rapidly. As the water reaches the boiling point of 100°C, the vapor pressure reaches 1 atm, known as 1 standard atmosphere. Which of the following models best describes the increase in vapor pressure as water is heated to its boiling point?

   A) Linear

   B) Quadratic

   C) Polynomial

   D) Exponential

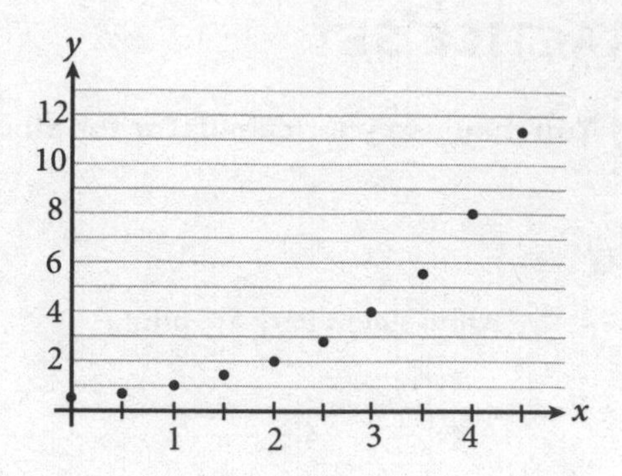

6. Which of the following equations best models the data shown in the scatterplot above?

   A) $y = \frac{1}{2}x$

   B) $y = \left(\frac{1}{2}\right)^x$

   C) $y = \frac{1}{2}(2^x)$

   D) $y = \frac{3}{2}x - 1$

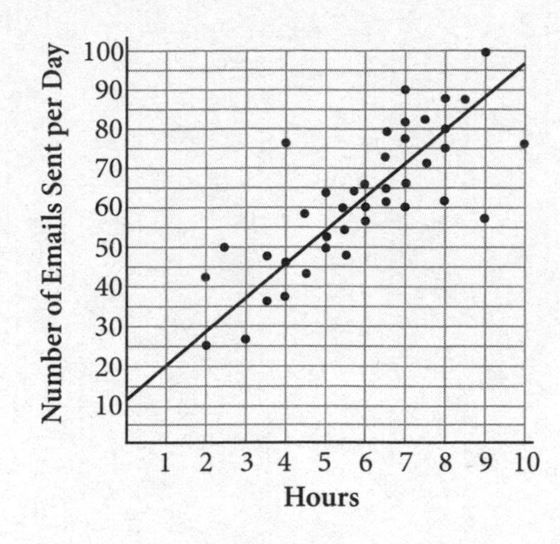

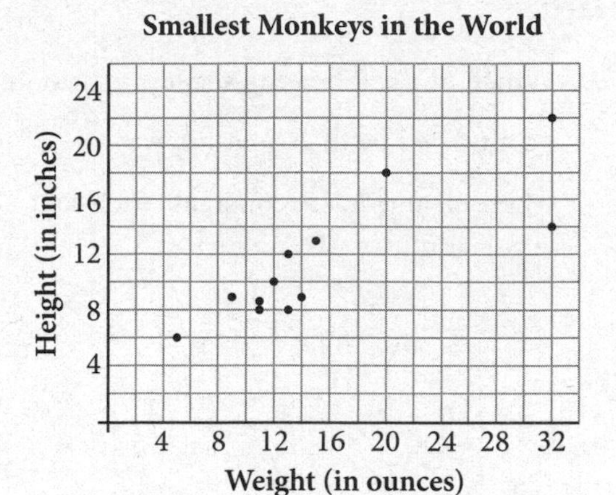

7. The figure above shows the number of emails sent per day plotted against the number of hours an adult works on a computer per day. Which of the following best estimates the average rate of change in the number of emails sent compared to the number of hours working on a computer?

   A) 0.12

   B) 1.5

   C) 8.5

   D) 12

8. There are more than 250 known species of monkeys in the world. The scatterplot above shows the average height and weight of 12 species of particularly small monkeys, most of which live in the Amazon Basin of South America. What is the height, in inches, of the monkey represented by the data point that is farthest from the line of best fit (not shown)?

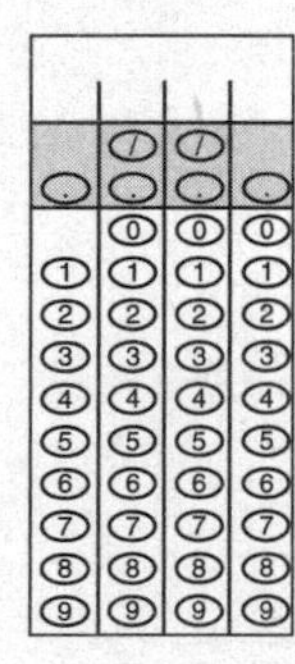

Hard

9. Which of the following scatterplots could be modeled by the equation $y = \frac{a}{b^x}$, where a and b are constants such that $a > 1$ and $x < -1$?

A) 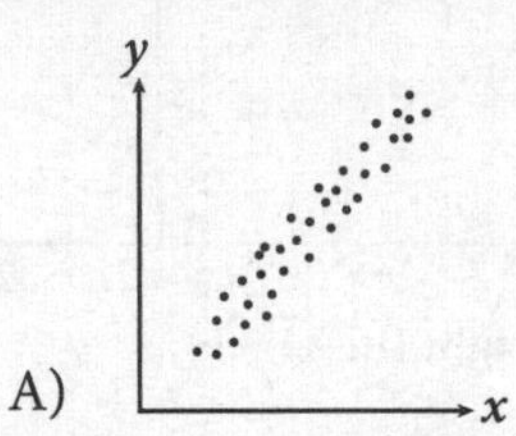

B) 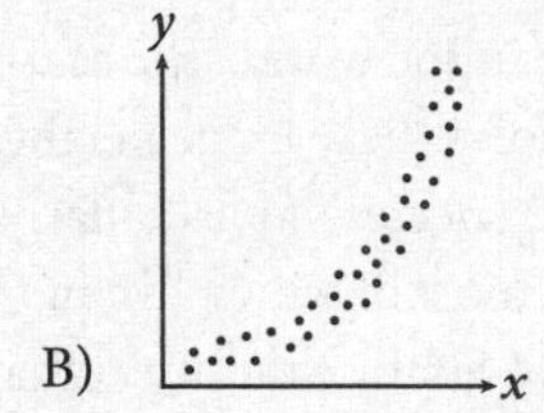

C) 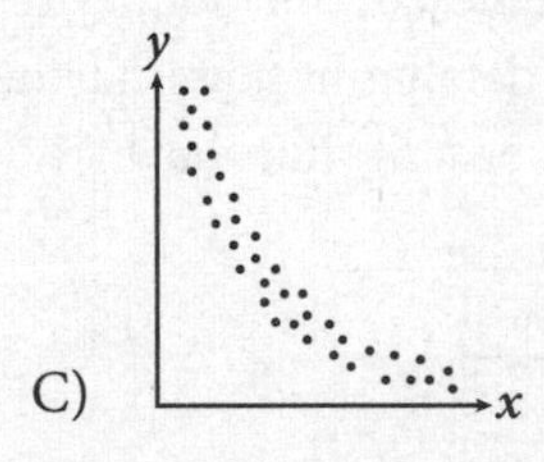

D) 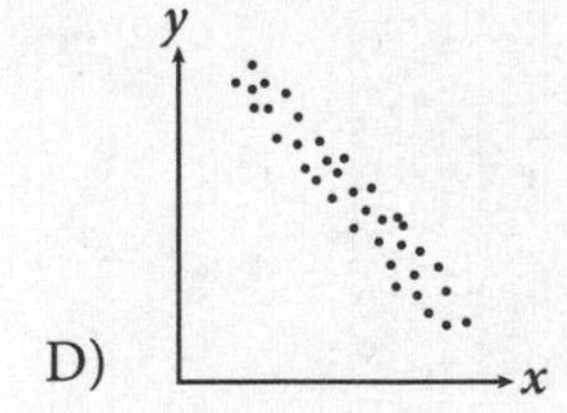

Multi-Part Math Questions

Questions 10-12 refer to the following information.

Sea ice extent is a measurement of the area of ocean with at least 15% sea ice. The graph below shows data for the extent of Arctic sea ice between 2000 and 2012 as reported by the National Snow and Ice Data Center. The line of best fit is also shown. The equation of the line is $A(t) = -0.187t + 380.5$, where $A(t)$ represents the area of measurable sea ice in the Arctic Ocean in the year t.

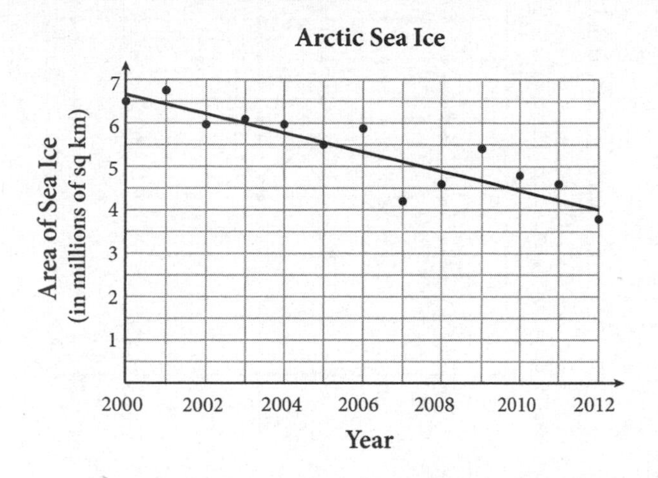

10. Based on the equation of the line of best fit, which of the following statements accurately describes the change in the amount of measurable Arctic sea ice during the given time period?

    A) The amount of sea ice increased approximately 187,000 square kilometers per year.

    B) The amount of sea ice increased approximately 187 million square kilometers per year.

    C) The amount of sea ice decreased approximately 187,000 square kilometers per year.

    D) The amount of sea ice decreased approximately 187 million square kilometers per year.

11. Assuming the trend of the data continues, what is the predicted area, in millions of square kilometers, of measurable Arctic sea ice in the year 2020?

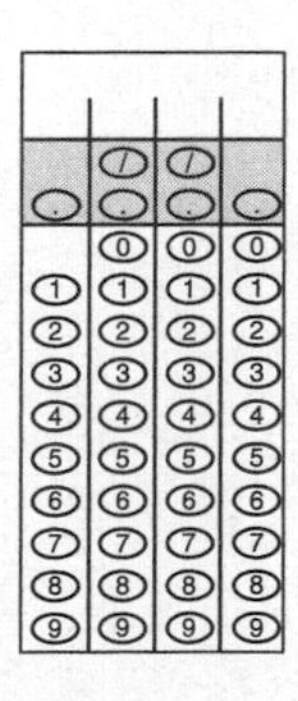

12. Assuming the trend of the data continues, in what year will the measurable sea ice in the Arctic Ocean cease to exist?

    A) 2026

    B) 2034

    C) 2066

    D) 2380

Questions 13 and 14 refer to the following information.

A dietician working in the children's ward at a hospital is monitoring the water intake of all the patients in the ward. The total water intake for each patient is recorded throughout the day and then averaged over the 24-hour period. The results are recorded in the scatterplot, where each dot represents the average hourly water intake for one child.

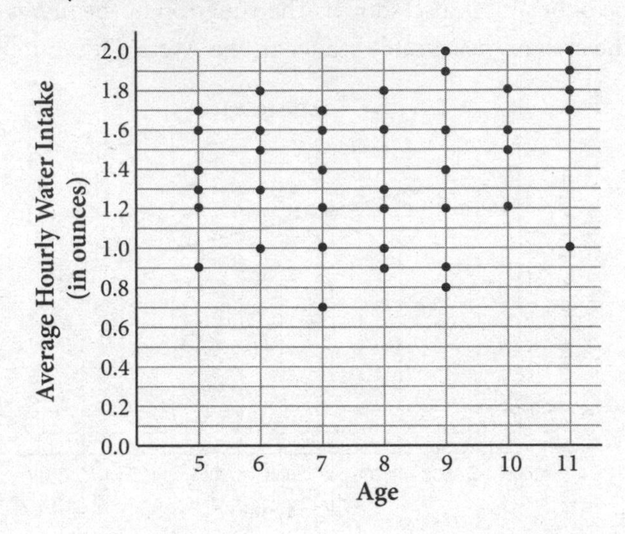

13. Which of the following ages has the greatest range of values for the average hourly water intake?

A) Age 5

B) Age 6

C) Age 8

D) Age 11

14. A child between the ages of 5 and 8 should consume at least 1 liter of water per day. Based on the data presented in the scatterplot, what percent of the children, ages 5-8, in the ward consumed less than the daily-recommended amount? (Note: There are 24 hours in one day and approximately 33.814 ounces in 1 liter.)

A) 35%

B) 60%

C) 70%

D) 85%

Questions 15 and 16 refer to the following information.

As the elevation above sea level increases, the corresponding air pressure decreases. A scatterplot comparing elevation, in feet, to air pressure, as a percent of sea level pressure, is shown below.

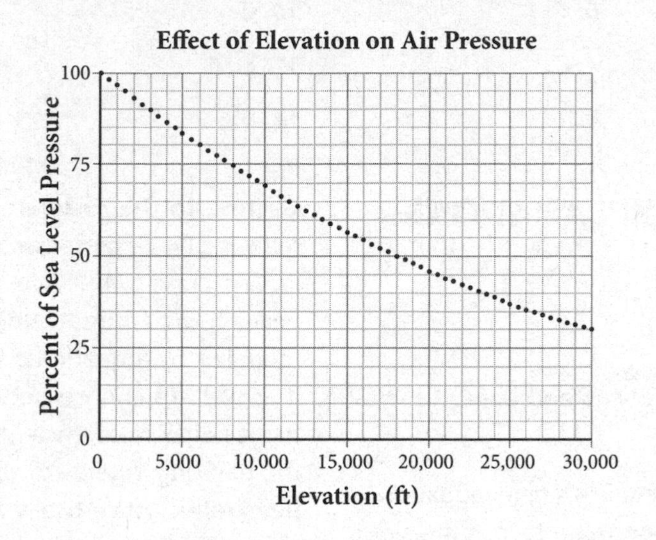

15. Elevations above 8,000 meters are said to be in the "death zone," a range of elevations in which oxygen levels are not high enough to sustain human life. What is the approximate air pressure (as a percent of sea level pressure) at the death zone threshold? (Note: 1 meter ≈ 3.28 feet)

    A) 35%

    B) 50%

    C) 74%

    D) 92%

16. Air pressure is sometimes measured in inches of mercury. If sea level air pressure measures as 29.2 inches of mercury, what is the approximate difference, in inches of mercury, between the pressure at sea level and the pressure at the death zone threshold?

    A) 7.6

    B) 10.2

    C) 19.0

    D) 21.6

ANSWER KEY

1. A	5. D	9. B	13. D
2. D	6. C	10. C	14. B
3. B	7. C	11. 2.76	15. A
4. C	8. 14	12. B	16. C

ANSWERS & EXPLANATIONS

1. A

Difficulty: Easy

Category: Problem Solving and Data Analysis / Scatterplots

Strategic Advice: When a linear equation is written in the form $y = mx + b$, m is the slope and b is the y-intercept. In a real-world scenario, the y-intercept represents a flat fee or a starting amount, and the slope represents a unit rate.

Getting to the Answer: The 3 in the equation is the coefficient of x, so it is the slope of the line of best fit. This means that the number 3 represents a unit rate. Read the axis labels carefully. The numbers of rides are plotted along the horizontal axis, so x represents the number of rides, which means 3 must be the cost *per* ride, or the cost of 1 ride, which is (A).

2. D

Difficulty: Easy

Category: Problem Solving and Data Analysis / Scatterplots

Strategic Advice: Data that can be modeled by a quadratic function should take the shape of a parabola (or part of a parabola), which looks like a U. Part of the data should increase from left to right, and part of the data should decrease (at some point, the data turns around).

Getting to the Answer: The data in choice A follows a linear pattern, so eliminate it. The data in C is curved, but it is always decreasing (it never turns around), which means it can't be modeled by a quadratic function, so eliminate it as well. To choose between B and (D), recall that a negative coefficient on the x^2 term means the parabola opens downward (because it has been reflected vertically across the x-axis). This means (D) is correct.

3. B

Difficulty: Easy

Category: Problem Solving and Data Analysis / Scatterplots

Strategic Advice: Skim the answer choices—they involve the shape of the data (linear or exponential) and the direction of the data (growth/increasing or decay/decreasing). Read the question carefully and then examine the data to determine which is a match.

Getting to the Answer: Read the data carefully—as the number of participants *eliminated* increases, the number of participants *remaining* decreases, which means as the rounds go up, the number of participants goes down. Eliminate A and C. According to the table, the total number of participants eliminated increases at varying rates (very fast at first, and then more slowly), which means the number of participants remaining decreases at varying rates. This indicates an exponential relationship. Choice (B) is the only answer that contains both traits.

4. C

Difficulty: Easy

Category: Problem Solving and Data Analysis / Scatterplots

Strategic Advice: The *domain* of a set of data points is the set of inputs, which corresponds to the x-values of the data points when plotted on a graph. The set of values that make up the *range* corresponds to the y-values. To remember this, think alphabetically: x-axis before y-axis, domain before range, so domain corresponds to x-values on a graph.

Getting to the Answer: You don't need to write each point to find the domain. Just take note of where along the x-axis each of the points is plotted—there is a point on each grid-line between 0 and 400, and the grid-lines increase by 50, so the domain is {0, 50, 100, 150, 200, 250, 300, 350, 400}, which is (C).

5. D

Difficulty: Medium

Category: Problem Solving and Data Analysis / Scatterplots

Strategic Advice: Don't let the scientific terminology in a question like this intimidate you. You don't need to know any chemistry to answer this question.

Getting to the Answer: The key to this question is understanding the shapes of various models. Linear models show a *constant* increase or decrease, so eliminate A. Quadratic models show data that either increases then decreases, or the reverse, so eliminate B. Polynomial models show data that change direction multiple times, so eliminate C as well. This means (D) must be correct. Exponential models show data that either increase slowly at first and then more rapidly (as in this question), or decrease rapidly at first and then more slowly.

6. C

Difficulty: Medium

Category: Problem Solving and Data Analysis / Scatterplots

Strategic Advice: Whenever you need to match data to a model, use the trend (shape) of the data to determine the *type* of model (linear, exponential, quadratic, or polynomial), and then use specific data points (if needed) to identify the best choice.

Getting to the Answer: The trend of the data shown in the scatterplot is increasing, slowly at first and then much more quickly, which indicates an exponential model. This means you can eliminate A and D (both are linear). To choose between B and (C), choose a data point (preferably one that lies on a grid-line and represents whole number values) and test it in each answer choice. Avoid $x = 0$ and $x = 1$ because those numbers have special properties when exponents are involved. Using the point (3, 4), the results are:

$$y = \left(\frac{1}{2}\right)^x \rightarrow \left(\frac{1}{2}\right)^3 = \frac{1}{2} \times \frac{1}{2} \times \frac{1}{2} = \frac{1}{8} \neq 4$$

Eliminate B, making (C) the correct choice. If you're still unsure, you can check (C):

$$y = \frac{1}{2} \cdot (2)^x \rightarrow \frac{1}{2} \cdot (2)^3 = \frac{1}{2} \cdot 8 = 4$$

7. C

Difficulty: Medium

Category: Problem Solving and Data Analysis / Scatterplots

Strategic Advice: When real-world data is presented in a scatterplot, the average rate of change is the same as the slope of the best-fit line.

Getting to the Answer: Find the slope of the best-fit line using the slope formula,

$m = \frac{y_2 - y_1}{x_2 - x_1}$, and any two points that lie on (or very close to) the line. Using the two points (4, 46) and (8, 80), the average rate of change is about $\frac{80-46}{8-4} = \frac{34}{4} = 8.5$, which is (C). Note that the answer choices are spread out enough that any two points along the line, even points for which you estimate the coordinates, will yield a slope that is close enough to one of the answer choices.

8. 14

Difficulty: Medium

Category: Problem Solving and Data Analysis / Scatterplots

Strategic Advice: When working with scatterplots, a line of best fit should be drawn so that about half the data points are above the line and half are below the line.

Getting to the Answer: Draw in the line of best fit. It doesn't have to be perfect—you just need to get an idea of where it should be.

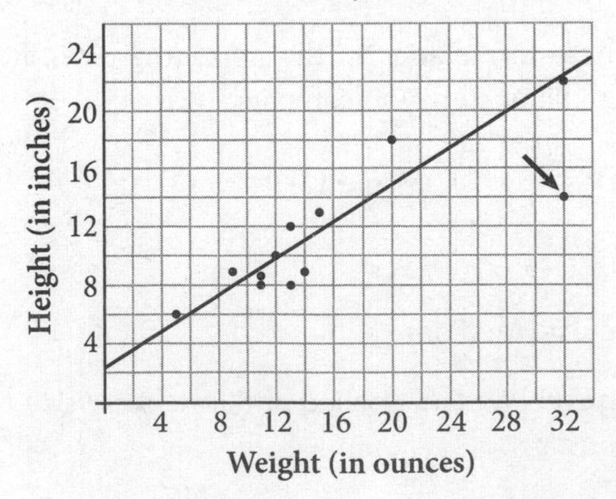

The data point that is farthest from the line is located at (32, 14). Height is plotted on the *y*-axis, so this data point represents a monkey that is 14 inches tall.

9. B

Difficulty: Hard

Category: Problem Solving and Data Analysis / Scatterplots

Strategic Advice: When matching a scatterplot (or any graph) to an equation, first identify the type of equation present to narrow down the choices.

Getting to the Answer: Because $x < -1$, the b term in the denominator will have a negative exponent. You can move the b term to the numerator and reverse the sign of its exponent to get $y = ab^x$, where $a > 1$ and $x > 1$. It is now easier to see that the equation is an exponential equation, which will graph as a curve. This means you can eliminate A and D. Because all the values are now greater than 1, the curve will increase from left to right (gradually at first and then more rapidly), which makes (B) the correct choice.

10. C

Difficulty: Medium

Category: Problem Solving and Data Analysis / Scatterplots

Strategic Advice: In a real-world scenario, the slope of a line represents a unit rate. In this scenario, the unit rate translates as the change in the area of the sea ice per year.

Getting to the Answer: The slope of the line of best fit is −0.187. The negative slope indicates a *decrease* in the amount of ice, so you can eliminate A and B. To decide between (C) and D, read the vertical axis label carefully—the amount of ice is given in millions of square kilometers. Therefore, a slope of −0.187 indicates a decrease per year of 0.187 million square kilometers. This is not one of the answer choices (read D carefully), so multiply 0.187 by 1,000,000 to find that this is equivalent to a decrease of 187,000 square kilometers per year, which is (C).

11. 2.76

Difficulty: Easy

Category: Problem Solving and Data Analysis / Scatterplots

Strategic Advice: Assuming the trend of the data continues, you can use the equation of the line of best fit to predict values outside the reported time period.

Getting to the Answer: Because t represents the year, substitute 2020 into the equation of the line of best fit for t and simplify:

$$A(t) = -0.187t + 380.5$$
$$A(2020) = -0.187(2020) + 380.5$$
$$A(2020) = 2.76$$

The predicted area of Arctic sea ice in 2020 is 2.76 million square kilometers.

12. B

Difficulty: Medium

Category: Problem Solving and Data Analysis / Scatterplots

Strategic Advice: You'll need to translate from English into math to answer this question. The words "cease to exist" tell you that you're looking for the year (t) in which the area of Arctic sea ice will be 0.

Getting to the Answer: Substitute 0 for $A(t)$ and use inverse operations to solve for t:

$$A(t) = -0.187t + 380.5$$
$$0 = -0.187t + 380.5$$
$$0.187t = 380.5$$
$$t = 380.5 \div 0.187 \approx 2034.76$$

Relative to time, this means about three-fourths of the way through the year 2034, so (B) is correct.

13. D

Difficulty: Medium

Category: Problem Solving and Data Analysis / Scatterplots

Strategic Advice: The range of a data set is the largest value minus the smallest value. In this question, you are comparing the ranges for each age, not for the whole data set.

Getting to the Answer: Read the axis labels carefully. Average hourly water intake is plotted along the vertical axis, and each grid-line represents 0.1. Estimate the y-coordinate of the lowest point and the highest point for each of the ages given in the answer choices. (Don't waste time calculating the ranges for all the ages plotted along the horizontal axis.) Then, subtract to find the range for each age:

Age 5: highest = 1.7; lowest = 0.9; difference = 0.8

Age 6: highest = 1.8; lowest = 1.0; difference = 0.8

Age 8: highest = 1.8; lowest = 0.9; difference = 0.9

Age 11: highest = 2.0; lowest = 1.0; difference = 1.0

The greatest difference is 1.0 (age 11), so (D) is correct.

14. B

Difficulty: Hard

Category: Problem Solving and Data Analysis / Scatterplots

Strategic Advice: Always pay careful attention to the units when examining a scatterplot. Here, the daily recommended amount of water is given in liters per day; the data is given in ounces per hour. You'll need to convert the units before you can answer the question. Also make sure you're only including data from ages 5-8.

Getting to the Answer: Start by converting 1 liter per day to ounces per hour. Use the factor-label method to keep your calculations organized:

$$\frac{1\ \cancel{L}}{1\ \cancel{day}} \times \frac{33.814\text{ oz}}{1\ \cancel{L}} \times \frac{1\ \cancel{day}}{24\text{ h}} \approx 1.41\text{ oz per hour}$$

To find the percent of the children, ages 5-8, in the ward that consumed less than this amount, divide the number of dots that are below 1.41 (including those right on the 1.4 line) for ages 5-8 by the total number of dots for ages 5-8. The result is $14 \div 23 \approx 0.6087$, which is closest to 60%. This means (B) is correct.

15. A
Difficulty: Medium

Category: Problem Solving and Data Analysis / Scatterplots

Strategic Advice: When information is presented in a graphic, always read the axis titles and labels carefully, including the units given.

Getting to the Answer: The death zone threshold is given in meters, but elevation is given in feet on the scatterplot. This means you'll need to convert the death zone threshold from meters to feet before doing anything else. Use the factor-label method to assemble a quick calculation:

$$8{,}000\ \cancel{m} \times \frac{3.28\text{ ft}}{1\ \cancel{m}} = 26{,}240\text{ ft}$$

Locate 26,240 ft on the x-axis of the scatterplot, and then identify the corresponding pressure on the y-axis, which is 35%. (A) is the correct answer.

16. C
Difficulty: Medium

Category: Problem Solving and Data Analysis / Scatterplots

Strategic Advice: When questions are presented in a "set," you can sometimes use your answer from the previous question to answer the next one. However, make sure you answer the *right* question.

Getting to the Answer: You need to find the difference between the two indicated pressures. You know that the air pressure at the death zone threshold is approximately 35% of sea level air pressure from the previous question. Find 35% of 29.2 by multiplying: $0.35 \times 29.2 = 10.2$. Be careful here, as the question asks for the difference between this pressure and sea level pressure, which is $29.2 - 10.2 = 19.0$. (C) is the correct answer.

PRACTICE SET 6: TWO-WAY TABLES, STATISTICS, AND PROBABILITY

Two-Way Tables

A two-way table is a table that is used to summarize data that pertains to two variables, sometimes referred to as bivariate data. The information given in a two-way table can be used to assemble ratios, make comparisons, find relative frequencies, and even determine possible relationships between sets of data. For example, if you have a two-way table that summarizes book sales for three local bookshops over the course of one week, you can compare Store 1's sales to total sales, Monday sales to Tuesday sales, and so on. The key to properly using a two-way table is in focusing only on the cell(s), row(s), and/or column(s) that pertain to the question being asked.

Statistics

Statistics is a part of almost every major in college and can be used in a variety of careers, which explains why there are entire high school and college courses devoted to the study of statistics. Fortunately, the SAT will only test you on a few basic statistical concepts. Suppose you took five tests in a world history class and earned scores of 85, 92, 85, 80, and 96. Descriptions of six common statistical measures that you can find for this data set are described below.

- *Mean* (also called average): The sum of the values divided by the number of values. For your history class, the mean of your test scores is $\frac{85+92+85+80+96}{5}=\frac{438}{5}=87.6$.
- *Median*: The value that is in the middle of the set *when the values are arranged in ascending order.* The history scores in ascending order are 80, 85, 85, 92, and 96, making the median 85. To find the median of a data set that contains an even number of terms, find the mean of the two middle terms.
- *Mode*: The value that occurs most frequently. The test score that appears more than any other is 85, so it is the mode. If more than one value appears the most often, that's fine. A set of data can have more than one mode.
- *Range*: The difference between the highest and lowest values. The highest and lowest scores are 96 and 80, so the range is $96 - 80 = 16$.
- *Standard deviation*: A measure of how far a typical data point is from the mean of the data. A low standard deviation indicates that most values in the set are fairly close to the mean; a high standard deviation indicates that the data is more spread out.
- *Margin of error*: A description of the maximum expected difference between a true value for a data pool (e.g., mean) and a random sampling from the data pool. A lower margin of error is achieved by increasing the size of the data pool.

Note: On the SAT, you will need to understand what the standard deviation and margin of error tell you about a set of data, but you won't have to calculate either of them.

Mean, median, and mode are referred to as *measures of central tendency* because they can be used to represent a typical value in the data set. Range, standard deviation, and margin of error are *measures of spread* because they show how much the data in a set vary.

You might also be asked to describe or analyze the shape of a data set, which can be either symmetric or skewed (asymmetric). Many data sets have a head, where many data points are clustered in one area, and tails, where the number of data points slowly decreases to 0. A data set is skewed in the direction of its longer tail.

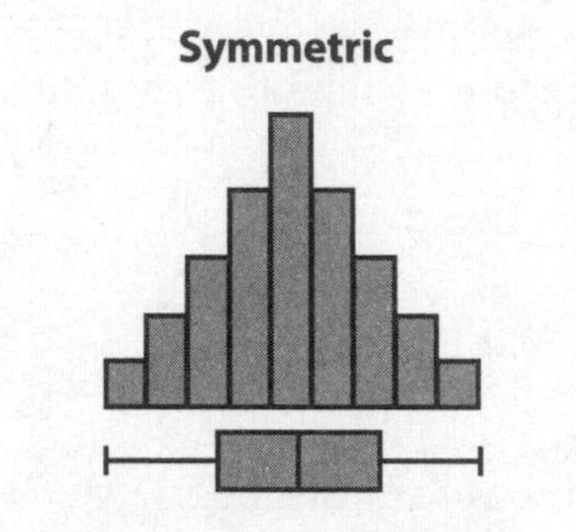

The data is evenly spread out.
mean ≈ median

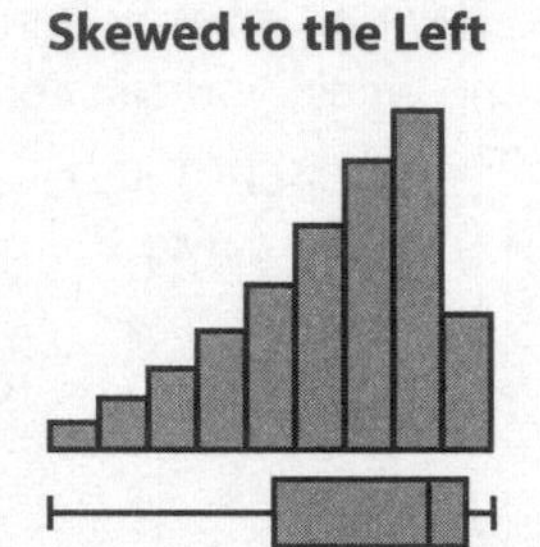

The tail is longer on the left.
mean < median

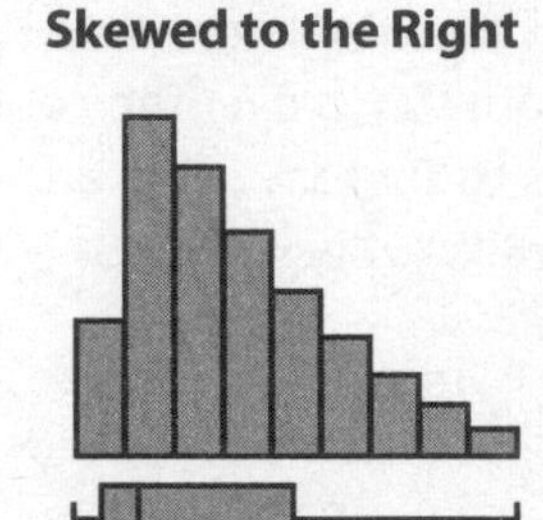

The tail is longer on the right.
mean > median

Probability

A concept closely linked to statistics is probability. Probability is a fraction or decimal comparing the number of desired outcomes to the number of total possible outcomes. From a two-way table, you can find the probability that a randomly selected data value (be it a person, object, etc.) will fit a certain profile (in a certain age range, married or single, etc.). You could also be asked to calculate a conditional probability. Conditional probability questions are easy to spot, as the word *given* is usually present. Be careful on these questions, as the total number of possible outcomes is dictated by the question stem and is not necessarily the grand total of items in the table.

Word Problems

You're already well versed in deciphering SAT word problems, but word problems involving statistics and/or probability sometimes require a different skill set. You'll need to use your analytical abilities to make inferences, justify conclusions, and evaluate the appropriateness of data collection techniques. Sometimes you'll need to use a two-way table or an infographic, and in other cases you'll need to study the question stem to gather relevant information. Perhaps the best part about these questions is the fact that the mathematical calculations involved are usually very straightforward and that you'll be able to use your calculator for all questions of this type.

PRACTICE SET

You may use your calculator for all questions in this practice set.

Easy

Number of Graduates by Focus and Year		
School Focus (District 1)	**2013**	**2014**
Career and Technical Education (CTE)	120	115
Fine and Performing Arts	146	151
International Studies	84	104
Science, Technology, Engineering, and Math (STEM)	163	163
World Languages	112	117

1. A magnet school is a free public school that has a focused theme. According to the table above, what was the increase in the total number of graduates in five of these magnet schools in District 1 from 2013 to 2014?

   A) 25
   B) 30
   C) 35
   D) 40

Rating	1	2	3	4	5
Frequency	1	3	5	5	6

2. Many people shop online primarily because they get access to product ratings submitted by fellow shoppers. The frequency table above gives the most recent 20 customer ratings for a certain product sold online. What is the mean rating for this product?

   A) 3.6
   B) 3.8
   C) 4
   D) 5

3. Phase I clinical trials are run to determine the safety of all new drugs, especially with respect to the severity and duration of side effects. A physician is overseeing a Phase I trial based on 800 healthy participants. Half of the participants are given the drug, and half are given an inert pill. The mean duration of side effects for those participants who were given the drug was 72 hours with a margin of error of 6 hours. The physician is planning to replicate the trial in an attempt to decrease this margin of error. Which of the following will most likely lead to a decrease in the margin of error for the mean duration of side effects?

   A) Decrease the trial size to 400 healthy participants.
   B) Decrease the trial size to 400 participants, and replace half the healthy participants with sick patients.
   C) Increase the trial size to 1,600 healthy participants.
   D) Increase the trial size to 1,600 participants, and replace half the healthy participants with sick patients.

4. In a survey of 1,600 adults in the United States, 48% indicated that television is their primary source of news. The margin of error for the survey is ±2.5%. Which value is outside the interval that is likely to contain the exact percent of all adults in the United States who get the news primarily from television?

   A) 45
   B) 47
   C) 49
   D) 50

5. A medical practice surveyed a random sample of 80 patients to determine whether their facility should open earlier in the morning or close later in the evening. Of the 80 patients surveyed, 37.5% preferred that the facility open earlier. Based on this information, about how many of the practice's 600 patients would be expected to prefer that the facility open earlier in the morning?

Medium

2015 Unemployment Rates (as a Percent)				
By Age and Gender	**Jan.**	**Feb.**	**Mar.**	**Apr.**
Men, 16+ years	5.9	5.7	5.6	5.5
Women, 16+ years	5.6	5.4	5.3	5.4
Total, 16+ years	5.7	5.5	5.5	5.4

6. Unemployment rates in the U.S. for the first four months of 2015 are given in the table above. These rates are based on people, 16 years and older, that are considered to be part of the workforce. In a certain district, there are 12,800 people eligible to be part of the workforce, of which 53.5% are men. Based on the information in the table, how many more men in this district were unemployed than women in April 2015 in this district?

A) 13
B) 56
C) 90
D) 896

7. Luge is a winter sport in which a person slides down an ice track feet first on a small sled. The luger lies supine (on his back) and uses his calf and shoulder muscles to steer the sled. Below is a summary of the times, in seconds, of three lugers' practice runs on a track in Utah.

Run	Marcelle	Aaron	Danielle
1	59.209	55.302	56.850
2	57.916	52.631	55.414
3	58.402	57.914	54.650
4	58.808	53.215	55.845
Mean	58.584	54.766	55.670
Std. Dev.	0.554	2.392	0.918

Which of the following conclusions can be drawn based on the data in the table?

A) Aaron performed the least consistently because his mean time is the lowest.
B) Marcelle performed the least consistently because his mean time is the highest.
C) Aaron performed the most consistently because his standard deviation is the highest.
D) Marcelle performed the most consistently because his standard deviation is the lowest.

8. Frederick is a business major and is conducting a study to determine the effect of having agendas for project meetings at large companies. He surveyed a randomly selected group of 500 project managers for large companies in Boston, MA, and found substantial evidence of an association between the duration of a project meeting and whether an agenda was followed. Which of the following can Frederick conclude from this information?

A) Use of a meeting agenda causes a decrease in meeting time at large companies in Boston, MA.

B) Use of a meeting agenda causes a decrease in meeting time at large companies in the United States.

C) There is an association between the use of a meeting agenda and the duration of meetings at large companies in Boston, MA.

D) There is an association between the use of a meeting agenda and the duration of meetings at large companies in the United States.

9. A study of gasoline prices across a particular state showed that the state's mean gas price was \$2.87 per gallon, and the median gas price was \$3.25 per gallon. Which of the following most likely accounts for the difference between the mean and median gas prices in this state?

A) A few areas of the state had substantially lower gas prices than the rest.

B) A few areas of the state had substantially higher gas prices than the rest.

C) The majority of areas in the study had gas prices between \$2.87 and \$3.25 per gallon.

D) An error was made during data collection or data analysis.

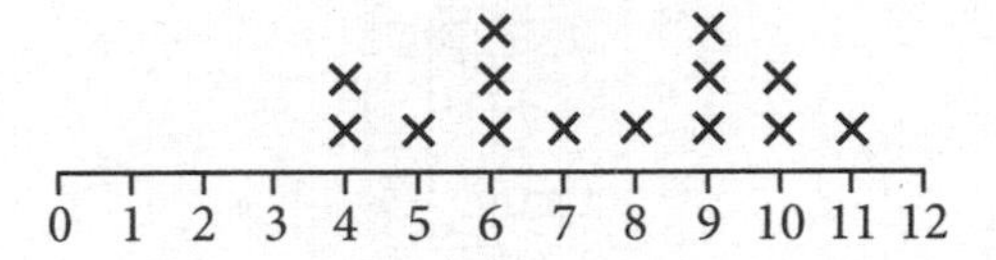

10. Brian and several of his work colleagues are on a kickball team together. The dot plot above shows the number of points his team scored during each of 14 games played so far this season. The league's record for highest average points scored per game during a season is 7.5. To break this record, a team's season average must be at least half a point higher. If Brian and his teammates are to set a new season record, how many points must they average during their final two games?

A) 9

B) 12

C) 14

D) 15

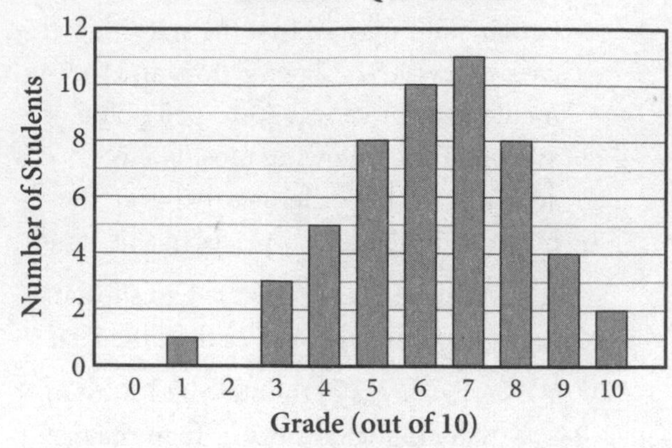

11. The bar graph above shows the results of a grammar quiz in a language arts class. What is the difference between the mean and median of the quiz scores? Round your answer to the nearest hundredth.

Hard

12. The average gas mileage for n vehicles on one car lot is 21 miles per gallon. The average gas mileage for p vehicles on another car lot is 25 miles per gallon. When the vehicles on both lots are combined, the average gas mileage is 22.5 miles per gallon. What is the value of $\frac{p}{n}$?

    A) $\frac{3}{8}$

    B) $\frac{1}{2}$

    C) $\frac{3}{5}$

    D) 1

13. Five numbers are given by the expressions x, $2x - 3$, $2x + 1$, $3x - 4$, and $3x + 1$. If the average (arithmetic mean) of these numbers is 10, what is the mode of the numbers?

    A) 10

    B) 11

    C) 12

    D) There is not enough information to determine the value of the mode.

Top Four East Coast Metro Areas for New Job Growth per Capita				
	Metro A	Metro B	Metro C	Metro D
Total jobs added in metro area	94,200	86,350	72,600	66,000
New jobs per 1,000 residents	245	375	250	200

14. According to the information in the table, which of the following correctly orders the populations of the metro areas from greatest to least?

    A) A, B, C, D

    B) D, C, B, A

    C) A, D, C, B

    D) A, D, B, C

Multi-Part Math Questions

Questions 15 and 16 refer to the following information.

	Tropical Storms	Hurricanes	Major Hurricanes (Cat. 3+)	Total
1990–1994	16	23	5	44
1995–1999	5	41	20	66
2000–2004	18	37	18	73
2005–2009	23	37	18	78
Total	91	175	74	340

15. The table above summarizes storm activity in the Atlantic Ocean between 1990 and 2009. What portion of the storms between 1990 and 1999 were classified as major hurricanes?

    A) 0.1136

    B) 0.2176

    C) 0.2273

    D) 0.8091

16. The students in an environmental science class will be randomly assigned a storm to research, with no two students researching the same storm. Given that the students will each be assigned a hurricane or a major hurricane for their reports, what is the probability that a student will be assigned a storm that occurred prior to 2000?

    A) 0.2618

    B) 0.3574

    C) 0.6426

    D) 0.7324

Questions 17 and 18 refer to the following information.

	A	B	C	D	E	F	G	H
Nonstop	n/a	$340.20	$340.20	$536.20	$380.19	$382.20	$383.95	$429.00
1 + stops	$332.20	$376.70	n/a	$363.19	$395.59	$386.20	$392.45	$493.99

17. The table above is a summary of prices for round-trip airplane tickets offered by eight airlines between New York City and San Francisco. What fraction of the nonstop flights are under $400?

18. Airlines C and E offer six nonstop flights each to San Francisco throughout the day at the prices shown in the table. Each flight on each airline can hold 170 passengers. If all six of Airline C's flights are filled to capacity and all six of Airline E's flights are filled to 90% capacity on a particular day, how much more revenue does Airline E generate than Airline C on that day?

A) $1,005.21

B) $2,010.42

C) $20,394.90

D) $40,789.80

Questions 19 and 20 refer to the following information.

	$m < 50{,}000$	$50{,}000 \le m \le 100{,}000$	$m > 100{,}000$	Total
2-Door Car	9	22	17	48
4-Door Car	16	48	34	98
SUV	19	35	40	94
Truck	12	27	21	60
Total	56	132	112	300

19. The table above shows the distribution of vehicles at a used car lot by type and by mileage. According to the data, which type of vehicle on the lot has the smallest percentage with less than 50,000 miles?

A) 2-door car

B) 4-door car

C) SUV

D) Truck

20. Based on the table, if a single vehicle is selected at random from all the vehicles on the lot to be given away in a raffle, what is the probability that it will be an SUV or a truck that has been driven between 50,000 and 100,000 miles?

A) $\frac{7}{60}$

B) $\frac{11}{25}$

C) $\frac{31}{66}$

D) $\frac{31}{150}$

ANSWER KEY

1. A	6. B	11. 29	16. B
2. A	7. D	12. C	17. 5/7
3. C	8. C	13. B	18. B
4. A	9. A	14. C	19. B
5. 225	10. B	15. C	20. D

ANSWERS & EXPLANATIONS

1. A
Difficulty: Easy

Category: Problem Solving and Data Analysis / Statistics and Probability

Strategic Advice: A question like this is as straightforward as it sounds—so be extra careful with your calculations and you are sure to get it correct.

Getting to the Answer: Don't try to keep track of the changes in your head. Instead, jot them down to the right of the last column. Be sure to include a positive or negative sign to indicate whether each change is an increase or a decrease. Then add the changes to find the total increase.

2013	2014	Change
120	115	−5
146	151	+5
84	104	+20
163	163	0
112	117	+5

To add quickly, notice that the first two changes cancel each other, so the total increase is $20 + 5 = 25$, which is (A).

2. A
Difficulty: Easy

Category: Problem Solving and Data Analysis / Statistics and Probability

Strategic Advice: To find the mean of a data set presented in a frequency table, multiply each data value (here, the rating) by the number of times it occurs (the frequency). Then, add the results and divide by the total number of data values.

Getting to the Answer: The mean rating means the average score given. The sum of the ratings is: $(1 \times 1) + (2 \times 3) + (3 \times 5) + (4 \times 5) + (5 \times 6) = 1 + 6 + 15 + 20 + 30 = 72$. Divide by the total number of ratings (20) to find that the mean is $72 \div 20 = 3.6$, which is (A).

3. C
Difficulty: Easy

Category: Problem Solving and Data Analysis / Statistics and Probability

Strategic Advice: Margin of error is a measure of data consistency. Think about how you could alter the sample pool for a study to decrease the margin of error.

Getting to the Answer: When conducting trials (or surveys), larger sample sizes lead to more reliable data. This, in turn, results in smaller margins of error. Therefore, the physician should increase the number of participants in the trial. Eliminate A and B. Replacing half the healthy participants with sick patients introduces another variable into the trial; consequently, doing this will prevent a valid comparison of the margins of error of the two trials. Choice (C) must be correct.

4. A
Difficulty: Easy

Category: Problem Solving and Data Analysis / Statistics and Probability

Strategic Advice: When the sample size of a random survey is sufficiently large, the margin of error provides a limit on how much the sample would differ from the responses of the population.

Getting to the Answer: To find the limit on how much the sample size would differ, add and subtract the margin of error:

$48\% - 2.5\% = 45.5\%$ and $48\% + 2.5\% = 50.5\%$

Based on this survey, the exact percent of all adults in the United States who get the news primarily from television should fall between 45.5% and 50.5%. The only answer choice that is outside this range is 45%, making (A) the correct answer.

5. 225
Difficulty: Easy

Category: Problem Solving and Data Analysis / Statistics and Probability

Strategic Advice: When a random sample is used to survey part of a population, the results are fairly representative of the entire population.

Getting to the Answer: Because the medical practice surveyed a random sample, the percent of all the patients expected to prefer that the facility open earlier can be estimated by the percent of patients who preferred that option in the sample, 37.5%. Therefore, of the practice's 600 patients, approximately $600 \times 0.375 = 225$ patients would be expected to prefer that the facility open earlier.

6. B
Difficulty: Medium

Category: Problem Solving and Data Analysis / Statistics and Probability

Strategic Advice: Whenever information is presented in a table, be sure to read the column and row headers carefully and only use the information that applies to the question.

Getting to the Answer: Before you can apply percentages from the table, you need to determine how many men and how many women there were in the workforce in the district. The number of men was 53.5% of 12,800, or $0.535 \times 12{,}800 = 6{,}848$, and the number of women was $12{,}800 - 6{,}848 = 5{,}952$. Now, use the information from the table for April:

Men: 5.5% of $6{,}848 = 0.055 \times 6{,}848 = 376.64$, or about 377 unemployed men

Women: 5.4% of $5{,}952 = 0.054 \times 5{,}952 = 321.408$, or about 321 unemployed women

This means there were $377 - 321 = 56$ more unemployed men than women in the district in April. Choice (B) is correct.

7. D
Difficulty: Medium

Category: Problem Solving and Data Analysis / Statistics and Probability

Strategic Advice: Skim through the answer choices—each one involves consistency, so you need to examine the statistic that measures consistency, which is standard deviation.

Getting to the Answer: The mean run times have nothing to do with consistency, so eliminate A and B. A lower standard deviation means the data is less spread out and therefore more consistent. Marcelle's standard deviation is the lowest, so he performed the most consistently. This matches (D).

8. C

Difficulty: Medium

Category: Problem Solving and Data Analysis / Statistics and Probability

Strategic Advice: You cannot make assumptions beyond what the question provides, so analyze the information (and the answer choices) carefully. For example, if you're given information about a group of teenagers, you can't assume the same information applies to adults as well.

Getting to the Answer: Only large companies in Boston were included in the study, so Frederick cannot draw conclusions about large companies throughout the United States. Eliminate B and D. Frederick's findings revealed a link (an association) between the use of an agenda and a meeting's duration. However, an association alone does not prove causation, so eliminate A. This leaves (C) as the correct answer.

9. A

Difficulty: Medium

Category: Problem Solving and Data Analysis / Statistics and Probability

Strategic Advice: Knowing when the mean and median of a data set are equal is the key to answering a question like this.

Getting to the Answer: The mean of a data set is approximately equal to the median when the data has a symmetrical distribution. In this scenario, the mean and median are not equal, which indicates that the data distribution is asymmetric. The difference between them is fairly large (compared to the values themselves), which could be explained by the presence of outliers (values that are significantly smaller or larger than the other values in the data). Outliers affect the mean of a data set but not the median. Because the mean is lower than the median, the outliers must also be on the low side, which means you can reasonably conclude that a few areas in the study had gas prices that were much lower than those elsewhere in the state. Choice (A) is correct.

10. B

Difficulty: Medium

Category: Problem Solving and Data Analysis / Statistics and Probability

Strategic Advice: The average of a data set is found by adding all the values and dividing the result by the number of values. Use this information in reverse to answer a question like this.

Getting to the Answer: The question implies that there are 16 games per season (Brian's team has already played 14 and has 2 left) and that Brian's team must average 8 points (the current record, 7.5, plus half a point) per game to beat the old record. To average 8 points per game, a team must score a total of $8 \times 16 = 128$ points during the season. Each '×' on the dot plot represents one game, so Brian's team has already scored a total of $(2 \times 4) + 5 + (3 \times 6) + 7 + 8 + (3 \times 9) + (2 \times 10) + 11 = 104$ points. They are currently 24 points short of the 8-point average with two games left to play, which means they must average at least 12 points per game during their last two games. (B) is the correct answer.

11. 29

Difficulty: Medium

Category: Problem Solving and Data Analysis / Statistics and Probability

Strategic Advice: Knowing basic statistics definitions is the key to answering a question like this. The calculations aren't difficult—you just have to know how to perform them.

Getting to the Answer: When data values are arranged from least to greatest, the median is the value in the middle. Add the bar heights to find that there are 52 students represented in the bar graph, which is an even number, so take the average of the two middle values to get the median. Half of 52 is 26, so the values in the middle are the 26th and 27th value. Both of these values are 6, so the median is 6. To find the mean, add all the quiz scores together and divide by the number of students, 52. To speed up the calculations, multiply each bar height by the corresponding score (mentally if possible):

$$[(1 \times 1) + (3 \times 3) + (5 \times 4) + (8 \times 5) + (10 \times 6) + (11 \times 7) + (8 \times 8) + (4 \times 9) + (2 \times 10)] \div 52 = \frac{327}{52} \approx 6.2885.$$

Rounded to the nearest hundredth, the difference between the median and the mean is $6.29 - 6 = 0.29$. Grid this in as .29.

12. C

Difficulty: Hard

Category: Problem Solving and Data Analysis / Statistics and Probability

Strategic Advice: In some questions, you can't find the value of individual variables, but you can find a comparison (or ratio) of the variables. In questions such as these, start with definitions and formulas that you know, and work toward the requested ratio.

Getting to the Answer: The average of a set of values is the sum of the values divided by the number of values. Use the information provided in the question to fill in as much information as you can about the two car lots. You might want to give the lots names, such as Lot A and Lot B:

Lot A

$$\text{avg gas mileage} = \frac{\text{sum of gas mileages on Lot A}}{\text{number of vehicles on Lot A}}$$

$$21 = \frac{sum_A}{n} \rightarrow 21n = sum_A$$

Lot B

$$\text{avg gas mileage} = \frac{\text{sum of gas mileages on Lot B}}{\text{number of vehicles on Lot B}}$$

$$25 = \frac{sum_B}{p} \rightarrow 25p = sum_B$$

Now, put the averages together:

Lot A + Lot B

$$\text{avg gas mileage} = \frac{\text{sum on Lot A } + \text{sum on Lot B}}{\text{vehicles on A } + \text{vehicles on B}}$$

$$22.5 = \frac{sum_A + sum_B}{n + p}$$

It may not seem like you're making progress, but you're actually very close. You're looking for the ratio of p over n, so substitute $21n$ for sum_A (found earlier) and $25p$ for sum_B and simplify the equation by cross-multiplying and combining like terms:

$$\frac{22.5}{1} = \frac{21n + 25p}{n + p}$$

$$21n + 25p = 22.5(n + p)$$

$$21n + 25p = 22.5n + 22.5p$$

$$2.5p = 1.5n$$

$$\frac{p}{n} = \frac{1.5}{2.5}$$

To remove the decimals from the fraction, multiply the numerator and dominator by 2 to get $\frac{p}{n} = \frac{3}{5}$, which matches (C).

13. B

Difficulty: Hard

Category: Problem Solving and Data Analysis / Statistics and Probability

Strategic Advice: Try not to overthink a question like this—instead, use what you know about finding an average and see where that takes you.

Getting to the Answer: The average of a set of values is the sum of the values divided by the number of values. Use this definition along with the average given in the question (10) to solve for x:

$$\text{Avg} = \frac{\text{sum of terms}}{\text{number of terms}}$$

$$10 = \frac{x+(2x-3)+(2x+1)+(3x-4)+(3x+1)}{5}$$

$$10 = \frac{11x-5}{5}$$

$$50 = 11x - 5$$

$$55 = 11x$$

$$x = 5$$

Now, substitute 5 for x in each of the expressions: 5, 2(5) − 3, 2(5) + 1, 3(5) − 4, and 3(5) + 1 give the numbers 5, 7, 11, 11, and 17. The *mode* of a data set is the number that occurs most often, which in this case is 11. This means (B) is correct.

14. C

Difficulty: Hard

Category: Problem Solving and Data Analysis / Statistics and Probability

Strategic Advice: This is a tricky question because the two numbers given for each metro area (total jobs added and jobs per 1,000 residents) can't be used alone to answer the question. Think conceptually before making calculations.

Getting to the Answer: The table shows the number of jobs added in total and the number added per 1,000 residents in each area. You need to set up and solve (using cross-multiplication) a proportion for each metro area that looks like $\frac{\text{new jobs}}{1{,}000} = \frac{\text{total new jobs}}{\text{total population}}$. The total population is the unknown, so call it p.

Metro A: $\frac{245}{1{,}000} = \frac{94{,}200}{p} \rightarrow$
$245p = 94{,}200{,}000 \rightarrow p \approx 384{,}490$

Metro B: $\frac{375}{1{,}000} = \frac{86{,}350}{p} \rightarrow$
$375p = 86{,}350{,}000 \rightarrow p \approx 230{,}267$

Metro C: $\frac{250}{1{,}000} = \frac{72{,}600}{p} \rightarrow$
$250p = 72{,}600{,}000 \rightarrow p \approx 290{,}400$

Metro D: $\frac{200}{1{,}000} = \frac{66{,}000}{p} \rightarrow$
$200p = 66{,}000{,}000 \rightarrow p \approx 330{,}000$

Ordering the results from greatest to least gives A, D, C, B, which matches (C).

15. C

Difficulty: Easy

Category: Problem Solving and Data Analysis / Statistics and Probability

Strategic Advice: When there is a lot of data presented in a table, be sure to focus only on the information that the question asks about (here, 1990 to 1999 and major hurricanes).

Getting to the Answer: According to the table, between 1990 and 1999 (which includes 1990-1994 and 1995-1999), there were a total of $44 + 66 = 110$ storms. Of these, $5 + 20 = 25$ were major hurricanes. Therefore, the portion you're looking for is $\frac{25}{110}$, which is approximately 0.2273. (C) is the correct answer.

16. B

Difficulty: Medium

Category: Problem Solving and Data Analysis / Statistics and Probability

Strategic Advice: In a probability question, the phrase "given that" indicates conditional probability, which means the total possible outcomes will be different from the total number of values in the table. Take this into account when setting up your calculations.

Getting to the Answer: You're told that all students will be assigned a hurricane or major hurricane, so the total possible outcomes is the sum of these categories: $175 + 74 = 249$. The question asks for the probability that a randomly selected storm occurred prior to 2000, so add together the number of hurricanes and major hurricanes that occurred before 2000 (which includes 1990–1994 and 1995–1999): $23 + 5 + 41 + 20 = 89$. This means the probability you're after is $\frac{89}{249} \approx 0.3574$. (B) is therefore correct.

17. 5/7
Difficulty: Easy

Category: Problem Solving and Data Analysis / Statistics and Probability

Strategic Advice: When dealing with straightforward ratio questions, pay careful attention to the wording.

Getting to the Answer: There are 7 nonstop flights total (Airline A does not offer any). Of these flights, 5 cost less than \$400. Therefore, the correct answer is $\frac{5}{7}$.

18. B
Difficulty: Medium

Category: Problem Solving and Data Analysis / Statistics and Probability

Strategic Advice: Always read carefully to ensure that you know all the facts, and verify that you are using the correct data from a table.

Getting to the Answer: Six flights of 170 passengers each is 1,020 passengers per day. Airline C collects \$340.20 from each passenger for this flight, which equals \$347,004 in revenue. Airline E's flights are only 90% full, which means they have $0.9 \times 1{,}020 = 918$ passengers. Each one pays \$380.19, so Airline E generates \$349,014.42 in revenue. The difference between these two amounts is \$2,010.42, which makes (B) the correct answer.

19. B
Difficulty: Medium

Category: Problem Solving and Data Analysis / Statistics and Probability

Strategic Advice: You rarely need to use all of the information presented in a table to answer a data analysis question. Read the question carefully to make sure you use only what you need.

Getting to the Answer: To calculate the percentage of the vehicles for each type that has less than 50,000 miles ($m < 50{,}000$), divide the number of vehicles of *that* type with less than 50,000 miles by the total number of vehicles of *that* type.

A: 2-door car $= 9 \div 48 = 0.1875$

B: 4-door car $= 16 \div 98 \approx 0.1633$

C: SUV $= 19 \div 94 \approx 0.2021$

D: Truck $= 12 \div 60 = 0.2$

Choice (B) is correct because 16.33% is a lower percentage than any of the other types of vehicle.

20. D
Difficulty: Hard

Category: Problem Solving and Data Analysis / Statistics and Probability

Strategic Advice: Most two-way table questions require that you first identify the rows and columns of interest. It may help to circle the row(s) and column(s) that you need to answer the question.

Getting to the Answer: The first criterion (the vehicle is an SUV or a truck) tells you to focus on rows 4 and 5. The second criterion (has been driven between 50,000 and 100,000 miles) tells you to focus on the center column. Of the 300 vehicles on the lot, there are 35 SUVs and 27 trucks with the indicated mileage, resulting in a total of $35 + 27 = 62$ out of 300 vehicles. The probability of randomly selecting one of these vehicles is $\frac{62}{300}$, which reduces to $\frac{31}{150}$, making (D) correct.

PRACTICE SET 7: EXPONENTS, RADICALS, POLYNOMIALS, AND RATIONAL EXPRESSIONS

Exponents

Questions involving exponents often look intimidating, but knowing the rules of exponents provides plenty of shortcuts. Make sure you're comfortable with the following rules and terminology before Test Day:

- Terminology: The *base* is the value being multiplied by itself, and the *exponent* (written as a superscript) tells you how many times to multiply ($3^5 = 3 \times 3 \times 3 \times 3 \times 3$).
- Multiplying terms with the same base: $a^b \times a^c = a^{(b+c)}$
- Dividing terms with the same base: $\frac{a^b}{a^c} = a^{(b-c)}$
- Raising a power to a power: $(a^b)^c = a^{bc}$
- Raising a product or quotient to a power: $(ab)^c = a^c \times b^c$ and $\left(\frac{a}{b}\right)^c = \frac{a^c}{b^c}$
- Raising a quantity to the zero power: $a^0 = 1$
- Raising a quantity to a negative power: $a^{-b} = \frac{1}{a^b}$ and $\frac{1}{a^{-b}} = a^b$

Radicals

Radicals (square roots, cube roots, etc.) can be written using fractional exponents (for example, $\sqrt{x} = x^{\frac{1}{2}}$, $\sqrt[3]{x} = x^{\frac{1}{3}}$, and so on). Fortunately, this means that radicals follow the same rules as exponents.

- Multiplying: $\sqrt{ab} = \sqrt{a} \times \sqrt{b}$ (because $\sqrt{ab} = (ab)^{\frac{1}{2}} = a^{\frac{1}{2}}b^{\frac{1}{2}} = \sqrt{a} \times \sqrt{b}$)
- Dividing: $\sqrt{\frac{a}{b}} = \frac{\sqrt{a}}{\sqrt{b}}$ (because $\sqrt{\frac{a}{b}} = \left(\frac{a}{b}\right)^{\frac{1}{2}} = \frac{a^{\frac{1}{2}}}{b^{\frac{1}{2}}} = \frac{\sqrt{a}}{\sqrt{b}}$)
- Powers: $a^{\frac{b}{c}} = \sqrt[c]{a^b}$ (because $a^{\frac{b}{c}} = (a^{\frac{1}{c}})^b = (\sqrt[c]{a})^b$)
- Rationalizing a denominator: When a fraction contains a radical in the denominator, multiply the numerator and denominator by the radical in the denominator.

Note: Radicals with different indices (square root versus cube root) can only be multiplied (or divided) by first writing the radicals using fractional exponents. For example, $\sqrt{7} \times \sqrt[3]{7} = 7^{\frac{1}{2}} \times 7^{\frac{1}{3}} = 7^{\frac{1}{2}+\frac{1}{3}} = 7^{\frac{5}{6}} = \sqrt[6]{7^5}$.

Solving Radical Equations

To solve a radical equation, follow these steps:

- Isolate the radical part of the equation.
- Remove the radical using an inverse operation. For example, to remove a square root, square both sides of the equation; to remove a cube root, cube both sides; and so on.
- Solve for the variable. Note: If $x^2 = 81$, then $x = \pm 9$, BUT $\sqrt{81} = 9$ only.
- Check for extraneous (invalid) solutions.

Polynomials

A *polynomial* is an expression or equation with one or more terms consisting of variables with non-negative integer exponents and coefficients, joined by addition, subtraction, and multiplication. You can combine like terms in polynomials as you did with linear expressions and equations. Adding and subtracting polynomials are straightforward operations—simply combine like terms, paying careful attention to negative signs. Multiplying polynomials is slightly more involved, requiring a careful distribution of terms followed by combining like terms if possible. You can use FOIL (First, Outer, Inner, Last) when you multiply two binomials.

When a polynomial is written in descending order, the term with the highest power (called the leading term) tells you the basic shape of its graph and how many x-intercepts (also called roots or zeros) its graph can have. To find the zeros of a polynomial equation, factor the equation and set each factor equal to 0. You can have simple zeros and/or multiple zeros. For example, in the equation $y = (x + 6)(x - 3)^2$, the factor $x + 6$ gives a simple zero of $x = -6$, while the factor $(x - 3)^2$ gives a double zero of $x = 3$ (because technically, the factor is $(x - 3)(x - 3)$). Graphically, when a polynomial has a simple zero (multiplicity 1) or any zero with an odd multiplicity, its graph will cross through the x-axis. When a polynomial has a double zero (multiplicity 2) or any zero with an even multiplicity, its graph just touches the x-axis, creating a turning point in the graph.

Rational Expressions

A rational expression is a fraction that contains one or more variables in the denominator (basically a polynomial over a polynomial). The rules that govern fractions and polynomials also govern rational expressions (e.g., finding common denominators and factoring to simplify). However, rational expressions have a few extra features that should be considered.

- Because rational expressions, by definition, have variables in the denominator, they are often undefined for certain values. Watch for expressions that could lead to a denominator of 0. For example, the expression $\frac{4x}{x - 11}$ is undefined at $x = 11$.
- Factors in a rational expression can be cancelled when simplifying, but under no circumstances can you do the same with individual terms. For example, $\frac{\cancel{(x+3)}(x - 4)}{2\cancel{(x+3)}}$ is allowed, but $\frac{x^2 + \cancel{2x} + 3}{\cancel{2x}}$ is NOT allowed.

PRACTICE SET

Easy

1. Which of the following is equivalent to the expression $\left(2x^4 - 5x^4\right)^2$?
   A) $-21x^8$
   B) $-6x^8$
   C) $9x^8$
   D) $9x^{16}$

2. Which of the following is equivalent to $\left(2b^3c^2 + b^2c - 4bc\right) - \left(b^3c^2 - b^2c - 4bc\right)$?
   A) 0
   B) b^3c^2
   C) $b^3c^2 + 2b^2c$
   D) $b^3c^2 + 2b^2c - 8bc$

3. When completely simplified, $\dfrac{25^4 \times 5^2}{25^5}$ has a value of:
   A) 0
   B) 1
   C) 5
   D) 25

4. Which of the following is equivalent to $x^{\frac{5}{7}}$, for all values of x?
   A) $\dfrac{5}{x^7}$
   B) $\dfrac{1}{x^2}$
   C) $\sqrt[5]{x^7}$
   D) $\sqrt[7]{x^5}$

5. Which of the following is the expanded form of $4(5x + 3)(2x - 1)$?
   A) $40x^2 + 12$
   B) $40x^2 - 12$
   C) $40x^2 - 4x + 12$
   D) $40x^2 + 4x - 12$

Medium

6. If $\dfrac{a^{x^2}}{a^{(x^2-y^2)}} = a^4$ and $y > 0$, what is the value of y?
   A) 0
   B) 1
   C) 2
   D) 4

7. Which sequence of steps correctly gives the value of $4^{\frac{3}{2}}$ and algebraically justifies the value?
   A) $4^{\frac{3}{2}} = (4^2)^{\frac{1}{3}} = \sqrt[3]{4^2} = \sqrt[3]{16}$
   B) $4^{\frac{3}{2}} = (4^2) \div 3 = 16 \div 3 = \dfrac{16}{3}$
   C) $4^{\frac{3}{2}} = (4^3) \div 2 = 64 \div 2 = 32$
   D) $4^{\frac{3}{2}} = (4^3)^{\frac{1}{2}} = \sqrt{4^3} = \sqrt{64} = 8$

8. What is the factored form of $16x^6 - 8x^3y^3 + y^6$?
   A) $(4x^3 - y^3)^2$
   B) $(4x^3 + y^3)^2$
   C) $(4x + y)^6$
   D) $(16x^2 + y)^3$

$$xy\left(\frac{x}{y} - y\right)$$

9. Which of the following is equivalent to the expression above?

A) $x^2 - y^2$

B) $1 - xy^2$

C) $x^2 - xy^2$

D) $2x - 2xy$

$$A = \frac{S}{360} \times \pi r^2$$

10. The area of a sector of a circle is given by the formula above, where S is the angle measure in degrees of the sector and r is the radius of the circle. Which of the following gives r in terms of A and S?

A) $r = \frac{360A\pi}{S}$

B) $r = \frac{360A}{S\pi}$

C) $r = \sqrt{\frac{360A\pi}{S}}$

D) $r = \sqrt{\frac{360A}{S\pi}}$

11. If $x^2 = a^{-\frac{1}{3}}$, where $x > 0$ and $a > 0$, which of the following gives a in terms of x?

A) $a = \frac{1}{x^6}$

B) $a = \frac{2}{x^3}$

C) $a = x^{\frac{3}{2}}$

D) $a = -x^6$

$$\frac{20u^3v^2 - 15u^2v}{10u^4v + 30u^3v^3}$$

12. Which of the following is the reduced form of the expression above?

A) $\frac{5uv}{40u^7v^4}$

B) $\frac{2v - 1}{u + 2uv^2}$

C) $\frac{4uv - 3}{2u^2 + 6uv^2}$

D) $\frac{2uv - 3uv^2}{u^2 + 6}$

13. If $x = 5$ and $\sqrt{2m + 11} - x = 0$, what is the value of m?

14. If $x = 8\sqrt{3}$ and $3x = \sqrt{3y}$, what is the value of y?

15. Given an account with interest compounded annually, the formula $A = P(1+r)^t$ can be used to calculate the total amount of money, A, in the account after t years, where P is the principal (the amount originally invested) and r is the interest rate (expressed as a decimal). Suppose Valeera invests \$5,000 in a savings account that pays 2% interest compounded annually. How much interest will Valeera earn in four years? Express your answer to the nearest whole dollar.

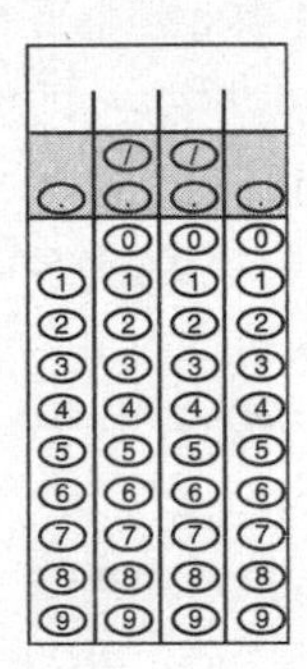

Hard

16. Which of the following gives $\frac{(x^2y)^3}{x^7y^2} \times \sqrt{xy}$ written in simplest form?

A) $\frac{y\sqrt{y}}{\sqrt{x}}$

B) $\frac{3\sqrt{y}}{\sqrt{x}}$

C) $\sqrt{\frac{3y}{x}}$

D) $\frac{y}{\sqrt{xy}}$

$$y - b = a - 2^{-x}$$

17. For the equation given, if $a > 0$ and $b > 0$, then which of the following statements is always true?

A) $y < a + b$

B) $y > a + b$

C) $y = a + b$

D) $y = -2(a + b)$

$$\frac{2x+6}{x^2+3x} - \frac{x+3}{x^2+x}$$

18. Which of the following is equivalent to the rational expression given above?

A) $-\frac{1}{x}$

B) $\frac{x+3}{2x}$

C) $\frac{x-1}{x(x+1)}$

D) $\frac{x+9}{2x^2+4x}$

$$E_n = -(2.18 \times 10^{-18})\frac{Z^2}{n^2}$$

19. Electrons follow paths, called orbits, around the nucleus of an atom. According to the Bohr model, the energy in joules of an atom's nth orbit containing a single electron is given by the formula shown above, where Z is the atomic number of the atom and n is the orbit number. Which of the following equations gives the atomic number of an atom given the energy of its nth orbit (assuming there is a single electron in that orbit)?

A) $Z = 10^9\sqrt{-\frac{n^2E_n}{2.18}}$

B) $Z = \frac{1}{10^9}\sqrt{-\frac{n^2E_n}{2.18}}$

C) $Z = 10^9\sqrt{-\frac{2.18E_n}{n^2}}$

D) $Z = \frac{1}{10^9}\sqrt{-\frac{2.18E_n}{n^2}}$

$$T = \frac{R^2}{r^2}\sqrt{\frac{2h}{g}}$$

20. Suppose an open cylindrical tank has a round drain with radius r in the bottom of the tank. When the tank is filled with water to a depth of h centimeters, the time it takes for all the water to drain from the tank is given by the formula above, where R is the radius of the tank (in centimeters) and $g = 980$ cm/s^2 is the acceleration due to gravity. Suppose such a tank has a radius of 2 meters and is filled to a depth of 4 meters. About how many minutes does it take to empty the tank if the drain has a radius of 5 centimeters? (1 meter = 100 centimeters)

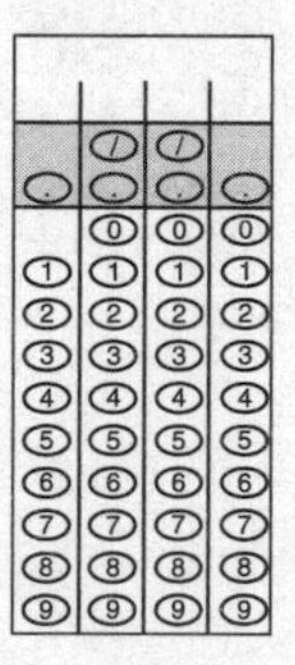

ANSWER KEY

1. C
2. C
3. B
4. D
5. D
6. C
7. D
8. A
9. C
10. D
11. A
12. C
13. 7
14. 576
15. 412
16. A
17. A
18. C
19. A
20. 24

ANSWERS & EXPLANATIONS

1. C

Difficulty: Easy

Category: Passport to Advanced Math / Exponents

Strategic Advice: When a quantity is squared, don't immediately starting FOILing. Look for like terms that you can combine first. Then follow the rules of exponents or use repeated multiplication to evaluate the power.

Getting to the Answer: If you examine the expression carefully, you'll see that you can combine the two terms inside the parentheses and then square the result:

Rules of Exponents	Repeated Multiplication
$(2x^4 - 5x^4)^2$	$(2x^4 - 5x^4)^2$
$= (-3x^4)^2$	$= (-3x^4)^2$
$= (-3)^2(x^4)^2$	$= (-3x^4)(-3x^4)$
$= 9 \cdot x^{(4\times2)}$	$= 9x^{4+4}$
$= 9x^8$	$= 9x^8$

The answer is (C).

2. C

Difficulty: Easy

Category: Passport to Advanced Math / Exponents

Strategic Advice: When you see negative signs, slow down and write everything out in your scratchwork. Even the best mathematicians make careless mistakes when they try to sort out negative signs in their heads.

Getting to the Answer: Start by distributing the negative sign to each term in the second set of parentheses. Then group and combine like terms, paying careful attention to the exponents on each variable:

$$(2b^3c^2 + b^2c - 4bc) - (b^3c^2 - b^2c - 4bc)$$
$$= 2b^3c^2 + b^2c - 4bc - b^3c^2 + b^2c + 4bc$$
$$= (2b^3c^2 - b^3c^2) + (b^2c + b^2c) + (-4bc + 4bc)$$
$$= b^3c^2 + 2b^2c$$

The answer is (C).

3. B

Difficulty: Easy

Category: Passport to Advanced Math / Exponents

Strategic Advice: Raising large numbers to high powers is very time-consuming without a calculator. Instead, start by using the rules of exponents to simplify expressions written in the same base. Then raise any remaining numbers to the resulting powers.

Getting to the Answer: The power of 25 is larger in the denominator, so subtract the exponents there. Then simplify further if possible: $\frac{25^4 \times 5^2}{25^5} = \frac{5^2}{25^{5-4}} = \frac{25}{25} = 1$, which is (B).

4. D
Difficulty: Easy

Category: Passport to Advanced Math / Exponents

Strategic Advice: If you forget the rules for fractional exponents on Test Day, you can remember them by thinking about the fact that $\sqrt{x} = x^{\frac{1}{2}} = \sqrt[2]{x^1}$. You can also use the saying "power over root."

Getting to the Answer: Think of the saying "power over root" for $x^{\frac{5}{7}}$. Think: 5 is the power of x and 7 is the root. This gives the seventh root of x raised to the fifth power, or $x^{\frac{5}{7}} = \sqrt[7]{x^5}$, choice (D).

5. D
Difficulty: Easy

Category: Passport to Advanced Math / Exponents

Strategic Advice: When the product of two binomials is multiplied by a constant, FOIL first, combine like terms, and then distribute the constant.

Getting to the Answer: Remember, FOIL stands for *first*, *outer*, *inner*, and *last*:

$$\begin{aligned} 4(5x+3)(2x-1) &= 4(10x^2 - 5x + 6x - 3) \\ &= 4(10x^2 + x - 3) \\ &= 40x^2 + 4x - 12 \end{aligned}$$

This matches choice (D).

6. C
Difficulty: Medium

Category: Passport to Advanced Math / Exponents

Strategic Advice: You can always rely on the rules of exponents, no matter how strange the exponents may look. Follow the rule for dividing powers of like bases: $\frac{a^b}{a^c} = a^{b-c}$.

Getting to the Answer: Although the left side of the equation seems strange, it melts away once you apply the exponent rule:

$$\frac{a^{x^2}}{a^{x^2-y^2}} = a^{x^2-(x^2-y^2)} = a^{x^2-x^2+y^2} = a^{y^2}$$

So $a^{y^2} = a^4$, which means that $y^2 = 4$ and $y = 2$ or -2. The question states that $y > 0$, so y must be 2, which is (C).

7. D
Difficulty: Medium

Category: Passport to Advanced Math / Exponents

Strategic Advice: Use rules of exponents to rewrite the expression. The saying "power over root" may help as well.

Getting to the Answer: Use the exponent rule $a^{mn} = (a^m)^n$ and the fact that $\frac{3}{2} = 3 \times \frac{1}{2}$ to rewrite the expression. Then simplify using the property $a^{\frac{1}{2}} = \sqrt{a}$:

$$4^{\frac{3}{2}} = (4^3)^{\frac{1}{2}} = \sqrt{4^3} = \sqrt{4 \times 4 \times 4} = \sqrt{64} = 8$$

This means (D) is correct.

8. A
Difficulty: Medium

Category: Passport to Advanced Math / Exponents

Strategic Advice: Whenever you see long polynomials with coefficients that are perfect squares, look for a factoring shortcut, such as recognizing the square of a sum or difference: $(a \pm b)^2 = a^2 \pm 2ab + b^2$. While that may not seem immediately applicable here, notice that $x^6 = (x^3)^2$ and $y^6 = (y^3)^2$.

Getting to the Answer: To confirm that you can use a factoring shortcut here, mentally "unsquare" the perfect squares on each end of the polynomial. Letting $a = 4x^3$ and $b = y^3$ (because $a^2 = 16x^6$ and $b^2 = y^6$), you get $2ab = 2(4x^3)(y^3) = 8x^3y^3$, which is indeed the middle term of the polynomial, so everything checks out. Thus, the expression equals $(a - b)^2$ or $(4x^3 - y^3)^2$, choice (A).

9. C
Difficulty: Medium

Category: Passport to Advanced Math / Exponents

Strategic Advice: When a product contains parentheses, and none of the terms inside the parentheses are like terms, chances are that you need to apply the distributive property. Then simplify using the rules of exponents.

Getting to the Answer: Rather than distributing mentally, rewrite the expression so that *xy* is multiplied by each term inside the parentheses:

$$xy\left(\frac{x}{y} - y\right) = xy\left(\frac{x}{y}\right) - xy(y) = x^2 - xy^2$$

The correct answer is (C).

10. D
Difficulty: Medium

Category: Passport to Advanced Math / Exponents

Strategic Advice: Don't let all the variables in a question like this intimidate you. Pretend they're simple numbers and use inverse operations to solve for the desired variable.

Getting to the Answer: To solve the equation for *r*, start by multiplying both sides of the equation by the only number that's in the denominator on the right-hand side (360). Then, divide by everything that *r* is being multiplied by ($S\pi$):

$$A = \frac{S}{360} \times \pi r^2$$

$$360(A) = 360\left(\frac{S}{360} \times \pi r^2\right)$$

$$360A = S\pi r^2$$

$$\frac{360A}{S\pi} = \frac{S\pi r^2}{S\pi}$$

$$\frac{360A}{S\pi} = r^2$$

Don't answer too quickly! The correct answer is not B. You're asked for *r* (not r^2), so take the square root of both sides of the equation to find that $r = \sqrt{\frac{360A}{S\pi}}$, which is (D).

11. A
Difficulty: Medium

Category: Passport to Advanced Math / Exponents

Strategic Advice: When a number (or a variable) is raised to a negative exponent, rewrite it as the reciprocal of the number with a positive exponent. You can also write fraction exponents as radicals to make using an inverse operation more recognizable.

Getting to the Answer: To write *a* in terms of *x*, you need to solve for *a*. To do this, you need to eliminate the fraction exponent, which also happens to be negative:

$$x^2 = a^{-\frac{1}{3}}$$

$$x^2 = \frac{1}{a^{\frac{1}{3}}}$$

$$x^2 = \frac{1}{\sqrt[3]{a}}$$

To eliminate the radical (the cube root), cube both sides of the equation:

$$\left(x^2\right)^3 = \left(\frac{1}{\sqrt[3]{a}}\right)^3$$
$$x^6 = \frac{1}{a}$$

Cross-multiply to get the a out of the denominator, and then divide both sides by x^6:

$$ax^6 = 1$$
$$a = \frac{1}{x^6}$$

Choice (A) is correct.

12. C

Difficulty: Medium

Category: Passport to Advanced Math / Exponents

Strategic Advice: When working with rational expressions, look for a common factor that can be divided out of the numerator and the denominator. If you find one, you can cancel it. It will save time if you look for the greatest common factor.

Getting to the Answer: Factor out the GCF of *both* the numerator and the denominator. Then cancel what you can. In this expression, the GCF is $5u^2v$.

$$\frac{20u^3v^2 - 15u^2v}{10u^4v + 30u^3v^3} = \frac{\cancel{5u^2v}\,(4uv - 3)}{\cancel{5u^2v}\left(2u^2 + 6uv^2\right)}$$
$$= \frac{4uv - 3}{2u^2 + 6uv^2}$$

This matches (C).

13. 7

Difficulty: Medium

Category: Passport to Advanced Math / Exponents

Strategic Advice: When you're given the value of a variable in an equation, start by plugging in the given value and see where that takes you.

Getting to the Answer: Substitute 5 for x, isolate the radical term by adding 5 to both sides of the equation, square both sides to get rid of the square root, and continue using inverse operations until you have m by itself:

$$\sqrt{2m+11} - 5 = 0$$
$$\sqrt{2m+11} = 5$$
$$\left(\sqrt{2m+11}\right)^2 = (5)^2$$
$$2m + 11 = 25$$
$$2m = 14$$
$$m = 7$$

One final step—when you solve an equation that involves a square root, you must check that an extraneous (invalid) root was not produced. To check, plug 7 in for m (and the given 5 for x) into the original equation to make sure the result is a true statement:

$$\sqrt{2(7)+11} - 5 = \sqrt{14+11} - 5$$
$$= \sqrt{25} - 5$$
$$= 5 - 5$$
$$= 0 \checkmark$$

14. 576

Difficulty: Medium

Category: Passport to Advanced Math / Exponents

Strategic Advice: You can "separate" radical expressions as long as the only operations involved are multiplication and division. For example, $\sqrt{3y} = \sqrt{3}\sqrt{y}$ because y is being *multiplied* by 3.

Getting to the Answer: First, substitute $8\sqrt{3}$ for x into the equation and simplify. Next, write $\sqrt{3y}$ as $\sqrt{3}\sqrt{y}$ so you can divide both sides by $\sqrt{3}$. Finally, square both sides to eliminate the radical:

$$3x = \sqrt{3y}$$
$$3(8\sqrt{3}) = \sqrt{3y}$$
$$24\sqrt{3} = \sqrt{3y}$$
$$24\sqrt{3} = \sqrt{3}\sqrt{y}$$
$$24 = \sqrt{y}$$
$$24^2 = y$$

Squaring 24, the final answer is 576.

15. 412
Difficulty: Medium

Category: Passport to Advanced Math / Exponents

Strategic Advice: Even though this question is wordy, if you read carefully you'll see that you're given a formula and the values of all the variables except the one you're looking for. This means you can simply plug in the values and simplify. Before filling in the grid, check that you answered the *right* question (how much interest was earned, not the total amount in the account).

Getting to the Answer: Jot down the formula $A = P(1+r)^t$, and then plug in the given values: $P = 5{,}000$, $r = 0.02$, and $t = 4$, to find that $A = 5{,}000(1+0.02)^4 \approx 5{,}412.16$. This is the total amount in the account after 4 years. The initial amount was \$5,000, so subtract to find that Valeera will earn \$5,412.16 – \$5,000 = \$412.16 or, rounded to the nearest whole dollar, \$412 in interest.

16. A
Difficulty: Hard

Category: Passport to Advanced Math / Exponents

Strategic Advice: Before you can combine radical expressions with non-radical expressions, you must write them in the same form. Here, you can rewrite the square root of *xy* as $(xy)^{\frac{1}{2}}$.

Getting to the Answer: First, simplify as much as possible using the rules of exponents. Then rewrite the radical as a power of *xy* and simplify further, using the rules of exponents to combine terms that have the same base:

$$\begin{aligned}
\frac{\left(x^2y\right)^3}{x^7y^2}\sqrt{xy} &= \frac{x^6y^3}{x^7y^2}\sqrt{xy} \\
&= \frac{y^{3-2}}{x^{7-6}}\sqrt{xy} \\
&= \frac{y}{x}\sqrt{xy} \\
&= x^{-1}y^1(xy)^{\frac{1}{2}} \\
&= x^{-1}x^{\frac{1}{2}}y^1y^{\frac{1}{2}} \\
&= x^{\left(-1+\frac{1}{2}\right)}y^{\left(1+\frac{1}{2}\right)} \\
&= x^{-\frac{1}{2}}y^{\frac{3}{2}}
\end{aligned}$$

The answer choices are given as radicals, so change the powers back to roots. The negative power on *x* moves it to the denominator, and for the *y* term, use the saying "power over root" to help you write the correct radical form:

$$x^{-\frac{1}{2}}y^{\frac{3}{2}} = \frac{y^{\frac{3}{2}}}{x^{\frac{1}{2}}} = \frac{\sqrt[2]{y^3}}{\sqrt{x}} = \frac{y\sqrt{y}}{\sqrt{x}}$$

The answer is (A).

17. A
Difficulty: Hard

Category: Passport to Advanced Math / Exponents

Strategic Advice: Sometimes it is necessary to rearrange an expression or equation to reveal properties about the quantity being defined.

Getting to the Answer: Each of the answer choices compares y to some expression, so rewrite the equation in terms of y by adding b to both sides. Then, group a and b so the equation looks more like the answer choices:

$$y - b = a - 2^{-x}$$
$$y = a - 2^{-x} + b$$
$$y = a + b - 2^{-x}$$

Now it's time to examine the information given in the question and some properties of exponents. It is given that a and b are both greater than 0, or in other words, positive numbers. This means $a + b$ must be a positive number as well. The term 2^{-x} is also a positive number because raising a number to a negative power simply means writing the reciprocal of the number $\left(\frac{1}{2^x}\right)$. When x is large, this number becomes very small, but it is still a positive number. Therefore, y is equal to a positive number $(a + b)$ minus another positive number. It follows that $a + b - 2^{-x} < a + b$, and therefore that $y < a + b$, which matches (A).

18. C

Difficulty: Hard

Category: Passport to Advanced Math / Exponents

Strategic Advice: When adding fractions, you need to find a common denominator, even when the fractions contain variables. Factor the denominators to find the least common denominator (LCD).

Getting to the Answer: The original expression, after factoring each of the denominators, is $\frac{2x+6}{x(x+3)} - \frac{x+3}{x(x+1)}$. The least common denominator is $x(x + 3)(x + 1)$. Multiply each numerator and each denominator by whatever is missing from the LCD in that particular term (the first term is missing the $x + 1$, and the second term is missing the $x + 3$). Then simplify each numerator as much as possible, and add the results by combining like terms:

$$\frac{2x+6}{x(x+3)} - \frac{x+3}{x(x+1)}$$
$$= \frac{(2x+6)(x+1)}{x(x+3)(x+1)} - \frac{(x+3)(x+3)}{x(x+1)(x+3)}$$
$$= \frac{2x^2+6x+2x+6}{x(x+3)(x+1)} - \frac{x^2+3x+3x+9}{x(x+1)(x+3)}$$
$$= \frac{2x^2+8x+6}{x(x+3)(x+1)} + \frac{-x^2-6x-9}{x(x+1)(x+3)}$$
$$= \frac{x^2+2x-3}{x(x+3)(x+1)}$$

This doesn't match any of the answer choices, so try to reduce further. Factor the numerator to get $\frac{(x+3)(x-1)}{x(x+3)(x+1)}$. You can divide the factor $(x + 3)$ from the numerator and denominator to get $\frac{x-1}{x(x+1)}$, which matches (C).

19. A

Difficulty: Hard

Category: Passport to Advanced Math / Exponents

Strategic Advice: A question like this requires patience and concentration. As with any equation, use inverse operations to solve for (isolate) the desired variable.

Getting to the Answer: To solve the equation for Z, start by multiplying both sides of the equation by the only expression that's in the denominator on the right-hand side (n^2). Then, divide by everything that Z is being multiplied by (-2.18×10^{-18}). To remove the negative exponent, you can move 10^{-18} to the numerator (but not 2.18 because it is not being raised to a negative power):

$$E_n = -(2.18 \times 10^{-18})\frac{Z^2}{n^2}$$

$$n^2(E_n) = \left[-(2.18 \times 10^{-18})\frac{Z^2}{n^2}\right]n^2$$

$$n^2E_n = -(2.18 \times 10^{-18})Z^2$$

$$\frac{n^2E_n}{-(2.18 \times 10^{-18})} = \frac{-(2.18 \times 10^{-18})Z^2}{-(2.18 \times 10^{-18})}$$

$$-\frac{n^2E_n \times 10^{18}}{2.18} = Z^2$$

You're asked for Z (not Z^2), so take the square root of both sides of the equation to find that $Z = \sqrt{-\frac{n^2E_n \times 10^{18}}{2.18}}$. This is not one of the answer choices, but you can take the square root of 10^{18} by writing it as $(10^9)^2$, which results in the equation $Z = 10^9\sqrt{-\frac{n^2E_n}{2.18}}$. This means (A) is correct.

Note: You may recall that you can't take the square root of a negative number, but electron energy is given as a negative number, which means you're multiplying a negative number by another negative number under the radical, resulting in a positive quantity.

20. 24

Difficulty: Hard

Category: Passport to Advanced Math / Exponents

Strategic Advice: The formula in this question looks quite complicated, but the value for each variable is given, which means your only task is to substitute the values into the formula and simplify.

Getting to the Answer: The variables in the formula are defined in terms of centimeters, so convert each of the tank's dimensions to centimeters: There are 100 centimeters in 1 meter, so 2 m = 200 cm, and 4 m = 400 cm. Substitute these values, the radius of the drain (5 cm), and the value given for g into the formula and simplify:

$$\begin{aligned} T &= \frac{R^2}{r^2}\sqrt{\frac{2h}{g}} \\ &= \frac{200^2}{5^2}\sqrt{\frac{2(400)}{980}} \\ &= \frac{40{,}000}{25}\sqrt{\frac{800}{980}} \\ &= 1{,}600 \times 0.903508 \\ &\approx 1{,}445.61 \end{aligned}$$

You should notice at this point that the result is not even close to the answer choices. The acceleration due to gravity was given in terms of seconds (seconds squared but then square rooted), so divide your final answer by 60 to find the number of minutes it takes to drain the tank: $1{,}445.61 \div 60 \approx 24.09$, or about 24 minutes.

PRACTICE SET 8: ADVANCED TECHNIQUES FOR POLYNOMIALS AND RATIONAL EQUATIONS

Division of Multi-Term Polynomials

Dividing polynomial expressions requires a fairly involved process called *polynomial long division*. Polynomial long division is just like regular long division except, as the name suggests, you use polynomials instead of numbers.

Suppose you wish to divide $x^3 + 3x + 7$ by $x + 4$. You can set this up as a long division problem.

$$x + 4 \overline{\smash{)}\, x^3 + 0x^2 + 3x + 7}$$

Notice that even though the dividend does not have an x^2 term, a placeholder is used to keep the terms organized. Because $0x^2$ is equal to 0, adding this placeholder term doesn't change the value of the polynomial. Start by dividing the first term of the dividend $(x^3 + 3x + 7)$ by the first term of the divisor $(x + 4)$ to get x^2. Multiply the entire divisor by x^2, and then subtract this product from the dividend. Bring down leftover terms as needed.

$$\begin{array}{r l}
 & x^2 \\
x + 4 & \overline{\smash{)}\, x^3 + 0x^2 + 3x + 7} \\
 & \underline{-(x^3 + 4x^2)} \\
 & \quad -4x^2 + 3x + 7
\end{array}$$

Continue by dividing the next term, $-4x^2$, by the first term of the divisor. Bring down leftover terms as needed. Multiply the quotient, $-4x$, by the entire divisor and subtract. Then repeat the process for the resulting x-term.

$$\begin{array}{r l}
 & x^2 - 4x + 19 \\
x + 4 & \overline{\smash{)}\, x^3 + 0x^2 + 3x + 7} \\
 & \underline{-(x^3 + 4x^2)} \\
 & \quad -4x^2 + 3x + 7 \\
 & \quad \underline{-(-4x^2 - 16x)} \\
 & \qquad\qquad 19x + 7 \\
 & \qquad\qquad \underline{-(19x + 76)} \\
 & \qquad\qquad\qquad -69
\end{array}$$

The quotient is $x^2 - 4x + 19$ with a remainder of -69; the remainder is written over the divisor in a separate term. The final answer is $x^2 - 4x + 19 - \dfrac{69}{x + 4}$.

Note: For a polynomial to be evenly divisible by another polynomial, the remainder must be 0.

Solving Rational Equations

Rational equations are just like rational expressions except for one difference: They have an equal sign. They follow the same rules as rational expressions and can be solved (just like linear equations) using inverse operations. However, there are often extra steps involved, which makes solving them more intimidating. Here are a few strategies for solving rational equations:

- If there are only two nonzero terms in the equation, separate the terms on opposite sides of the equal sign and cross-multiply.
- If there are more than two nonzero terms, find a common denominator, multiply all the terms in the equation by the common denominator (which "clears" the fractions), and then solve the resulting equation.
- Beware of extraneous solutions (solutions that don't actually satisfy the original equation). When a solution to a rational equation produces a 0 denominator in *any* of the terms in the equation, the solution is invalid (because division by 0 is not possible).

You might also be asked to match a graph to a rational equation or vice versa. To do this, make note of any x-values that cause the equation to be undefined, which will appear as open dots on the graph or, in more complicated equations, dashed vertical lines (called *asymptotes*). Then, plot a few key points to see how the graph behaves on each side of the undefined value(s).

Work Scenarios

A real-world scenario involving rational equations that you may see on Test Day is a work question (a question about completing a job or task). Recall that distance is the product of rate and time ($d = rt$), which can be rewritten in the form $r = \frac{d}{t}$ or $t = \frac{d}{r}$ (which are both rational equations). In questions involving work, you can replace d with W (for work), which allows you to determine how long it will take to complete some kind of work or a specific task.

You can also combine rates by rewriting the equation $W = rt$ as $r = \frac{W}{t}$ for each person (or machine) working on a job and then adding the rates together. Here's a typical example:

Suppose machine A can complete a job in 2 hours, and machine B can do the same job in 4 hours. You want to know how long it will take to complete the job if both machines work together. There is one job to complete, so let $W = 1$, resulting in two rates: $r_A = \frac{W_A}{t_A} = \frac{1}{2}$ job per hour and $r_B = \frac{W_B}{t_B} = \frac{1}{4}$ job per hour. The combined rate is $\frac{1}{2} + \frac{1}{4} = \frac{3}{4}$ job per hour, which means $t_{total} = \frac{W_{total}}{r_{total}} = \frac{1}{\frac{3}{4}} = \frac{4}{3}$ hours. Thus, it will take $\frac{4}{3}$ hours to complete the job if machines A and B work together.

PRACTICE SET

Easy

1. If a function $p(x)$ has four distinct zeros, which of the following could represent the entire graph of p in a standard coordinate plane?

A) 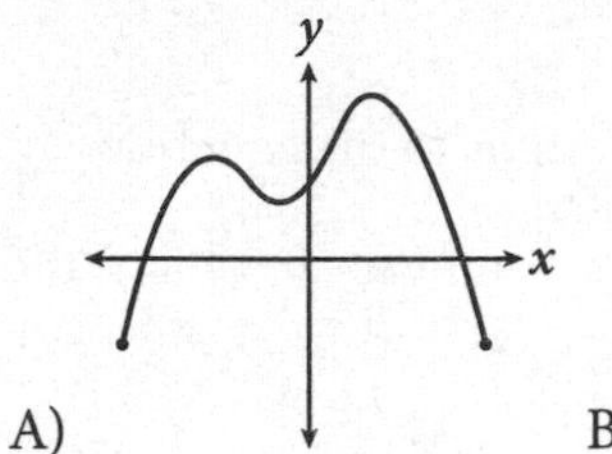

B) 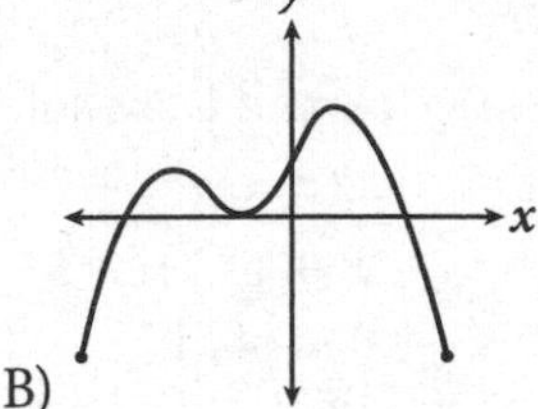

C) 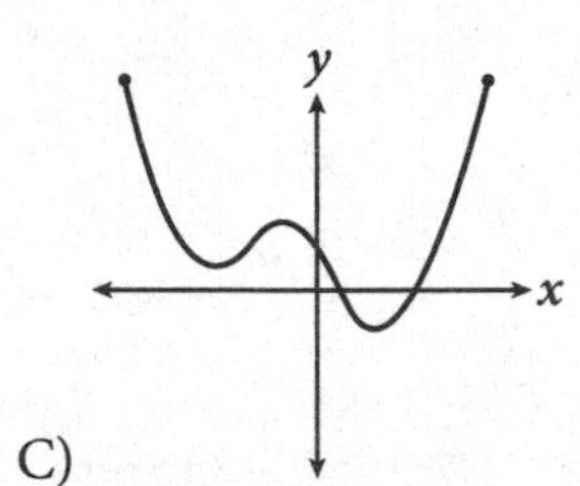

D) 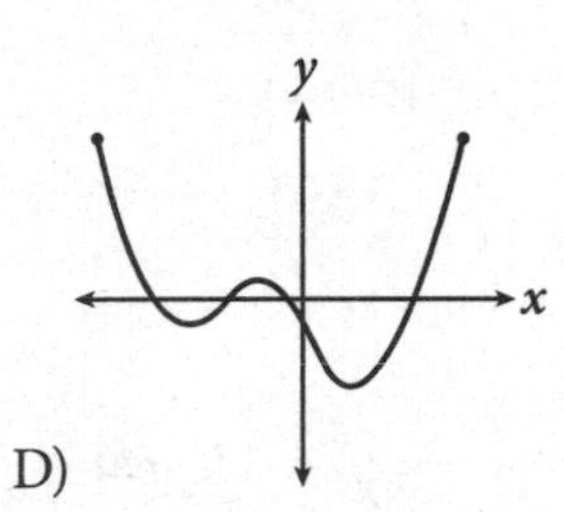

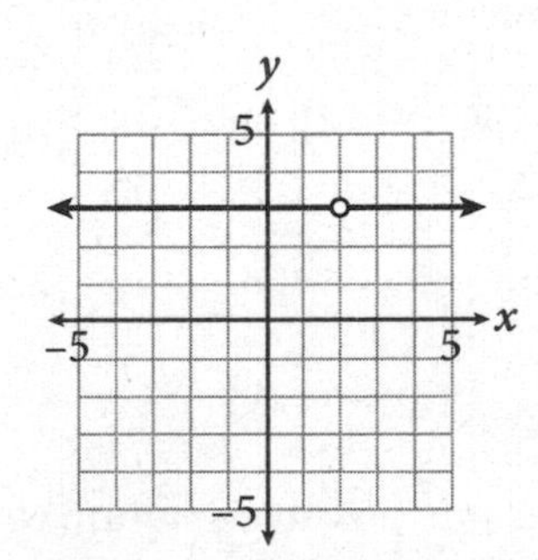

2. Which of the following rational functions could represent the graph shown?

A) $r(x) = 3$

B) $r(x) = \dfrac{x-3}{x-2}$

C) $r(x) = \dfrac{2}{x-3}$

D) $r(x) = \dfrac{3x-6}{x-2}$

3. For which values of x is the expression $\dfrac{3x+6}{3x(4x+8)(x-5)}$ undefined?

A) -2

B) $-2, 5$

C) $0, -2, 5$

D) $0, 2, -5$

4. If $y \neq z$, then $\dfrac{xy - zx}{z - y} =$

A) $-x$

B) -1

C) 1

D) x

5. If Q is the quotient when $(x^2 - 10x - 24)$ is divided by $(x + 2)$ and $x \neq -2$, which of the following represents Q?

A) $x - 22$

B) $x - 12$

C) $x + 12$

D) $x + 22$

Medium

6. Given that $a \neq \pm\dfrac{1}{2}$, which of the following is equivalent to $\dfrac{2a^2 + 5a - 3}{4a^2 - 1}$?

A) $\dfrac{2a-3}{2a+1}$

B) $\dfrac{a-3}{2a-1}$

C) $\dfrac{2a+3}{2a-1}$

D) $\dfrac{a+3}{2a+1}$

$$\frac{1}{\frac{1}{x-4}-\frac{1}{x+6}}$$

7. Which of the following is equivalent to the expression above?

A) 10

B) $\frac{10(x-4)}{x+6}$

C) $\frac{10}{x^2+2x-24}$

D) $\frac{x^2+2x-24}{10}$

$$\frac{2}{x+1}-\frac{x}{6}=0$$

8. If x_1 and x_2 are valid solutions to the rational equation given, what is the sum of $x_1 + x_2$?

A) −1

B) 0

C) 1

D) 2

9. Given that $\frac{2m+n}{2n}=\frac{3}{4}$, which of the following must also be true?

A) $m=\frac{1}{8}$

B) $m=\frac{3}{4}$

C) $\frac{m}{n}=\frac{1}{4}$

D) $\frac{m}{n}=\frac{5}{4}$

10. What is/are the solution(s) to the equation

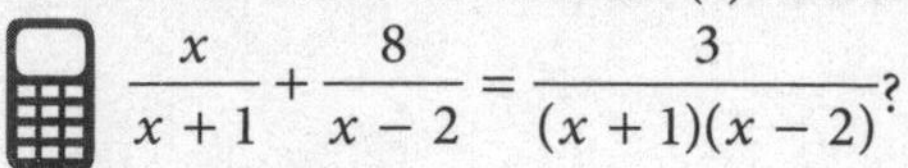

$\frac{x}{x+1}+\frac{8}{x-2}=\frac{3}{(x+1)(x-2)}$?

A) −5 only

B) −1 and −5

C) 1 and 5

D) No real solution

11. If a and b are solutions to the equation $\frac{2x}{2x+1}+\frac{8}{x-3}=0$, and $a<b$, what is the value of a?

A) −4

B) −1

C) 1

D) 4

12. If $g(x)=-2(x^2-6x-12)+3(k-x)$ and $g(x)$ is evenly divisible by x, then what is the value of k?

A) −8

B) −3

C) 2

D) 3

13. What is one possible solution to the rational equation $\frac{1}{x}-\frac{2}{x-2}=3$?

14. When $-2x^3 - 2x^2 + 27x - 30$ is divided by $2 - x$, what is the remainder?

Hard

15. Professors Lambert and Stassen team-teach an advanced psychology class. Their students recently took their final exams, and the professors are preparing to grade them. Professor Stassen predicts that, if she graded all the exams, she would take 50% more time than Professor Lambert would if he graded all the exams. They estimate that it will take them 40 hours to grade all the exams if they grade together. Which of the following equations best represents the situation described?

A) $x + \frac{3}{2}x = 40$

B) $\frac{1}{x} + \frac{2}{3x} = \frac{1}{40}$

C) $\frac{1}{x} + \frac{50}{x} = \frac{1}{40}$

D) $\frac{2}{x} + \frac{3}{x} = \frac{1}{40}$

16. Using a riding lawnmower, Joe can cut his lawn in 45 minutes. Using a push mower, Joe's son can cut the lawn in 3 hours. If they work together, how long will it take them to cut the lawn?

A) 24 minutes

B) 30 minutes

C) 36 minutes

D) 40 minutes

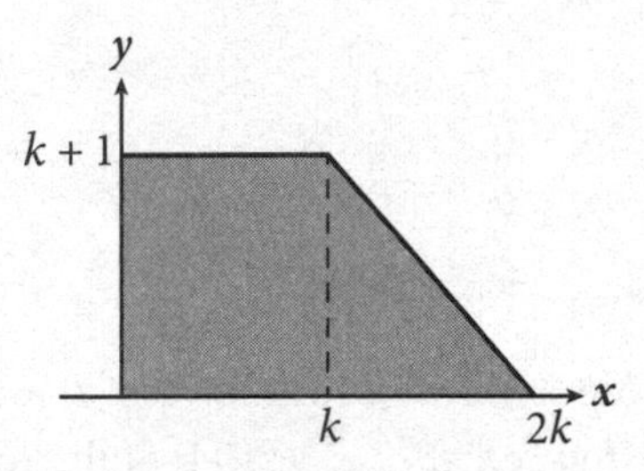

17. Which of the following polynomial equations could be used to represent the area of the trapezoid shown in the figure?

A) $A = k^2 + k$

B) $A = \frac{3k^2 + 3k}{2}$

C) $A = \frac{3k^2 + k}{2}$

D) $A = \frac{3k^3 + k^2}{2}$

$$p(x) = 12x^4 + 13x^3 - 35x^2 - 16x + 20$$

18. Which of the following is a factor of the polynomial given above?

A) $2x + 3$

B) $3x - 2$

C) $4x + 5$

D) $5x - 3$

19. If the expression $\frac{9x^2}{3x - 2}$ is rewritten in the form $A + \frac{4}{3x - 2}$, which of the following gives A in terms of x?

A) $\frac{9x^2}{4}$

B) $3x + 2$

C) $3x - 2$

D) $9x^2 - 4$

$$p(x) = 6x^2 + 7x - 2$$
$$q(x) = 2x^2 + 7x - 1$$

20. Two polynomial functions $p(x)$ and $q(x)$ are defined above. Which of the following polynomial functions is divisible by $2x + 5$?

A) $r(x) = p(x) + q(x)$

B) $r(x) = 2p(x) + q(x)$

C) $r(x) = p(x) + 3q(x)$

D) $r(x) = 2p(x) + 3q(x)$

ANSWER KEY

1. D	6. D	11. A	16. C
2. D	7. D	12. A	17. B
3. C	8. A	13. 2/3 or 1	18. B
4. A	9. C	14. 0	19. B
5. B	10. A	15. B	20. C

ANSWERS & EXPLANATIONS

1. D
Difficulty: Easy

Category: Passport to Advanced Math / Exponents

Strategic Advice: The *zeros* of a function are the same as its roots or solutions. Graphically, the zeros are the *x*-intercepts. *Distinct zeros* mean different values of *x*.

Getting to the Answer: You are looking for the graph that has four different *x*-intercepts, or in other words, crosses through the *x*-axis four times, which is the graph in (D). Note that the graphs in A and C have two distinct zeros, and the graph in B has three.

2. D
Difficulty: Easy

Category: Passport to Advanced Math / Exponents

Strategic Advice: When matching a graph to an equation, look for key features of the graph, particularly when there is a feature that stands out.

Getting to the Answer: There is a "hole" in this graph (the open dot), which means that the function is not defined at a particular value of *x*, in this case $x = 2$ (the *x*-coordinate of the open dot). Think about what could cause that: A denominator that equals 0 when $x = 2$ is the correct response. Now look at the answer choices—you can eliminate A (because there is no denominator), and you can eliminate C (because the denominator is equal to 0 at $x = 3$, not $x = 2$). To choose between B and (D), simplify the functions to see which results in a horizontal line at $r(x) = 3$ (because the line is at $y = 3$). The function in B doesn't simplify at all, so skip it for now. For (D):

$$r(x) = \frac{3x - 6}{x - 2}$$

$$r(x) = \frac{3(x - 2)}{(x - 2)}$$

$$r(x) = 3$$

The equation in (D) is indeed a horizontal line at $y = 3$ with a "hole" at $x = 2$.

3. C
Difficulty: Easy

Category: Passport to Advanced Math / Exponents

Strategic Advice: An expression (or equation) is undefined for any value or values of the variable that result in a denominator equal to 0 or a negative quantity inside a square root. Having the expression already written in factored form makes finding these values very straightforward.

Getting to the Answer: There are three distinct factors in the denominator of the expression, so set each factor equal to 0, and solve for x:

$$3x = 0 \rightarrow x = 0$$

$$4x + 8 = 0 \rightarrow 4x = -8 \rightarrow x = -2$$

$$x - 5 = 0 \rightarrow x = 5$$

The expression is undefined when $x = 0, -2,$ and 5. Choice (C) is correct.

4. A

Difficulty: Easy

Category: Passport to Advanced Math / Exponents

Strategic Advice: Whenever you need to simplify a fraction that involves binomials, your first thought should be: factor!

Getting to the Answer: Because x is in both terms of the numerator, factor out x to get:

$$\frac{xy - zx}{z - y} = \frac{x(y - z)}{z - y}$$

Next, rewrite $(z - y)$ as $-1(y - z)$ to get:

$$\frac{xy - zx}{z - y} = \frac{x(y - z)}{(-1)(y - z)}$$

The numerator and denominator now share a common factor, $y - z$, which can be cancelled:

$$\frac{x\cancel{(y - z)}}{(-1)\cancel{(y - z)}} = \frac{x}{-1} = -x$$

The result matches (A).

5. B

Difficulty: Easy

Category: Passport to Advanced Math / Exponents

Strategic Advice: When you see a polynomial being divided by another polynomial, don't automatically assume that you need to use long division (unless you just happen to like long division). Factoring may also be an option, especially when one of the polynomials is quadratic.

Getting to the Answer: Write the division as a fraction. Then, factor the numerator to see if anything cancels nicely. To factor the numerator, look for factors of -24 that add up to -10 (which are -12 and $+2$).

$$\frac{x^2 - 10x - 24}{x + 2} = \frac{(x - 12)\cancel{(x + 2)}}{\cancel{x + 2}} = x - 12$$

The quotient is $x - 12$, which is (B).

6. D

Difficulty: Medium

Category: Passport to Advanced Math / Exponents

Strategic Advice: When dividing polynomials, you have two choices: factor and cancel, or long division. Choose whichever method you're more comfortable with.

Getting to the Answer: Factoring is most likely the faster route. The denominator is easier to factor, so start there: $4a^2 - 1$ is a difference of squares, which becomes $(2a + 1)(2a - 1)$. The expression in the numerator will factor into two binomials, one of which must be either $2a + 1$ or $2a - 1$ so that the fraction can be reduced. Using factoring by grouping, find the factors of $2(-3) = -6$ that add up to $+5$, split the middle term using those numbers, and then factor by grouping:

$$\begin{aligned} 2a^2 + 5a - 3 &= 2a^2 + 6a - 1a - 3 \\ &= 2a(a+3) - 1(a+3) \\ &= (2a-1)(a+3) \end{aligned}$$

The factor $2a - 1$ appears in the numerator and the denominator, so cancel it to leave $\frac{a+3}{2a+1}$, which is (D).

7. D
Difficulty: Medium

Category: Passport to Advanced Math / Exponents

Strategic Advice: Questions that involve "complex" fractions (fractions within fractions) require patience and strategy. Ignore the big "1 over" part initially. If you rewrite the messy denominator as a single fraction, then you can simply "flip it" to get the final answer.

Getting to the Answer: To combine the two terms in the denominator, find the least common denominator, $(x - 4)(x + 6)$, and write each term as a fraction with that denominator. Then, simplify as needed by using FOIL and combining like terms:

$$\begin{aligned} \frac{1}{x-4} - \frac{1}{x+6} &= \left(\frac{x+6}{x+6}\right)\frac{1}{x-4} - \left(\frac{x-4}{x-4}\right)\frac{1}{x+6} \\ &= \frac{(x+6)-(x-4)}{(x+6)(x-4)} \\ &= \frac{x-x+6-(-4)}{x^2-4x+6x-24} \\ &= \frac{10}{x^2+2x-24} \end{aligned}$$

Don't answer too quickly! You still need to take the reciprocal of the simplified expression to perform the final "1 over" part of the original expression, making (D) the correct answer.

8. A
Difficulty: Medium

Category: Passport to Advanced Math / Exponents

Strategic Advice: When solving a rational equation (an equation with variables in the denominator) with exactly two terms, it is almost always quicker to set one term equal to the other (separate the terms on opposite sides of the = sign) and cross-multiply. You could also find a common denominator and go from there, but this usually takes more time.

Getting to the Answer: Add $\frac{x}{6}$ to both sides of the equation, and then cross-multiply:

$$\begin{aligned} \frac{2}{x+1} - \frac{x}{6} &= 0 \\ \frac{2}{x+1} &= \frac{x}{6} \\ x(x+1) &= 6(2) \\ x^2 + x &= 12 \end{aligned}$$

You now have a quadratic equation to solve. Start by subtracting 12 from both sides so that the equation equals 0. The answer choices are all integers, so this is a clue that you should be able to factor the equation:

$$\begin{aligned} x^2 + x - 12 &= 0 \\ (x+4)(x-3) &= 0 \end{aligned}$$

Set each factor equal to 0 and solve to find that $x = -4$ and $x = 3$. The sum of these values is $-4 + 3 = -1$, so (A) is correct.

9. C
Difficulty: Medium

Category: Passport to Advanced Math / Exponents

Strategic Advice: Whenever you see only one equation, but two variables, chances are that you can't solve for either variable. Rather, you're most likely looking for an answer that somehow compares the two variables. Notice that the first two answer choices give a solution for m, which is probably not possible to find. The last two choices give a ratio of $\frac{m}{n}$, so try to write the original equation with at least one $\frac{m}{n}$ in it.

Getting to the Answer: Start by breaking the fraction on the left side of the equation into two parts. Then, simplify and see what's left:

$$\frac{2m+n}{2n} = \frac{3}{4}$$

$$\frac{2m}{2n} + \frac{n}{2n} = \frac{3}{4}$$

$$\frac{m}{n} + \frac{1}{2} = \frac{3}{4}$$

$$\frac{m}{n} = \frac{3}{4} - \frac{2}{4}$$

$$\frac{m}{n} = \frac{1}{4}$$

This matches (C).

10. A

Difficulty: Medium

Category: Passport to Advanced Math / Exponents

Strategic Advice: When solving a rational equation that has more than two terms, a good strategy is to eliminate the fractions first by multiplying both sides of the equation by the least common denominator. Don't forget to check for extraneous solutions (solutions that result in a denominator equal to 0) before selecting your final answer.

Getting to the Answer: Multiply both sides of the equation by the common denominator, $(x+1)(x-2)$, to eliminate the fractions. Once complete, solve for x by factoring:

$$\left[\frac{(x+1)(x-2)}{1}\right]\left[\frac{x}{x+1} + \frac{8}{x-2} = \frac{3}{(x+1)(x-2)}\right]$$

$$x(x-2) + 8(x+1) = 3$$

$$x^2 - 2x + 8x + 8 = 3$$

$$x^2 + 6x + 5 = 0$$

$$(x+1)(x+5) = 0$$

Setting each factor equal to 0 and solving gives $x = -1$ and $x = -5$. Now, take another look at the denominators in the original equation. The equation is undefined when x is -1 or 2, so one of the solutions you found (-1) is extraneous. Therefore, -5 is the only valid solution, making (A) the correct answer.

11. A

Difficulty: Medium

Category: Passport to Advanced Math / Exponents

Strategic Advice: When a rational equation has two terms (not counting 0), move one term to the other side of the equation and cross-multiply.

Getting to the Answer: Move the second term to the right-hand side of the equal sign by subtracting it (mentally) from both sides. Then cross-multiply. The result gives a quadratic equation, which you can solve by factoring:

$$\frac{2x}{2x+1} + \frac{8}{x-3} = 0$$

$$\frac{2x}{2x+1} = -\frac{8}{x-3}$$

$$2x(x-3) = -8(2x+1)$$

$$2x^2 - 6x = -16x - 8$$

$$2x^2 + 10x + 8 = 0$$

$$2(x^2 + 5x + 4) = 0$$

$$(x+1)(x+4) = 0$$

Setting each factor equal to 0 gives $x = -1$ and $x = -4$. Neither solution is extraneous, and the question states that $a < b$, so a is the smaller of the two values, which is –4, or (A).

12. A
Difficulty: Medium

Category: Passport to Advanced Math / Exponents

Strategic Advice: If a polynomial is evenly divisible by x, then x has to be a factor of the polynomial, which means that when simplified, every term must have an x in it. In other words, the polynomial cannot contain a constant term.

Getting to the Answer: Start by simplifying $g(x)$. Be careful when distributing the –2:

$$\begin{aligned} g(x) &= -2(x^2 - 6x - 12) + 3(k - x) \\ &= -2x^2 + 12x + 24 + 3k - 3x \\ &= -2x^2 + 9x + 3k + 24 \end{aligned}$$

To ensure that there is no constant term, set $3k + 24$ equal to 0 and solve for k.

$$\begin{aligned} 3k + 24 &= 0 \\ 3k &= -24 \\ k &= -8 \end{aligned}$$

The correct answer is (A).

13. 2/3 or **.666** or **.667** or **1**
Difficulty: Medium

Category: Passport to Advanced Math / Exponents

Strategic Advice: Solving a rational equation is sometimes easier than just adding two rational expressions. You still may need to find a common denominator, but once you multiply both sides of the equation by that denominator, you're left with a much nicer looking equation to solve.

Getting to the Answer: Multiplying both sides of the equation by the common denominator $(x)(x - 2)$ yields the following:

$$\left[\frac{x(x-2)}{1}\right] \cdot \left[\frac{1}{x} - \frac{2}{x-2} = 3\right]$$

$$\begin{aligned} 1(x-2) - 2(x) &= 3(x)(x-2) \\ x - 2 - 2x &= 3(x^2 - 2x) \\ -x - 2 &= 3x^2 - 6x \\ 0 &= 3x^2 - 5x + 2 \end{aligned}$$

Now, factor the equation (or use the quadratic formula), set each factor equal to 0, solve for x, and check that the solutions don't result in a zero denominator.

$$\begin{aligned} 0 &= 3x^2 - 5x + 2 \\ 0 &= (3x - 2)(x - 1) \end{aligned}$$

The solutions are $x = \frac{2}{3}$ and $x = 1$, both of which are valid solutions, so grid in either value.

14. 0
Difficulty: Medium

Category: Passport to Advanced Math / Exponents

Strategic Advice: When you are looking for a remainder, long division is definitely the way to go. Just work carefully, particularly with the negative signs.

Getting to the Answer: To reduce the chance of careless errors, start by factoring –1 out of both expressions to remove some of the negative signs. The math will still work if you omit this step, but you'll have more negative signs to track.

$$\begin{array}{r} 2x^2 + 6x - 15 \\ x - 2 \overline{\big)\, 2x^3 + 2x^2 - 27x + 30} \\ \underline{-(2x^3 - 4x^2)} \\ 6x^2 - 27x + 30 \\ \underline{-(6x^2 - 12x)} \\ -15x + 30 \\ \underline{-(-15x + 30)} \\ 0 \end{array}$$

The remainder is 0.

15. B
Difficulty: Hard

Category: Passport to Advanced Math / Exponents

Strategic Advice: When building an equation that describes how much work two people can do together versus independently, remember the additive nature of rates and the equation $r = \frac{W}{t}$.

Getting to the Answer: Let x represent Professor Lambert. If W (the work) represents the job of grading the exams and there is one job to complete, then Professor Lambert's grading speed (his rate) is $\frac{1}{x}$. Professor Stassen's solo grading time is 50% greater than Professor Lambert's, so her rate is $\frac{1}{1.5x}$. Working together, the two would take 40 hours to grade all the exams, which is a rate of $\frac{1}{40}$ exams per hour. Add the individual rates together and set them equal to the combined rate: $\frac{1}{x} + \frac{1}{1.5x} = \frac{1}{40}$. None of the answer choices contain a decimal in the denominator, so multiply the second fraction by $\frac{2}{2}$ to eliminate the 1.5. The result is $\frac{1}{x} + \frac{2}{3x} = \frac{1}{40}$, which matches (B).

16. C
Difficulty: Hard

Category: Passport to Advanced Math / Exponents

Strategic Advice: This is a *work* question, so use the formula $W = rt$. When two people are working together, start by finding their combined rate and go from there. Don't forget to convert Joe's time to hours to match the units given for his son's time.

Getting to the Answer: Manipulate the work formula to find each person's rate. If $W = rt$, then $r = \frac{W}{t}$. Use 1 for W because there is 1 lawn to mow. Joe can mow the lawn in 45 minutes (or $\frac{3}{4}$ of an hour), so he can mow $\frac{1}{\frac{3}{4}} = \frac{4}{3}$ of the lawn in 1 hour. Joe's son can mow the lawn in 3 hours, so he can mow $\frac{1}{3}$ of the lawn in 1 hour. Working together, they can mow $\frac{4}{3} + \frac{1}{3} = \frac{5}{3}$ of the lawn in 1 hour. Substitute this rate back into the work formula to find the amount of time it will take them to mow the lawn together:

$$W = rt$$
$$1 = \frac{5}{3}t$$
$$\frac{3}{5}(1) = \frac{3}{5}\left(\frac{5}{3}t\right)$$
$$\frac{3}{5} = t$$

Working together, they can mow the lawn in $\frac{3}{5}$ of an hour, which is equal to $\frac{3}{5} \times 60 = 36$ minutes, making (C) correct.

17. B
Difficulty: Hard

Category: Passport to Advanced Math / Exponents

Strategic Advice: The area of a trapezoid is given by the formula $A = \frac{1}{2}(b_1 + b_2)h$, where b_1 and b_2 represent the lengths of the bases and h represents the height of the trapezoid.

Getting to the Answer: Although this question asks about a *polynomial* equation, focus on what you're looking for—an equation that represents the area of the trapezoid. The polynomial part will take care of itself. Examine the figure along with the labels along the axes to determine the dimensions of the trapezoid (which happen to be variables instead of numbers). The height of the trapezoid is the vertical distance from the

x-axis to the top base, which is given (labeled on the y-axis) as $k + 1$. The length of the top base is the distance from the y-axis to the dotted line inside the trapezoid, which is k. Finally, the length of the bottom base is the distance from the y-axis to the right-most point along the x-axis, which is $2k$. Substitute these values into the area formula and simplify:

$$\begin{aligned} A &= \frac{1}{2}(b_1 + b_2)h \\ &= \frac{1}{2}(k + 2k)(k + 1) \\ &= \frac{1}{2}(3k)(k + 1) \\ &= \frac{1}{2}(3k^2 + 3k) \\ &= \frac{3k^2 + 3k}{2} \end{aligned}$$

This matches (B).

18. B
Difficulty: Hard

Category: Passport to Advanced Math / Exponents

Strategic Advice: You will not be expected to factor a fourth-degree polynomial like the one in this question on Test Day, so there must be another way to approach the solution. This is a *long* polynomial, so think *long* division.

Getting to the Answer: Factors of polynomials must divide evenly into the polynomial, which means there is no remainder. In polynomial long division, the only time there is no remainder is when the *constant* in the factor divides evenly into the *constant* in the polynomial. This means you can eliminate A and D (because 3 does not divide evenly into 20). To decide between (B) and C, use long division. Start with (B):

$$\begin{array}{r} 4x^3 + 7x^2 - 7x - 10 \\ 3x - 2 \overline{) 12x^4 + 13x^3 - 35x^2 - 16x + 20} \\ -(12x^4 - 8x^3) \quad\quad\quad\quad\quad\quad\quad \\ \hline 21x^3 - 35x^2 - 16x + 20 \\ -(21x^3 - 14x^2) \quad\quad\quad\quad\; \\ \hline -21x^2 - 16x + 20 \\ -(-21x^2 + 14x) \quad\;\; \\ \hline -30x + 20 \\ -(-30x + 20) \\ \hline 0 \end{array}$$

When $p(x)$ is divided by $3x - 2$, the remainder is 0, so $3x - 2$ is a factor of the polynomial. This means (B) is correct.

19. B
Difficulty: Hard

Category: Passport to Advanced Math / Exponents

Strategic Advice: When a question involves *equivalent expressions*, compare how the original expression looks to how the desired result looks. Here, the original numerator has a higher power than the denominator, but the desired result doesn't. This is a clue that you need to divide.

Getting to the Answer: Set up the fraction as a long division problem and divide. Don't forget that anything left over at the end is the remainder and is written over the original denominator:

$$\begin{array}{r} 3x + 2 \quad + \frac{4}{3x - 2} \\ 3x - 2 \overline{) 9x^2 + 0x + 0} \quad\quad\quad \\ -(9x^2 - 6x) \quad\quad\quad\quad \\ \hline 6x + 0 \quad\quad\quad \\ -(6x - 4) \quad\quad\quad \\ \hline 4 \quad\quad\quad \end{array}$$

Next, compare the result to the desired expression. Because the remainder 4 matches the numerator in the fraction part of the expression, A must be the quotient $3x + 2$. Choice (B) is correct.

20. C

Difficulty: Hard

Category: Passport to Advanced Math / Exponents

Strategic Advice: Take a moment to think about this question logically before you start multiplying functions by constants and adding functions. Understanding what it means for a polynomial to be *divisible* by an expression can cut your work in half.

Getting to the Answer: If a polynomial is divisible by an expression, then the remainder when divided will be 0. The only way the remainder can be 0 is if the constant term of the expression divides evenly into the constant term of the polynomial. Based only on the constant terms:

A: $p(x)+q(x)=\ldots-3$ is not divisible by 5.

B: $2p(x)+q(x)=\ldots-5$ is divisible by 5.

C: $p(x)+3q(x)=\ldots-5$ is divisible by 5.

D: $2p(x)+3q(x)=\ldots-7$ is not divisible by 5.

The answer must be B or (C). Try B:

$$2p(x)=2(6x^2+7x-2)=12x^2+14x-4$$

$$2p(x)+q(x)=12x^2+14x-4+2x^2+7x-1$$
$$=14x^2+21x-5$$

Now use long division:

$$\begin{array}{r} 7x-7 \\ 2x+5\overline{)14x^2+21x-5} \\ -(14x^2+35x) \\ \hline -14x-5 \\ -(-14x-35) \\ \hline 30 \end{array}$$

The remainder is not 0 (it's 30), so $2p(x)+q(x)$ is not divisible by $2x+5$. This means (C) must be the correct answer. (For extra practice, work it out using the steps above to confirm that the remainder is indeed 0, but keep in mind that this is not necessary on Test Day.)

$$\begin{array}{r} 6x-1 \\ 2x+5\overline{)12x^2+28x-5} \\ -(12x^2+30x) \\ \hline -2x-5 \\ -(-2x-5) \\ \hline 0 \end{array}$$

PRACTICE SET 9: FUNCTIONS AND FUNCTION NOTATION

Functions and Function Notation

Functions act as rules that transform inputs into outputs, and they differ from equations in that each input can have only one corresponding output. The *domain* of a function is the set of input values (typically x) for which the function is defined, and the *range* of a function is the set of corresponding output values (typically y). The notation $f(3)$ is read "f of 3," and it means the output value when 3 is substituted into the function's equation.

Graphical Representation of Functions

Graphically, an input and corresponding output create ordered pairs of the form $(x, f(x))$. If given the graph of a function and asked to evaluate $f(3)$, locate 3 along the x-axis, trace up (or down) to the function's graph, and find the corresponding y-value. Conversely, suppose you're given a graph and asked to find the value of x for which $f(x) = 3$. Because $f(x)$ represents the output value, or range, translate this as "When does the y-value equal 3?" To answer the question, find 3 on the y-axis this time, and trace over to the function's graph. The corresponding x-value is your answer.

Working with Multiple Functions

There are several ways in which the SAT might ask you to evaluate multiple functions simultaneously. Fortunately, the rules governing what to do are easy to understand. Combining functions simply involves adding, subtracting, multiplying, and/or dividing the individual functions. The table below describes how to combine functions with the four basic operations.

When you see ...	convert it to:
$(f+g)(x)$	$f(x)+g(x)$
$(f-g)(x)$	$f(x)-g(x)$
$(fg)(x)$	$f(x) \cdot g(x)$
$\left(\frac{f}{g}\right)(x)$	$\frac{f(x)}{g(x)}$

A more challenging type of function combination that you're likely to see is a composition of functions (sometimes called nested functions). Questions involving a composition of functions require you to find the output of a function at a given input value, and then plug the result into another function to get the final answer. A composition could be written as $f(g(x))$ or $(f \circ g)(x)$. Regardless of which form you're given, always start with the function that is closest to the x, and work your way outward. Keep in mind that $f(g(x))$ is very rarely the same as $g(f(x))$, so order definitely matters.

Relationships between Variables

When analyzing the graph of a function or an interval (a specific segment) of the function, the relationship between the x- and y-values can be described as follows:

- Increasing: The y-values *increase* as the corresponding x-values increase.
- Decreasing: The y-values *decrease* as the corresponding x-values increase.
- Constant: The y-values *remain the same* as the x-values increase. The graph of a constant function is a horizontal line.

Transformations

A transformation occurs when a change is made to a function's equation and/or graph. Transformations include translations (moving a graph up/down, left/right), reflections (flips about an axis or other line), and expansions/compressions (stretching or squashing horizontally or vertically). The following table provides transformation rules for altering a given function $f(x)$.

Algebraic Change	**Corresponding Graphical Change**
$f(x)$	N/A—original function
$f(x) + a$	$f(x)$ moves up a units
$f(x) - a$	$f(x)$ moves down a units
$f(x + a)$	$f(x)$ moves left a units
$f(x - a)$	$f(x)$ moves right a units
$-f(x)$	$f(x)$ is reflected vertically over the x-axis (top-to-bottom)
$f(-x)$	$f(x)$ is reflected horizontally over the y-axis (left-to-right)
$af(x)$ for $0 < a < 1$	$f(x)$ undergoes vertical compression
$af(x)$ for $a > 1$	$f(x)$ undergoes vertical expansion
$f(ax)$ for $0 < a < 1$	$f(x)$ undergoes horizontal expansion
$f(ax)$ for $a > 1$	$f(x)$ undergoes horizontal compression

Note: One way to think about transformations is that if the transformation is "with the x," then a horizontal change occurs. Keep in mind that horizontal changes are the opposite of what they look like. If the transformation is not "with the x," a vertical change occurs (vertical changes are exactly what they look like).

PRACTICE SET

Easy

1. If $f(x) = x^3 + 6$, what is the value of $f(3)$?

   A) 15

   B) 33

   C) 39

   D) 729

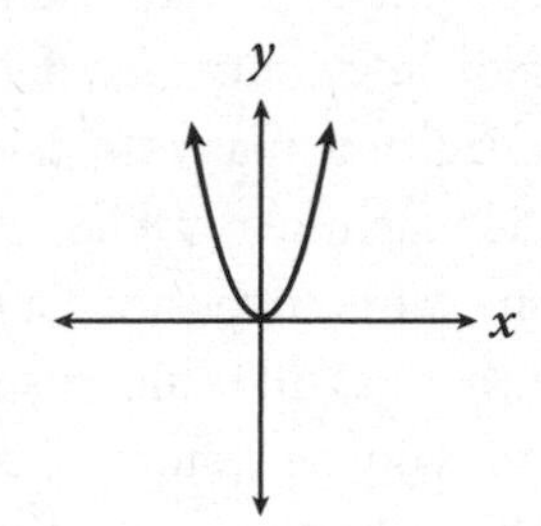

2. If the above graph shows $f(x) = 7x^2$, which of the following is the graph of $f(x) = 7x^2 + 1$?

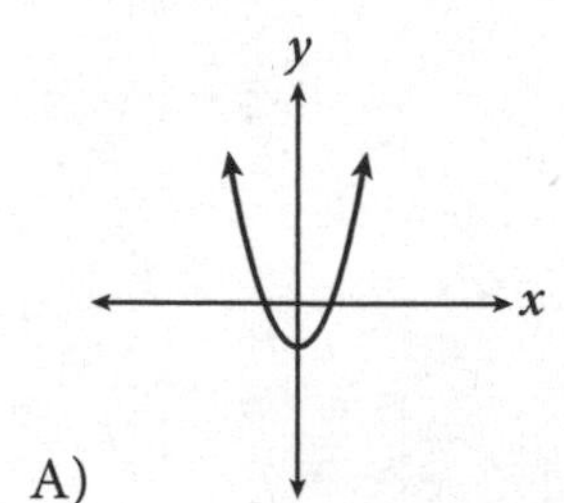

A)

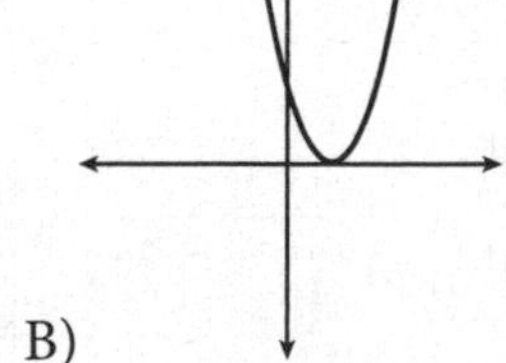

B)

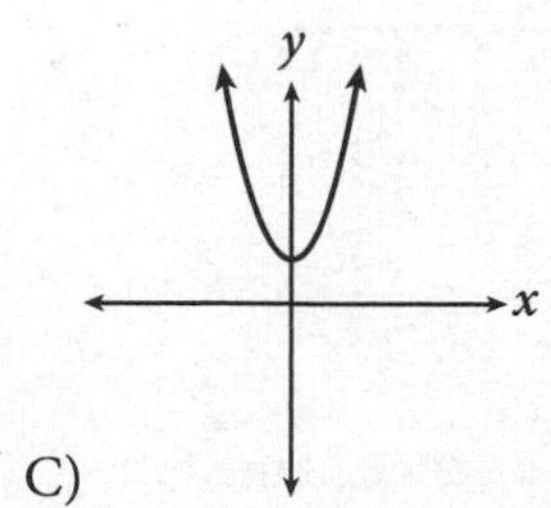

C)

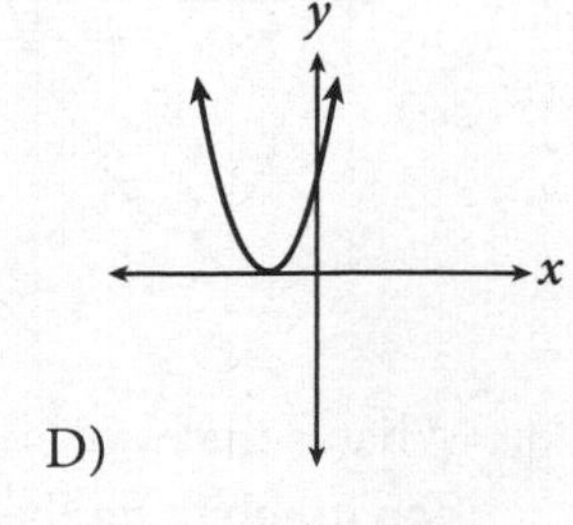

D)

3. If $g(x)$ represents a quadratic function, which of the following best describes the relationship between the graphs of $g(x)$ and $g(x - 3)$?

   A) The graph of $g(x - 3)$ is 3 units to the left of the graph of $g(x)$.

   B) The graph of $g(x - 3)$ is 3 units to the right of the graph of $g(x)$.

   C) The graph of $g(x - 3)$ is 3 units lower than the graph of $g(x)$.

   D) The graph of $g(x - 3)$ is 3 units higher than the graph of $g(x)$.

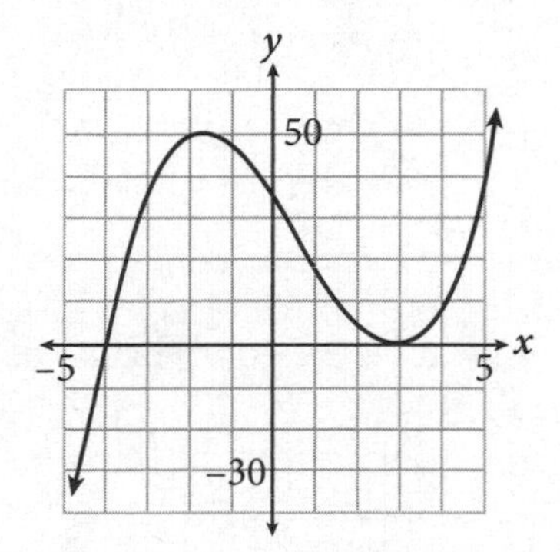

4. If $p(x)$ is the polynomial function shown in the graph above, which of the following could be the factored form of $p(x)$?

   A) $p(x) = (x - 4)(x + 3)$

   B) $p(x) = (x - 4)(x + 3)^2$

   C) $p(x) = (x + 4)(x - 3)$

   D) $p(x) = (x + 4)(x - 3)^2$

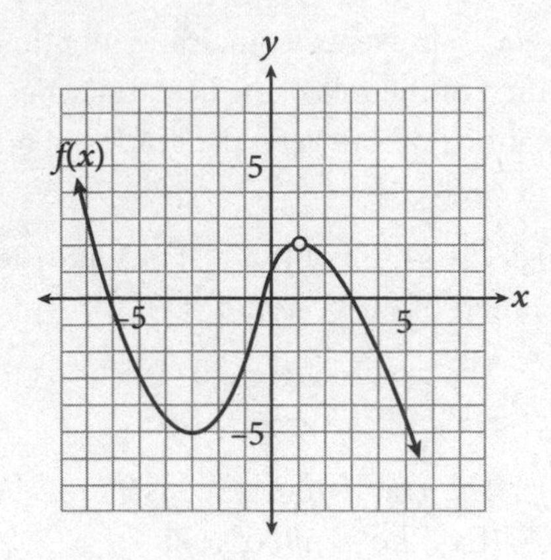

5. The graph of $y = f(x)$ is given above. For what value of x is $y = f(x - 4)$ undefined?

A) −3

B) 1

C) 2

D) 5

6. If $f(x) = \left|x^2 + 2x + 1\right|$, what is the value of $f(-4)$?

Medium

$$f(s) = \frac{s}{25} + 1$$

7. A college campus has vending machines, which must be periodically restocked. The restocking frequency depends primarily on how many students have classes near the machine each semester. The vending machine company uses the function shown above, where s is the number of students estimated to have classes within the immediate vicinity on a daily basis, to determine how many times per semester the machine must be restocked. How many more times must a vending machine that has 400 students in the immediate vicinity be restocked compared to one that has 300 students in the immediate vicinity?

A) 2

B) 4

C) 7

D) 13

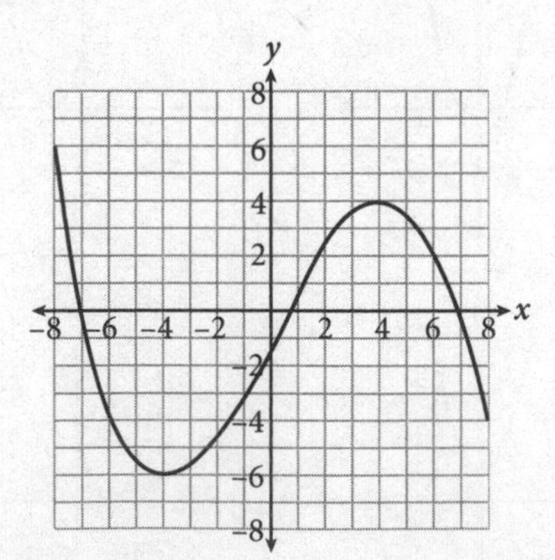

8. What is the maximum value of the function graphed on the coordinate plane above, over the interval $-8 \leq x \leq 8$?

A) 4

B) 6

C) 8

D) ∞ (infinity)

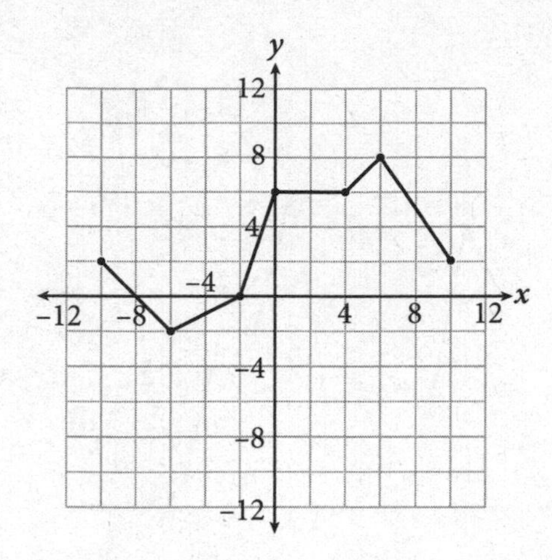

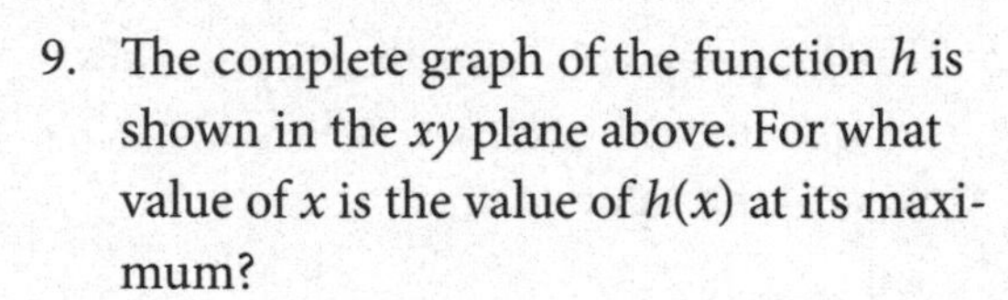

9. The complete graph of the function h is shown in the xy plane above. For what value of x is the value of $h(x)$ at its maximum?

A) 5

B) 6

C) 8

D) 10

$$g(x) = \frac{x}{x-2} \text{ and } h(x) = \sqrt{9-x}$$

10. Given the functions defined above, what is the value of $(g \circ h)(5)$?

A) 0

B) $\frac{8}{7}$

C) 2

D) Undefined

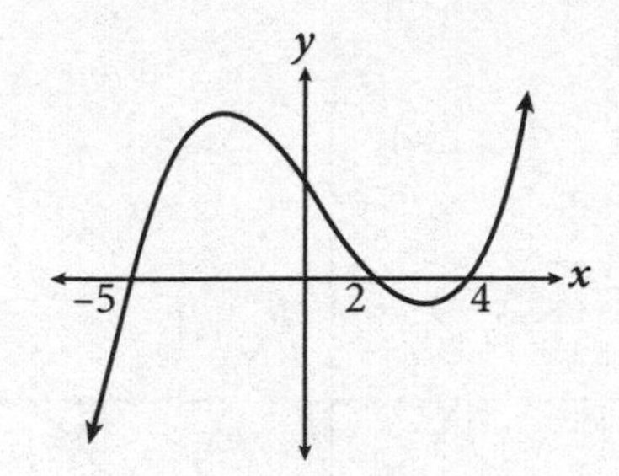

11. The polynomial function shown in the graph crosses the x-axis at $x = -5$, $x = 2$, and $x = 4$. If the equation for this polynomial is written in the form $y = ax^3 + bx^2 + cx + d$, with $a = 1$, which of the following could be the equation?

A) $y = x^3 - x^2 - 38x - 40$

B) $y = x^3 + x^2 - 22x - 40$

C) $y = x^3 - x^2 - 22x + 40$

D) $y = x^3 + 5x^2 + 8x + 40$

12. If $f(x) = \frac{1}{\sqrt{x}}$, what is the domain of $f(x)$?

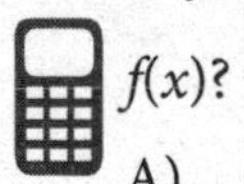

A) All real numbers

B) All integers except zero

C) All positive real numbers

D) All real numbers except zero

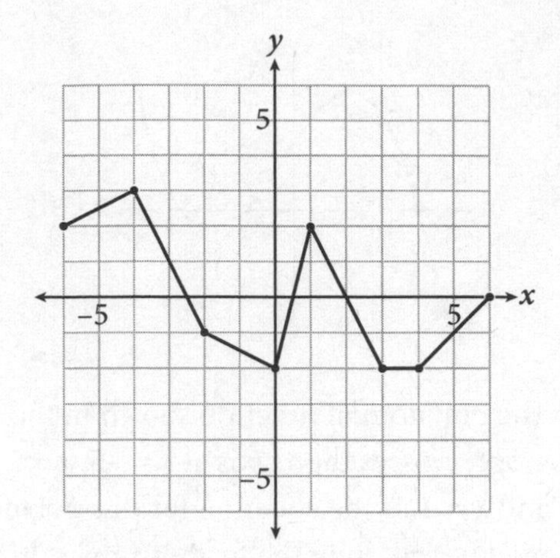

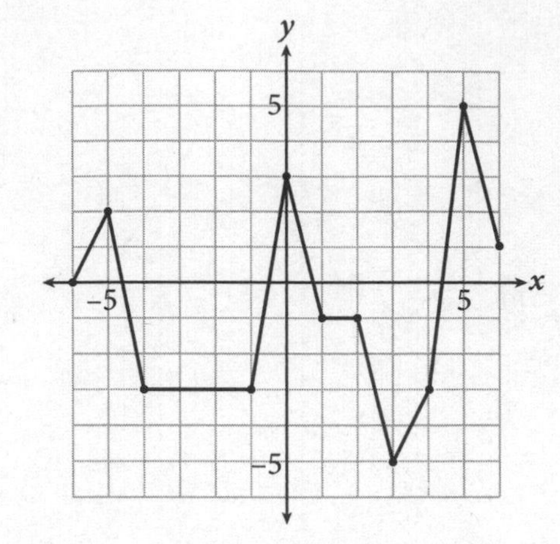

13. The complete graph of the function f is shown in the xy plane above. Which of the following is/are equal to −2?

    I. $f(-1)$

    II. $f(0)$

    III. $f(\frac{7}{2})$

    A) I only

    B) II only

    C) II and III only

    D) I, II, and III

14. The polynomial function p is defined by $p(x) = 4x^3 + bx^2 + 41x + 12$, where b is a constant. When graphed on a standard coordinate plane, p intersects the x-axis at (−0.25, 0), (3, 0), and (k, 0). What is the value of b?

    A) −27

    B) $-\sqrt{207}$

    C) 4

    D) 27

15. The figure above shows the graph of $h(x)$ on the interval $-6 \leq x \leq 6$. If $h(3) = a$, what is the value of $h(a)$?

    A) −6

    B) −5

    C) 0

    D) 2

16. A function g satisfies $g(5) = 3$ and $g(3) = 0$, and a function h satisfies $h(3) = -2$ and $h(0) = 5$. What is the value of $h(g(3))$?

	/	/	
.	.	.	.
	0	0	0
1	1	1	1
2	2	2	2
3	3	3	3
4	4	4	4
5	5	5	5
6	6	6	6
7	7	7	7
8	8	8	8
9	9	9	9

Hard

17. If $f(x + 2) = x^2 - x + 9$, then what is the value of $f(-6)$?

    A) -47

    B) 29

    C) 65

    D) 81

18. If $g(x + 3) = x^2 + 2x + 1$, then which of the following gives $g(x)$?

    A) $g(x) = x^2 + 8x + 16$

    B) $g(x) = x^2 - 4x + 4$

    C) $g(x) = x^2 + 2x + 4$

    D) $g(x) = x^2 + 2x - 2$

19. A commercial airline has calculated that the approximate fuel mileage for its 600-passenger airplane is 0.2 miles per gallon when the plane travels at an average speed of 500 miles per hour. Flight 818's fuel tank has 42,000 gallons of fuel at the beginning of an international flight. If the plane travels at an average speed of 500 miles per hour, which of the following functions f models the number of gallons of fuel remaining in the tank t hours after the flight begins?

    A) $f(t) = 42{,}000 - \dfrac{500t}{0.2}$

    B) $f(t) = 42{,}000 - \dfrac{0.2}{500t}$

    C) $f(t) = \dfrac{42{,}000 - 500t}{0.2}$

    D) $f(t) = \dfrac{42{,}000 - 0.2}{500t}$

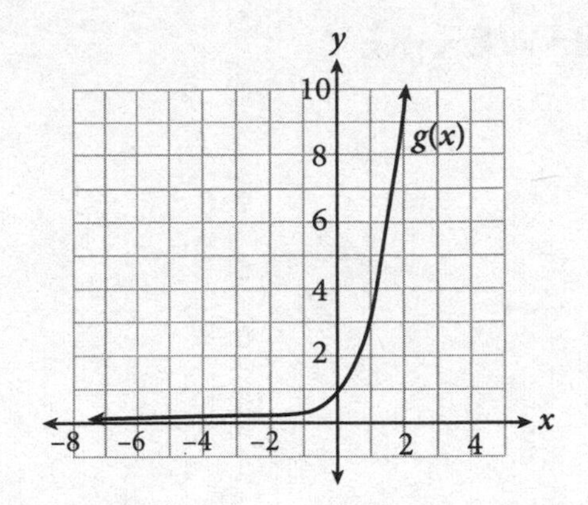

20. An exponential function $g(x)$ is shown in the figure above. What is the exact value of $g(-4)$?

ANSWER KEY

1. B	**6. 9**	**11. C**	**16. 5**
2. C	**7. B**	**12. C**	**17. D**
3. B	**8. B**	**13. C**	**18. B**
4. D	**9. B**	**14. A**	**19. A**
5. D	**10. D**	**15. D**	**20. 1/81**

ANSWERS & EXPLANATIONS

1. B
Difficulty: Easy

Category: Passport to Advanced Math / Functions

Strategic Advice: When asked to find the value of *f*(a number), just plug in the correct value and simplify.

Getting to the Answer: Substitute 3 for x and simplify: $f(3) = 3^3 + 6 = 27 + 6 = 33$. Choice (B) is the correct answer.

2. C
Difficulty: Easy

Category: Passport to Advanced Math / Functions

Strategic Advice: Understanding function notation and how transformations work will earn you points on Test Day.

Getting to the Answer: Adding 1 to the original function results in a vertical shift. Here, the new function is simply the old function moved up 1 unit, which is (C).

3. B
Difficulty: Easy

Category: Passport to Advanced Math / Functions

Strategic Advice: When working with function notation, transformations that are grouped with the x (here, inside the parentheses) represent horizontal shifts, which are always the opposite of how they appear. For example, the transformation $f(x + a)$ represents a shift to the *left* while $f(x - a)$ represents a shift to the *right*.

Getting to the Answer: The -3 is grouped with the x, so the graph of $g(x - 3)$ is a horizontal shift that is 3 units to the right of the original graph, making (B) correct.

4. D
Difficulty: Easy

Category: Passport to Advanced Math / Functions

Strategic Advice: The graph of a polynomial function crosses the x-axis when there is a *single root* and just touches the x-axis when there is a *double root*. A double root occurs when a factor is repeated, or in other words, is raised to the second power. For example, the factor $(x - a)$ produces a single root at $x = a$, while $(x - b)^2$ produces a double root at $x = b$.

Getting to the Answer: Examine the graph. The function crosses the x-axis at $x = -4$, which means one factor is $(x - (-4))$, or $(x + 4)$. This means you can eliminate A and B. The function just touches the x-axis at $x = 3$, which means the other factor (which should be raised to the second power) is $(x - 3)$. The factored form of the polynomial could be $p(x) = (x + 4)(x - 3)^2$, which is (D).

5. D
Difficulty: Easy

Category: Passport to Advanced Math / Functions

Strategic Advice: Algebraically, a function is undefined when a value (or values) in its domain cannot be evaluated (for example, taking the square root of a negative number or dividing by 0). Graphically, this produces a "hole" (or an asymptote) in the graph.

Getting to the Answer: The graph of $f(x)$, which is shown, is undefined at $x = 1$ because there is a hole in the graph, meaning that the function cannot be evaluated at $x = 1$. The question asks where the transformed (shifted) function is undefined. The notation $y = f(x - 4)$ represents a horizontal shift of $f(x)$ to the *right* 4 units (remember, horizontal shifts are always the opposite of how they appear), so imagine moving the hole shown in the graph to the right 4 units. The result is (5, 2), so the shifted function is undefined at $x = 5$, which is (D).

6. 9
Difficulty: Easy

Category: Passport to Advanced Math / Functions

Strategic Advice: Don't let an absolute value function intimidate you. Just treat the function as you would any other, and remember that when you take the absolute value, the end result must be positive.

Getting to the Answer: Substitute −4 for each x in the equation and simplify:

$$f(x) = |x^2 + 2x + 1|$$
$$f(-4) = |(-4)^2 + 2(-4) + 1|$$
$$f(-4) = |16 - 8 + 1|$$
$$f(-4) = 9$$

Grid in 9.

7. B
Difficulty: Medium

Category: Passport to Advanced Math / Functions

Strategic Advice: Some word problems can be fairly straightforward, but you must be careful to check that you answered the right question (here, how many more times a vending machine with 400 nearby students must be restocked than a vending machine with 300 nearby students).

Getting to the Answer: Start by evaluating the function at $s = 400$ and at $s = 300$. Make sure you follow the correct order of operations as you simplify:

$$f(s) = \frac{s}{25} + 1$$
$$f(400) = \frac{400}{25} + 1$$
$$f(400) = 16 + 1$$
$$f(400) = 17$$

$$f(300) = \frac{300}{25} + 1$$
$$f(300) = 12 + 1$$
$$f(300) = 13$$

The question asks how many more times the vending machine must be restocked, so find the difference in the amounts to get $17 - 13 = 4$ more times, which is (B).

8. B
Difficulty: Medium

Category: Passport to Advanced Math / Functions

Strategic Advice: The maximum value of a function that is graphed over a given interval is the highest y-value of all the points on the graph in that interval. Be careful, it's easy to mistake a *relative maximum* (a hump in the graph) for the actual maximum over the entire interval.

Getting to the Answer: For the graph shown, the maximum occurs at the left endpoint of the graph, (–8, 6). The y-value of the point is 6, so (B) is correct. Note that if the domain (x-values) had not been restricted to –8 to 8, and the graph had continued indefinitely upward on the left side, then there would be no maximum value (i.e., the maximum would be ∞).

9. B

Difficulty: Medium

Category: Passport to Advanced Math / Functions

Strategic Advice: The maximum of a function is the point at which the range (the y-values) of the function is at its highest. Identify that point on the graph, and then determine the corresponding x-value. Pay careful attention to the axis labels—they are not given in increments of 1.

Getting to the Answer: The maximum value of $h(x)$ (the highest point h reaches on the graph) is 8. Drop down from the maximum to the x-axis to find that the corresponding x-value is 6, which matches (B).

10. D

Difficulty: Medium

Category: Passport to Advanced Math / Functions

Strategic Advice: The notation $(g \circ h)(x)$ indicates a composition of two functions, which can also be written as $g(h(x))$ and read as "g of h of x." It means that the output when x is substituted in $h(x)$ becomes the input for $g(x)$. Just remember that when working with compositions, you always start with the function that is closest to x.

Getting to the Answer: Substitute 5 for x in $h(x)$, simplify, and then substitute the result into $g(x)$:

$$h(5) = \sqrt{9-5} = \sqrt{4} = 2$$

$$g(2) = \frac{2}{2-2} = \frac{2}{0}$$

Be careful—a number divided by 0 is not 0! In fact, division by 0 is not defined, so $(g \circ h)(5)$ is undefined, which is (D).

11. C

Difficulty: Medium

Category: Passport to Advanced Math / Functions

Strategic Advice: You can use the x-intercepts shown in a graph to write the factors of the corresponding polynomial—here, there are three x-intercepts, so you should have three factors.

Getting to the Answer: You are given that $a = 1$, which tells you that each factor is of the form $(x - a)$. To write the equation in the form requested, FOIL the first two factors, then multiply the result by the third factor. The order in which you multiply the factors doesn't matter. The function crosses the x-axis at $x = -5$, which means one factor is $(x - (-5))$, or $(x + 5)$. The function also crosses the x-axis at $x = 2$ and $x = 4$, which translate to factors of $(x - 2)$ and $(x - 4)$. Start by FOILing $(x + 5)$ and $(x - 2)$. Then multiply the result by the remaining factor, $(x - 4)$:

$$\begin{aligned}(x+5)(x-2) &= x^2 - 2x + 5x - 10 \\ &= x^2 + 3x - 10\end{aligned}$$

$$\begin{aligned}(x^2+3x-10)(x-4) &= x^3 - 4x^2 + 3x^2 - 12x - 10x + 40 \\ &= x^3 - x^2 - 22x + 40\end{aligned}$$

This matches (C).

12. C

Difficulty: Medium

Category: Passport to Advanced Math / Functions

Strategic Advice: The domain of a function is the set of *x*-values for which the function is defined (exists). For basic functions, there are only two rules that can't be broken—the quantity under a square root can't be negative, and the denominator of a fraction can't equal 0.

Getting to the Answer: The square root in the denominator tells you that only numbers greater than or equal to 0 are in the function's domain. However, because the *x* is in the denominator, it can't be equal to 0. Putting these two restrictions together results in a domain of all real numbers that are strictly greater than 0, or in other words, all positive real numbers, which is (C).

13. C

Difficulty: Medium

Category: Passport to Advanced Math / Functions

Strategic Advice: When a function is graphed on a coordinate plane, the points take on the form $(x, f(x))$, so $f(x)$ is just another way of saying *y*.

Getting to the Answer: Read the graph to find the *y*-coordinates that correspond to the given *x*-coordinates (the numbers inside the parentheses) in I, II, and III. The points on the graph of *f* with *x*-coordinates –1, 0, and $\frac{7}{2}$ (or 3.5) are $(-1, -1.5)$, $(0, -2)$, and $(3.5, -2)$. Therefore, of the values given, only 0 (II) and $\frac{7}{2}$ (III) are equal to –2, making (C) correct.

14. A

Difficulty: Medium

Category: Passport to Advanced Math / Functions

Strategic Advice: A function's *solutions* are the point or points at which its graph crosses the *x*-axis. You're given two *x*-intercepts (two solutions), which you can use to find the value of *b*.

Getting to the Answer: The solution (3, 0) is the simplest of the intercepts given, so plug in 3 for each *x* and 0 for $p(x)$, and solve for *b*:

$$p(x) = 4x^3 + bx^2 + 41x + 12$$
$$0 = 4(3)^3 + b(3)^2 + 41(3) + 12$$
$$0 = 108 + 9b + 123 + 12$$
$$0 = 243 + 9b$$
$$-9b = 243$$
$$b = -27$$

(A) is the correct answer.

15. D

Difficulty: Medium

Category: Passport to Advanced Math / Functions

Strategic Advice: When presented with a function in equation form (here, $h(3) = a$), 3 is the input (*x*) and *a* is the output (*y*). Use this information and the graph to first find *a*, and then $h(a)$.

Getting to the Answer: If $h(3) = a$, then *a* is the *y*-coordinate of the point on the graph when $x = 3$; the point is (3, –5), so $a = -5$. You're not done yet—you're asked for the value of $h(a)$, or $h(-5)$. Find the point on the graph that has an *x*-coordinate of –5; the point is (–5, 2), so $h(a) = 2$. This matches (D).

16. 5

Difficulty: Medium

Category: Passport to Advanced Math / Functions

Strategic Advice: When asked for the value of a composition of functions, start with the innermost set of parentheses and work outward from there.

Getting to the Answer: The innermost set of parentheses dictates that you should find $g(3)$ first, which the question tells you is equal to 0. Substitute 0 for $g(3)$ in the requested expression to get $h(0)$. The question states that $h(0) = 5$, so 5 is your final answer.

17. D

Difficulty: Hard

Category: Passport to Advanced Math / Functions

Strategic Advice: In a question like this, where a *shifted* function is given $(x + 2)$ but a value for the *original* function is requested, try to write the input (the number inside the parentheses) in terms of the shift.

Getting to the Answer: Start by writing −6 in terms of $x + 2$ (in other words, some number plus 2): −6 can be written as −8 + 2, so x is −8. Now, substitute this value into the function:

$$f(x + 2) = x^2 - x + 9$$
$$f(-8 + 2) = (-8)^2 - (-8) + 9$$
$$f(-6) = 64 + 8 + 9 = 81$$

The correct answer is (D).

18. B

Difficulty: Hard

Category: Passport to Advanced Math / Functions

Strategic Advice: The key to answering a question like this is having a conceptual understanding of function notation. Here, the input $(x + 3)$ has already been substituted and simplified in the given function. Your job is to determine what the function would have looked like had x been the input instead.

Getting to the Answer: To keep things organized, let $u = x + 3$, the old input. This means $x = u - 3$. Substitute this into $g(x + 3)$ and simplify:

$$g(x + 3) = x^2 + 2x + 1$$
$$g(u) = (u - 3)^2 + 2(u - 3) + 1$$
$$= u^2 - 6u + 9 + 2u - 6 + 1$$
$$= u^2 - 4u + 4$$

This means $g(u) = u^2 - 4u + 4$. Now, when working with function notation, you evaluate the function by substituting a given input value for the variable in the parentheses. Here, if the input value is x, then $g(x) = x^2 - 4x + 4$, which is (B).

19. A

Difficulty: Hard

Category: Passport to Advanced Math / Functions

Strategic Advice: Sometimes, the units on the variables can help you piece together a function that represents a real-world scenario. In this particular function, the independent variable is time (in hours), and the dependent variable is remaining fuel (in gallons).

Getting to the Answer: The airplane is traveling at an average speed of 500 miles per hour, and the plane's fuel mileage is 0.2 miles per gallon. You can calculate the number of gallons of fuel used each hour using the factor-label method:

$\frac{500\ \cancel{mi}}{1\ h} \times \frac{1\ gal}{0.2\ \cancel{mi}} = \frac{500}{0.2}$ gal/h. This means that, in t hours, the plane uses $\frac{500t}{0.2}$ gallons of fuel. The plane's tank has 42,000 gallons of fuel at the beginning of the flight, so the amount of fuel left after t hours is $42{,}000 - \frac{500t}{0.2}$. This makes (A) the correct answer.

20. 1/81

Difficulty: Hard

Category: Passport to Advanced Math / Functions

Strategic Advice: When you're told that a function is exponential, start with the form $g(x) = a \cdot b^x$ (with $a = 1$) to write an equation that represents the graph.

Getting to the Answer: The notation $g(-4)$ means that you are looking for the y-value when x is -4. Because you cannot tell what the exact value is from the graph, you need to find the equation of the function. To find the base, b, which is being raised to a power, make a list of several points on the graph and look for a pattern: (0, 1), (1, 3), and (2, 9). The y-values are all powers of 3, so try $b = 3$. Notice that $3^0 = 1$, $3^1 = 3$, and $3^2 = 9$. The equation of the function is $g(x) = 3^x$, so substitute -4 for x and simplify:

$$g(-4) = 3^{-4} = \frac{1}{3^4} = \frac{1}{3 \times 3 \times 3 \times 3} = \frac{1}{81}$$

Grid in 1/81.

PRACTICE SET 10: QUADRATIC EQUATIONS AND THEIR GRAPHS

Introduction to Quadratic Equations

A quadratic equation or expression is one that contains a squared variable (x^2) as the highest-order term. In standard form, a quadratic equation is written as $y = ax^2 + bx + c$, where a, b, and c are constants. A quadratic equation can have zero, one, or two real solutions, which are also called roots, x-intercepts (because solutions are where the graph of an equation crosses the x-axis), or zeros.

Solving Quadratic Equations

In most cases, before you can solve a quadratic equation, you must set the equation equal to 0. In other words, move everything to one side of the equal sign so that 0 is the only thing remaining on the other side. Once complete, you can use a variety of algebraic techniques, the quadratic formula, or graphing to find the solutions.

Factoring

Factoring, also known as reverse-FOIL, allows you to go from a quadratic equation written in standard form to a product of two binomials. Once you have a pair of binomials, set each factor equal to 0 and solve for the variable. Here are some general rules for factoring:

- Factoring is easiest when $a = 1$, so whenever possible, try to simplify your expression so that this is the case.
- If you see nice-looking numbers (integers or simple fractions) in the answer choices, this is a clue that factoring is possible.
- To factor a quadratic expression that is written in the form $x^2 + bx + c$, look for the factors of c that add up to b.
- To factor a quadratic expression that is written in the form $ax^2 + bx + c$, multiply a times c, look for the factors of ac that add up to b, use those factors to break the middle term into two pieces, and then factor by grouping.
- If you're ever not sure that you've factored correctly, use FOIL to check your work. You should get the equation you started with.

Square Rooting and Completing the Square

Occasionally, a quadratic equation is already set up perfectly to use *square rooting*, which simply means taking the square root of both sides. For example, $(x + 3)^2 = 49$ is ready to square root because both sides of the equation are perfect squares. For more difficult quadratic equations, you can *complete the square* and then use square rooting. To complete the square, start by putting the quadratic equation in standard form. Once there, divide b by 2 and square the result. Then, add the result to both sides of the equation, factor, and solve by square rooting.

The Quadratic Formula

The quadratic formula can be used to solve any quadratic equation. However, because the math can often get complicated, use it as a last resort or when you need to find exact solutions (or when you see radicals in the answer choices). Be sure to write the equation in standard form before plugging the values of a, b, and c into the formula. Memorize the formula before Test Day:

$$x = \frac{-b \pm \sqrt{b^2 - 4ac}}{2a}$$

The expression under the radical ($b^2 - 4ac$) is called the *discriminant*, and its sign dictates the number of real solutions that the equation has. If this quantity is positive, there are two distinct real solutions; if it is equal to 0, there is just one distinct real solution; and if it is negative, there are no real solutions.

Finding Solutions by Graphing

To solve a quadratic equation in the Calculator section on Test Day, you might also consider graphing the equation and finding its x-intercepts. Be sure you are familiar with your calculator's graphing capabilities in advance.

Graphing Parabolas

All quadratic equations and functions graph as parabolas (U-shaped), opening either up or down. The graph of a quadratic equation written in standard form will open down when $a < 0$ (negative) and up when $a > 0$ (positive). You should have a foundational knowledge of the structure of a parabola for Test Day. Some of the basic pieces you could be asked about are shown in the following figure.

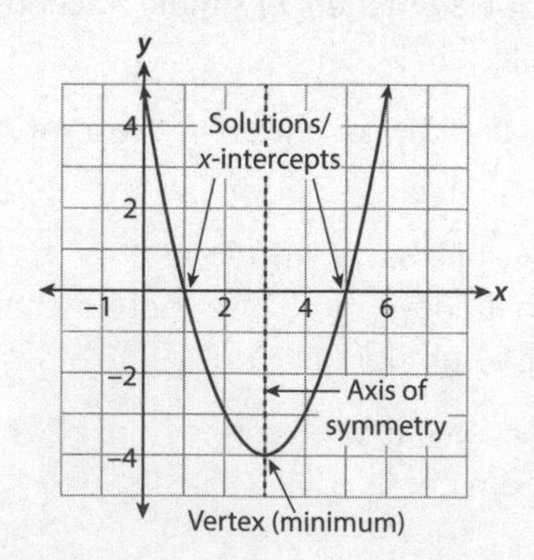

Along with the standard and factored forms of quadratic equations, you might also see them written in vertex form, $y = a(x - h)^2 + k$, where h and k are the x- and y-coordinates of the parabola's vertex, respectively, and the equation $x = h$ gives the axis of symmetry. From standard form, you can find the x-coordinate of the vertex (and therefore the location of the axis of symmetry) by plugging the appropriate values into the equation $x = \frac{-b}{2a}$ (the quadratic formula without the square root part). To find the y-coordinate of the vertex (which tells you the maximum or minimum value of the parabola), plug the x-coordinate into the original equation and solve for y.

PRACTICE SET

Easy

1. Etienne is graphing the quadratic equation $y = x^2 - 8x - 48$. He substitutes 0 for x and finds that the y-intercept of the graph is -48. Next, he wants to plot the x-intercepts of the graph, so he rewrites the equation in a different form. Assuming he rewrote the equation correctly and the equation reveals the x-intercepts, which of the following is Etienne's new equation?

   A) $y = (x - 8)^2 - 16$

   B) $y = (x - 8)^2 + 16$

   C) $y = (x - 8)(x + 6)$

   D) $y = (x - 12)(x + 4)$

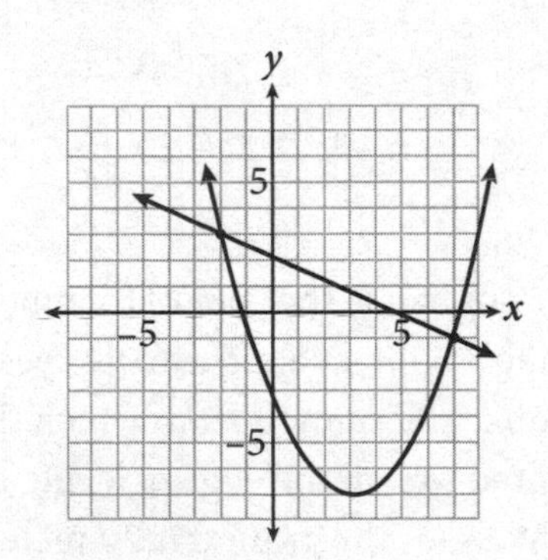

2. If (x_1, y_1) and (x_2, y_2) are solutions to the system of equations shown on the graph, what is the value of $x_1 + x_2$?

   A) −3

   B) 0

   C) 2

   D) 5

$$\begin{cases} x^2 + y^2 = 17 \\ 2x^2 + y = 6 \\ 3x - 2y = -5 \end{cases}$$

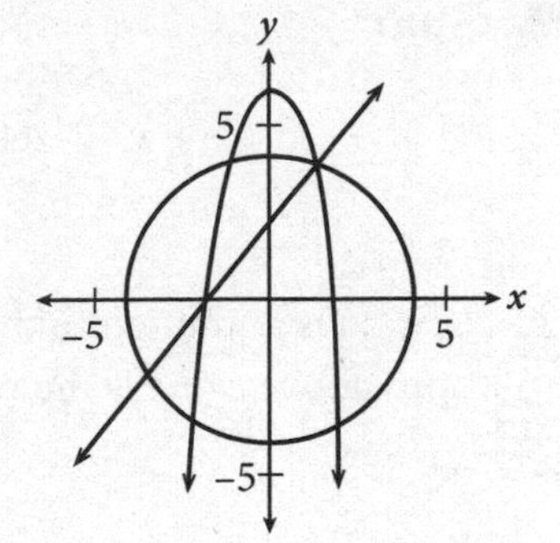

3. A system of three equations and their graphs are shown above. How many solutions does the system of equations have?

   A) None

   B) One

   C) Two

   D) Three

4. If $x^2 + 10x = 24$ and $x > 0$, what is the value of $x + 5$?

5. If −7 is one solution of the equation $y^2 + cy - 35 = 0$, what is the value of c?

Medium

$$x^2 + y^2 = 160$$
$$y = 3x$$

6. If (x, y) is a solution to the system of equations above, what is the value of y^2?

A) 12

B) 16

C) 120

D) 144

7. Which of the following expressions is not a factor of $16x^4 - 81y^4$?

A) $2x - 3y$

B) $2x + 3y$

C) $4x - 9y$

D) $4x^2 + 9y^2$

8. What is the range of the quadratic function whose equation is $q(x) = -(x + 4)^2 + 3$?

A) $y \geq -4$

B) $y \leq 3$

C) $y \geq 3$

D) $y \leq 4$

9. If $r < 0$ and $(4r - 4)^2 = 36$, what is the value of r?

A) −6

B) −2

C) −1

D) $-\frac{1}{2}$

10. The graph of $y = ax^2 + bx + c$ passes through the points (1, −8), (2, −1), (3, 4), and (5, 8). If the maximum value of y occurs at $x = 5$, through which other point must the graph of y pass?

A) (4, 6)

B) (6, 4)

C) (8, −1)

D) (10, −8)

11. For what values of x is the equation $2x^2 = 4x + 3$ true?

A) $\frac{2 \pm \sqrt{10}}{2}$

B) $\frac{2 \pm 5\sqrt{2}}{2}$

C) $1 \pm \sqrt{5}$

D) $1 \pm \sqrt{10}$

$$y = -5t^2 + 2t + 20$$
$$x = 2t$$

12. A cannonball is shot out of a cannon at a 45° angle with an approximate speed of 283 m/s. The cannon from which the ball was fired sits on the edge of a cliff, and its height above the ground is 20 meters. The equations given above represent the cannonball's height above the ground (y) and its horizontal distance (x) from the face of the cliff, t seconds after it was fired, where $t \geq 0$. How many seconds after the ball was fired does its vertical height above the ground equal its horizontal distance from the cliff?

A) 1

B) 2

C) 3

D) 4

Object	Acceleration Due to Gravity (m/s²)
Earth	9.8
Moon	1.6
Mars	3.7
Venus	9.5
Jupiter	24.5

13. The approximate height h of an object launched vertically upward after an elapsed time t is represented by the equation $h = -\frac{1}{2}at^2 + v_0t + h_0$, where a is acceleration due to gravity, v_0 is the object's initial velocity, and h_0 is the object's initial height. The table above gives the acceleration due to gravity on Earth's moon and several planets. If two identical objects are launched vertically upward at an initial velocity of 40 m/s from the surfaces of Mars and Earth's moon, approximately how many more seconds will it take for the moon projectile to return to the ground than the Mars projectile?

A) 21

B) 28

C) 36

D) 50

14. If $x = -4$ when $x^2 + 2xr + r^2 = 0$, what is the value of r?

Hard

$$x^2 + 2kx = \frac{j}{3}$$

15. In the quadratic equation above, j and k are constants. What are the solutions for x?

A) $x = -k \pm \frac{\sqrt{3(3k^2 + j)}}{3}$

B) $x = -6k \pm \frac{\sqrt{3(3k^2 + j)}}{3}$

C) $x = -k \pm \frac{\sqrt{3(3k^2 + j)}}{6}$

D) $x = -6k \pm \left(k + \frac{\sqrt{3j}}{3}\right)$

16. If $x^2 - 2x - 15 = (x + r)(x + s)$ for all values of x, what is one possible value of $r - s$?

A) −3

B) −2

C) 2

D) 8

17. If k is a positive integer less than 17, what is the total number of possible integer solutions for the equation $x^2 + 8x + k = 0$?

A) 5

B) 6

C) 7

D) 8

18. All 50 states have legislation regarding pool safety, the majority of which includes a requirement for a safety fence around the perimeter of the top of an in-ground pool. If a rectangular in-ground pool has a length that is 2 feet less than twice its width, and the area of the top of the pool is 480 square feet, how many linear feet of fencing are required, assuming the fence is placed 1 foot from the edge of the water on all sides?

A) 92

B) 100

C) 108

D) 116

19. If $x - 4$ is a factor of $x^2 - kx + 2k$, where k is a constant, what is the value of k?

A) −4

B) 4

C) 8

D) 12

$$\frac{1}{3}(x-2)^2 - \frac{1}{2} = \frac{5}{2}$$

20. What is the sum of the roots of the quadratic equation given above?

ANSWER KEY

1. D	**6. D**	**11. A**	**16. D**
2. D	**7. C**	**12. B**	**17. C**
3. B	**8. B**	**13. B**	**18. B**
4. 7	**9. D**	**14. 4**	**19. C**
5. 2	**10. C**	**15. A**	**20. 4**

ANSWERS & EXPLANATIONS

1. D

Difficulty: Easy

Category: Passport to Advanced Math / Quadratics

Strategic Advice: Quadratic equations can be written in several different forms, each of which reveals something special about the graph. For example, the vertex form of a quadratic equation ($y = a(x - h)^2 + k$) gives the minimum or maximum value of the function (k), while the standard form ($y = ax^2 + bx + c$) reveals the y–intercept (c).

Getting to the Answer: The factored form of a quadratic equation reveals the solutions to the equation, which graphically represent the x-intercepts. You can eliminate A and B because the equations are not written in factored form. To choose between C and (D), you need to determine which is the correctly factored form of the given equation. The factors of –48 that add up to –8 are –12 and 4, so the factors are $(x - 12)$ and $(x + 4)$, which means (D) is correct.

2. D

Difficulty: Easy

Category: Passport to Advanced Math / Quadratics

Strategic Advice: The solutions to a system of equations presented graphically are the points where the graphs of the equations intersect.

Getting to the Answer: Read the graph carefully to find the points of intersection: (–2, 3) and (7, –1). The question asks for the sum of the x-coordinates of these points, and $-2 + 7 = 5$, so (D) is correct.

3. B

Difficulty: Easy

Category: Passport to Advanced Math / Quadratics

Strategic Advice: Graphically speaking, the solution to a system of equations is the point or points where ALL the graphs intersect.

Getting to the Answer: There is only one point on the graph where all three equations intersect, (1, 4). Therefore, (B) is the correct answer.

4. 7

Difficulty: Easy

Category: Passport to Advanced Math / Quadratics

Strategic Advice: When the coefficient of the x^2 term in a quadratic equation is 1, the fastest way to solve it is usually by factoring. If that doesn't work, try the quadratic formula.

Getting to the Answer: Regardless of which method you choose, the equation must be set equal to 0 before you begin, so start by subtracting 24 from both sides: $x^2 + 10x - 24 = 0$. To factor the equation, look for the factors of −24 that add up to 10; the numbers are −12 and 2, so the equation in factored form is $(x + 12)(x - 2) = 0$. Therefore, the solutions are −12 and 2. The question states that $x > 0$, so x must equal 2, making $x + 5 = 2 + 5 = 7$.

5. 2
Difficulty: Easy

Category: Passport to Advanced Math / Quadratics

Strategic Advice: The *solution* to any equation is a value that satisfies the equation, or in other words, makes the equation true.

Getting to the Answer: Substitute −7 for y and solve for c:

$$\begin{aligned}(-7)^2 + (-7)c - 35 &= 0\\ 49 - 7c - 35 &= 0\\ -7c &= -14\\ c &= 2\end{aligned}$$

Grid in 2.

6. D
Difficulty: Medium

Category: Passport to Advanced Math / Quadratics

Strategic Advice: When you encounter a system of equations in which one of the equations is already solved for a variable, substitute its value into the other equation and solve.

Getting to the Answer: The second equation tells you that y is equal to $3x$, so substitute $3x$ for y in the first equation. Then solve for x^2 (not x because x^2 will be more useful when solving for y^2):

$$\begin{aligned}x^2 + (3x)^2 &= 160\\ x^2 + 9x^2 &= 160\\ 10x^2 &= 160\\ x^2 &= 16\end{aligned}$$

Now, substitute 16 for x^2 in the first equation and solve for y^2:

$$\begin{aligned}16 + y^2 &= 160\\ y^2 &= 144\end{aligned}$$

The correct answer is (D).

7. C
Difficulty: Medium

Category: Passport to Advanced Math / Quadratics

Strategic Advice: Being able to rewrite an expression as an equivalent expression (for example, by factoring it) will certainly earn you points on Test Day.

Getting to the Answer: Examine each of the terms in the expression: Each term is a perfect square and the terms are joined by a minus sign, so the expression is a difference of two squares. Use the factoring rule $a^2 - b^2 = (a + b)(a - b)$ to factor the expression:

$$\begin{aligned}16x^4 - 81y^4 &= (4x^2)^2 - (9y^2)^2\\ &= (4x^2 + 9y^2)(4x^2 - 9y^2)\end{aligned}$$

You can eliminate D because $(4x^2 + 9y^2)$ *is* one of the factors given. Next, notice that the second factor is another difference of two squares, so you can factor it further using the same rule:

$$\begin{aligned}4x^2 - 9y^2 &= (2x)^2 - (3y)^2\\ &= (2x + 3y)(2x - 3y)\end{aligned}$$

You can now eliminate A and B, leaving (C) as the expression that is not a factor of $16x^4 - 81y^4$.

8. B
Difficulty: Medium

Category: Passport to Advanced Math / Quadratics

Strategic Advice: The range of a function is the set of outputs that the function takes on, or in other words, the set of *y*-values through which the graph of the function passes.

Getting to the Answer: Use your knowledge of parabolas to draw a quick sketch of the function. You are only interested in the *y*-values of the graph, so plot the vertex and sketch the direction in which the parabola opens. Don't worry about the *x*-intercepts or any other specific points on the graph. The equation is written in vertex form, $y = a(x - h)^2 + k$, so you know the vertex is (−4, 3). Because a is negative (−1), the parabola opens downward, so your sketch should look like the following:

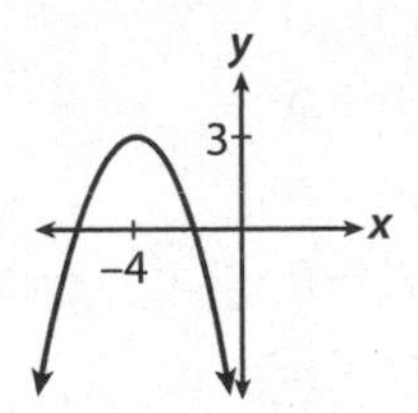

The *y*-values of the graph start below the *x*-axis and go up to the *y*-value of the vertex (3), so the range of the function is given by $y \leq 3$, which is (B).

9. D
Difficulty: Medium

Category: Passport to Advanced Math / Quadratics

Strategic Advice: When both sides of a quadratic equation are perfect squares, use square rooting to solve for the variable. Just remember that you'll end up with two possible solutions (and you'll need to review the question to see which is correct).

Getting to the Answer: Take the square root of both sides of the equation to remove the exponent and get $4r - 4 = \pm 6$. You're looking for a value of r that is less than 0, so start by solving the equation with −6:

$$4r - 4 = -6$$
$$4r = -2$$
$$r = -\frac{1}{2}$$

The value of r is indeed negative, so no need to try +6. Choice (D) is the correct answer. (Don't fall for choice A: $4r - 4 = -6$, but you're looking for the value of r, not $4r - 4$.)

10. C
Difficulty: Medium

Category: Passport to Advanced Math / Quadratics

Strategic Advice: The graph of every quadratic equation is a parabola (a symmetric U-shape), and its maximum (or minimum) value occurs at the *x*-coordinate of its vertex.

Getting to the Answer: Examine the points that are given—they increase from left to right. You also know that the maximum *y*-value (the highest point on the graph) occurs at $x = 5$, which means the vertex of the parabola is (5, 8). This tells you that the points will begin to decrease immediately to the right of that point. Using symmetry, you can find the point that corresponds to each of the given points. The first point, (1, −8), is 4 units to the left of the vertex ($5 - 1 = 4$), so the point that is 4 units to the right of the vertex ($x = 5 + 4$) will have the same *y*-coordinate—the graph must pass through (9, −8). This means you can eliminate D. The second point, (2, −1), is 3 units to the left of the vertex, so the point that is 3 units to the right of the vertex is (8, −1). This means (C) is correct.

11. A

Difficulty: Medium

Category: Passport to Advanced Math / Quadratics

Strategic Advice: Take a peek at the answer choices—they contain radicals, so your best bet is using the quadratic formula to solve the equation.

Getting to the Answer: First, write the equation in the form $ax^2 + bx + c = 0$:

$$2x^2 = 4x + 3$$
$$2x^2 - 4x - 3 = 0$$

Next, jot down the values that you'll need: $a = 2$, $b = -4$, and $c = -3$. Then, substitute these values into the quadratic formula and simplify:

$$\begin{aligned} x &= \frac{-b \pm \sqrt{b^2 - 4ac}}{2a} \\ &= \frac{-(-4) \pm \sqrt{(-4)^2 - 4(2)(-3)}}{2(2)} \\ &= \frac{4 \pm \sqrt{16 + 24}}{4} \\ &= \frac{4 \pm \sqrt{40}}{4} \end{aligned}$$

This isn't one of the answer choices, so you'll need to simplify the radical:

$$\begin{aligned} x &= \frac{4 \pm \sqrt{(4)(10)}}{4} \\ &= \frac{4 \pm 2\sqrt{10}}{4} \\ &= \frac{2(2 \pm \sqrt{10})}{2(2)} \\ &= \frac{2 \pm \sqrt{10}}{2} \end{aligned}$$

The correct answer is (A).

12. B

Difficulty: Medium

Category: Passport to Advanced Math / Quadratics

Strategic Advice: In real-world scenarios, a good bit of the text is probably just providing the context of the question. Be sure to focus on what the question is asking (here, find t when x and y are equal).

Getting to the Answer: Because both equations are given in terms of t, set the two equations equal to each other and simplify:

$$-5t^2 + 2t + 20 = 2t$$
$$-5t^2 + 20 = 0$$

The resulting equation is quadratic, so you can either factor and solve, or use square rooting:

Factoring	*Square rooting*
$-5t^2 + 20 = 0$	$-5t^2 + 20 = 0$
$-5(t^2 - 4) = 0$	$-5t^2 = -20$
$-5(t - 2)(t + 2) = 0$	$t^2 = 4$
$(t - 2)(t + 2) = 0$	$t = \pm\sqrt{4} = \pm 2$
$t = 2$ or $t = -2$	

Time can't be negative, so t must equal 2. This matches (B).

13. B

Difficulty: Medium

Category: Passport to Advanced Math / Quadratics

Strategic Advice: For a question like this, you'll need to interpret what is being asked for. Here, "to reach the ground" means that the height, h, equals 0, so your goal is to find the times (solve for t) when $h = 0$ for the two projectiles, and then find the difference between them.

Getting to the Answer: You know both projectiles begin at their respective surfaces and will be launched vertically at 40 m/s, so $h_0 = 0$ and $v_0 = 40$ for both. According to the table, acceleration due to gravity on Mars is 3.7 m/s^2, so the final equation for the height of the Mars projectile is $h_{Mars} = -1.85t^2 + 40t$. Similarly, acceleration due to gravity on the moon is 1.6 m/s^2, which makes the equation for the height of the moon projectile $h_{moon} = -0.8t^2 + 40t$. Substitute 0 for h in each equation, and solve for t via factoring as follows:

Mars	Earth's moon
$0 = -1.85t^2 + 40t$	$0 = -0.8t^2 + 40t$
$0 = t(-1.85t + 40)$	$0 = t(-0.8t + 40)$
$t = 0,\ -1.85t + 40 = 0$	$t = 0,\ -0.8t + 40 = 0$
$-1.85t = -40$	$-0.8t = -40$
$t = 21.62$	$t = 50$

At $t = 0$, the objects have not yet been launched, so disregard this value of t. Subtract the other values you found to get to the difference between the landing times, $50 - 21.62 = 28.38$ seconds. The closest answer choice is (B).

You could also graph the equations in a graphing calculator and use the *trace* or *zero* commands to determine when $h = 0$.

14. 4
Difficulty: Medium

Category: Passport to Advanced Math / Quadratics

Strategic Advice: Resist autopilot! Always look for a way to simplify an equation before plugging any numbers in.

Getting to the Answer: The equation is a special product of binomials (the square of a sum), so when factored, it becomes $(x + r)^2 = 0$ (you could also factor normally if you didn't recognize this). Substitute -4 for x, and then solve for r by taking the square root of both sides:

$$(-4 + r)^2 = 0$$
$$r - 4 = 0$$
$$r = 4$$

Grid in 4.

15. A
Difficulty: Hard

Category: Passport to Advanced Math / Quadratics

Strategic Advice: When an equation involves fractions, it is usually best to clear the fractions first to avoid messy calculations later. Also, when the answer choices involve square roots, you'll need to use the quadratic formula, $x = \frac{-b \pm \sqrt{b^2 - 4ac}}{2a}$.

Getting to the Answer: Multiply both sides of the equation by 3 to eliminate the fraction on the right side, and then subtract j from both sides to yield $3x^2 + 6kx - j = 0$. This means $a = 3$, $b = 6k$, and $c = -j$. Substitute these values into the quadratic formula, and then simplify as shown below:

$$x = \frac{-(6k) \pm \sqrt{(6k)^2 - 4(3)(-j)}}{2(3)}$$
$$= \frac{-6k}{6} \pm \frac{\sqrt{36k^2 + 12j}}{6}$$
$$= -k \pm \frac{\sqrt{4(9k^2 + 3j)}}{6}$$
$$= -k \pm \frac{2\sqrt{9k^2 + 3j}}{6}$$
$$= -k \pm \frac{\sqrt{3(3k^2 + j)}}{3}$$

Choice (A) is correct.

16. D
Difficulty: Hard

Category: Passport to Advanced Math / Quadratics

Strategic Advice: Whenever you are given two expressions and told that they are equal, look for ways to manipulate one so that it resembles the other.

Getting to the Answer: Notice that the expressions on both sides of the equal sign are quadratics—one is written in factored form and the other in standard form. Begin by factoring the left side of the equation to make it look like the right side. The factors of –15 that add up to –2 are 3 and –5, so the factored form of the left side is $(x + 3)(x - 5)$. Although you can't tell which constant represents r and which represents s, remember that the question asks for a possible value of $r - s$. The two possible differences are $3 - (-5) = 8$ and $-5 - 3 = -8$. The first combination is a match for (D).

17. C
Difficulty: Hard

Category: Passport to Advanced Math / Quadratics

Strategic Advice: When a quadratic equation is factored, the value of the constant (here, k) is equal to the product of the two constants in the binomial factors. Determine how many combinations there are, and you'll have your answer.

Getting to the Answer: Imagine factoring the given equation: You would need to find the factors of k (which you're told is less than 17) that add up to 8. List all possible integer combinations whose sum is 8 and whose product is less than 17:

$(x + 1)(x + 7) = 0$: 2 solutions

$(x + 2)(x + 6) = 0$: 2 solutions

$(x + 3)(x + 5) = 0$: 2 solutions

$(x + 4)(x + 4) = 0$: 1 solution

There are 7 different integer solutions for this equation, which matches (C).

18. B
Difficulty: Hard

Category: Passport to Advanced Math / Quadratics

Strategic Advice: In a question like this, translate from English into math to write an equation that represents the scenario. Drawing a diagram will also be very helpful.

Getting to the Answer: First, write expressions to represent the dimensions of the pool. If you let w represent the width, and the length is 2 feet less than twice the width, then the length is $2w - 2$. To find the area of a rectangle, multiply its length times its width. Here, the area is $(2w - 2) \times w$. Set this equal to the given area, 480, and solve for w. The equation is quadratic, so solve it by factoring or by using the quadratic formula:

$$\begin{aligned} w(2w - 2) &= 480 \\ 2w^2 - 2w &= 480 \\ 2w^2 - 2w - 480 &= 0 \\ 2(w^2 - w - 240) &= 0 \\ 2(w - 16)(w + 15) &= 0 \end{aligned}$$

The solutions for w are 16 and –15. Because the width of the pool can't be negative, the width must be 16 feet. This means the length must be $2(16) - 2 = 30$ feet. Now, add the 1 extra foot around all the edges to represent the fence. Drawing a diagram like the following one may help:

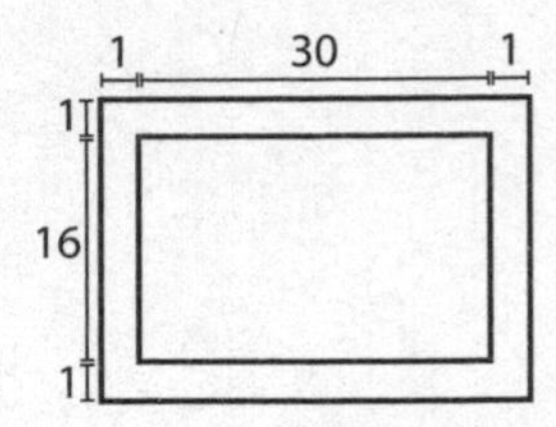

The new dimensions (for the fence) are 18 by 32, so the perimeter, and therefore the number of linear feet of fencing required, is $18 + 18 + 32 + 32 = 100$, which is (B).

19. C

Difficulty: Hard

Category: Passport to Advanced Math / Quadratics

Strategic Advice: You could substitute each of the answer choices for k and factor the resulting expression, but this will use up valuable time on Test Day. Instead, think about what it means for $x - 4$ to be a *factor* of the expression.

Getting to the Answer: If $x - 4$ is a factor of $x^2 - kx + 2k$, then $x^2 - kx + 2k$ can be written as the product $(x - 4)(x - a)$ for some real number a. Expanding the product $(x - 4)(x - a)$ yields $x^2 - 4x - ax + 4a$, which can be rewritten as $x^2 - (4 + a)x + 4a$. Substituting this for the factored form of the original expression results in the equation $x^2 - (4 + a)x + 4a = x^2 - kx + 2k$. Two quadratic equations are equal if and only if the coefficients of their like terms are equal, so $4 + a = k$ and $4a = 2k$. You now have a system of equations to solve. The first equation is already solved for k, so substitute $4 + a$ into the second equation for k and solve for a:

$$4a = 2(4 + a)$$
$$4a = 8 + 2a$$
$$2a = 8$$
$$a = 4$$

The questions asks for the value of k, and $k = 4 + a$, so $k = 4 + 4 = 8$, which is (C).

20. 4

Difficulty: Hard

Category: Passport to Advanced Math / Quadratics

Strategic Advice: *Roots* are the same as *solutions* to an equation, so you need to solve the equation for x. Don't let the fractions intimidate you, and don't jump to the conclusion that you need to find common denominators.

Getting to the Answer: Notice that adding $\frac{1}{2}$ to both sides of the equation will actually eliminate two of the fractions. Then, you can multiply both sides of the equation by 3 to eliminate the remaining fraction:

$$\frac{1}{3}(x - 2)^2 - \frac{1}{2} = \frac{5}{2}$$
$$\frac{1}{3}(x - 2)^2 = \frac{6}{2}$$
$$\frac{1}{3}(x - 2)^2 = 3$$
$$(x - 2)^2 = 9$$

Both sides of the equation are now perfect squares, so take their square roots. Then solve the resulting equations. Remember, there will be two equations to solve.

$$(x - 2)^2 = 9$$
$$\sqrt{(x - 2)^2} = \pm\sqrt{9}$$
$$x - 2 = \pm 3$$

Now, simplify each equation: $x - 2 = -3$ yields $x = -1$, and $x - 2 = 3$ yields $x = 5$. The question asks for the sum of the roots, so add $-1 + 5$ to get 4.

PRACTICE SET 11: IMAGINARY NUMBERS

Introduction to Imaginary Numbers

Until you reached more advanced math classes like Algebra 2, Pre-Calculus, or Trigonometry, you were likely taught that it is impossible to take the square root of a negative number. There is some truth to this, as the result isn't a real number. However, it is mathematically possible, and what you'll get is an imaginary number.

To take the square root of a negative number, it is necessary to use i, which is defined in math as the square root of –1. For example, to find $\sqrt{-49}$ rewrite it as $\sqrt{-1 \times 49}$, take the square root of –1 (which is by definition i), and then take the square root of 49, which is 7. The end result is $7i$.

Powers of i

When an imaginary number is raised to a power, you can use the pattern shown below to determine what the resulting term will be. Knowing the cycles of i will save you time on Test Day. Here are the first several in the pattern:

- $i^1 = i$
- $i^2 = -1$ (by definition because $\sqrt{-1} = i$)
- $i^3 = i^2 \times i = -1 \times i = -i$
- $i^4 = i^2 \times i^2 = -1 \times -1 = 1$
- $i^5 = i^4 \times i = 1 \times i = i$
- $i^6 = i^4 \times i^2 = 1 \times -1 = -1$

If we continued the pattern indefinitely, you would see that it repeats over and over in cycles of four ($i, -1, -i, 1, i, -1, -i, 1, \ldots$). To simplify a large power of i, divide the exponent by 4 (the number of terms in the cycle) and look at the remainder. The remainder tells you which term in the cycle the number is equivalent to. For example, to evaluate i^{35}, divide 35 by 4 to get 8, remainder 3. This means that i^{35} is equivalent to i^3, which is $-i$. (Note that graphing calculators are programmed to include the definition of i, so they can also successfully take the square root of a negative number and simplify powers of i.)

Multiplying Radicals

Be particularly careful when multiplying two radicals that both contain negative numbers. The first step is *always* to rewrite each quantity as the square root of the product of –1 and a positive number. Take the square root of –1, and then multiply the resulting expressions together. For example, if asked to multiply $\sqrt{-16} \times \sqrt{-25}$, you must first rewrite the expression as $i\sqrt{16} \times i\sqrt{25}$, which becomes $4i \times 5i = 20i^2 = 20(-1) = -20$. Combining the two radicals into a single one and canceling the negative signs to give $\sqrt{16 \times 25}$ is incorrect and will likely lead to a trap answer.

Complex Numbers

When a number is written in the form $a + bi$, where a is the real component and bi is the imaginary component, it is referred to as a *complex number.* You can add, subtract, multiply, and divide complex numbers. Here are a few rules:

- To add (or subtract) complex numbers, simply add (or subtract) the real parts and then add (or subtract) the imaginary parts.
- To multiply complex numbers, treat them as binomials and use FOIL. To simplify the product, use the definition $i^2 = -1$ and combine like terms.
- To divide complex numbers, write them in fraction form, and then *rationalize the denominator* (just as you would a fraction with a radical in the denominator) by multiplying top and bottom by the conjugate of the complex number in the denominator. For example, if asked to divide $21 \div (3 + 5i)$, start by writing it as $\frac{21}{3+5i}$. Then multiply top and bottom by the conjugate of $3 + 5i$, which is $3 - 5i$.

Quadratic Equations with Imaginary Solutions

In addition to real solutions, quadratic equations can have imaginary solutions, which you can find by using the quadratic formula. Recall from the previous practice set that the quadratic formula is $x = \frac{-b \pm \sqrt{b^2 - 4ac}}{2a}$ and that the sign of the discriminant $(b^2 - 4ac)$ dictates the nature of the solutions. When $b^2 - 4ac < 0$, the equation will have two imaginary solutions because you are taking the square root of a negative quantity.

Graphically, a quadratic equation has imaginary solutions when its vertex is above the x-axis and the parabola opens upward, or when its vertex is below the x-axis and the parabola opens downward. In either case, the graph does not cross the x-axis and therefore has no real solutions.

PRACTICE SET

Easy

1. Which of the following are zeros of the quadratic equation $x^2 + 9 = 0$? (Note: $i = \sqrt{-1}$)

   A) ± 3

   B) $\pm 3i$

   C) ± 81

   D) $\pm 81i$

2. Which of the following is equivalent to $(4 + 7i) - (3 - 2i)$?

   A) $1 + 5i$

   B) $1 + 9i$

   C) $7 + 5i$

   D) $7 + 9i$

3. Which of the following shows $(2 + 6i)(3i - 4)$ written as a complex number in the form $a + bi$? (Note: $i^2 = -1$)

   A) $-26 - 18i$

   B) $-18 + 10i$

   C) $10 - 18i$

   D) $30 + 10i$

4. What is the value of $-i^{48}$?

   A) $-i$

   B) i

   C) -1

   D) 1

Medium

5. Which of the following is a solution to the equation $4x^5 + 4x^3 = 360x$? (Note: $i = \sqrt{-1}$)

   A) -10

   B) $-i\sqrt{10}$

   C) $10i$

   D) $\sqrt{10}$

6. What is $\sqrt{-18} \times \sqrt{-50}$ written in simplest form? (Note: $i = \sqrt{-1}$)

   A) -30

   B) $-30i$

   C) 30

   D) $30i$

7. If u and v are complex numbers such that $u = 3 - 5i$ and $v = -6 + i$, which of the following is equivalent to $(u + v)^2$? (Note: $i^2 = -1$)

   A) $-7 + 24i$

   B) $9 + 8i$

   C) $9 + 16i$

   D) $25 - 24i$

8. Which of the following is equivalent to $\dfrac{4 + 6i}{10 - 5i}$? (Note: $i^2 = -1$)

   A) $\dfrac{2}{25} + \dfrac{16}{25}i$

   B) $\dfrac{2}{25} - \dfrac{16}{25}i$

   C) $\dfrac{2}{5} + \dfrac{6}{5}i$

   D) $\dfrac{2}{5} - \dfrac{6}{5}i$

$$\frac{2-i}{5-2i}$$

9. If the expression above is rewritten in the form $a + bi$, where a and b are real numbers, what is the value of $-b$ written as a fraction? (Note: $i^2 = -1$)

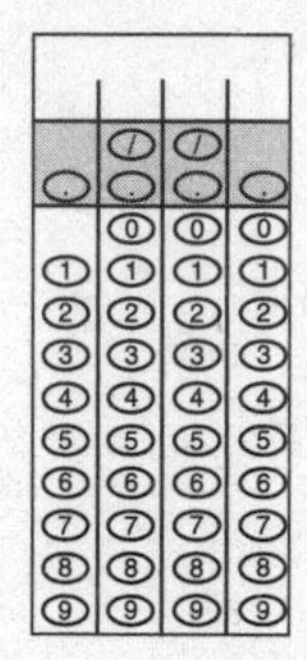

11. If w and z represent two complex numbers such that $w = 3 - 2i$ and $z = -2 + 4i$, which of the following gives $\frac{w}{z}$ written in the form $a + bi$? (Note: $i^2 = -1$)

A) $-\frac{3}{2} - \frac{1}{2}i$

B) $-\frac{7}{10} - \frac{2}{5}i$

C) $\frac{13}{20} - \frac{1}{2}i$

D) $\frac{7}{6} + \frac{2}{3}i$

Hard

10. Which of the following is equivalent to $\frac{10-\sqrt{-12}}{1-\sqrt{-27}}$? (Note: $i = \sqrt{-1}$)

A) $-\frac{2}{7}$

B) $\frac{28}{3}$

C) $-\frac{2}{7} + i\sqrt{3}$

D) $1 + i\sqrt{3}$

12. Two complex numbers, A and B, are defined as $A = (3 - 4i)$ and $B = (9 + ki)$, where k is a constant. If $AB - 15 = 60$, what is the value of k? (Note: $i^2 = -1$)

ANSWER KEY

1. B	5. B	9. 1/29
2. B	6. A	10. D
3. A	7. A	11. B
4. C	8. A	12. 12

ANSWERS & EXPLANATIONS

1. B

Difficulty: Easy

Category: Additional Topics in Math / Imaginary Numbers

Strategic Advice: The *zeros* of an equation are the same as its solutions. Take a peek at the answers. The *i*'s in some of the choices are a clue that solving by factoring may not be possible for an equation like this.

Getting to the Answer: Don't be tempted to use a factoring shortcut here—this is *not* a difference of two squares. The left side of the equation is a *sum* of two squares, so you cannot factor it. Instead, subtract 9 from both sides of the equation and use square rooting. Keep in mind that $\sqrt{-1} = i$:

$$x^2 + 9 = 0$$
$$x^2 = -9$$
$$\sqrt{x^2} = \pm\sqrt{-9}$$
$$\sqrt{x^2} = \pm\sqrt{(-1) \cdot 9}$$
$$x = \pm i\sqrt{9}$$
$$x = \pm 3i$$

This means (B) is correct.

2. B

Difficulty: Easy

Category: Additional Topics in Math / Imaginary Numbers

Strategic Advice: The rules of algebra still apply when working with imaginary numbers; as you work, be careful of negative signs when distributing and combining like terms.

Getting to the Answer: Start by distributing the negative sign to the expression in the second set of parentheses to get $4 + 7i - 3 + 2i$. Combining like terms yields $1 + 9i$, which matches (B).

3. A

Difficulty: Easy

Category: Additional Topics in Math / Imaginary Numbers

Strategic Advice: You can treat complex numbers as binomials and use FOIL to multiply them. Also remember that $i^2 = -1$.

Getting to the Answer: Use FOIL to multiply the two complex numbers, and then combine like terms (the real parts and the imaginary parts):

$$(2 + 6i)(3i - 4)$$
$$= (2)(3i) + (2)(-4) + (6i)(3i) + (6i)(-4)$$
$$= 6i - 8 + 18i^2 - 24i$$
$$= -8 + 18(-1) + 6i - 24i$$
$$= -26 - 18i$$

The correct answer is (A).

4. C
Difficulty: Easy

Category: Additional Topics in Math / Imaginary Numbers

Strategic Advice: Powers of i cycle in the pattern $i, -1, -i, 1$. Memorizing this pattern may save you valuable time on Test Day. To determine the value of a high power of i, divide the exponent by 4 (the number of terms in the cycle). The remainder tells you which term in the pattern is the answer. For example, $i^{11} = i^3$ because $11 \div 4 = 2 \text{ R } 3$.

Getting to the Answer: When 48 is divided by 4, the remainder is 0; therefore, $i^{48} = i^4$, which is 1. Don't forget about the negative sign in front of it, though. This makes the correct answer −1, which is (C).

5. B
Difficulty: Medium

Category: Additional Topics in Math / Imaginary Numbers

Strategic Advice: Plugging each choice into the equation will take far too long. Instead, factor the trinomial to find the solutions. Because this question does not allow the use of a calculator, look for ways to simplify the equation first, and then try obvious factors.

Getting to the Answer: Set the equation equal to 0, and then factor $4x$ out of the trinomial to get $4x(x^4 + x^2 - 90) = 0$. Factoring the new trinomial inside the parentheses (by finding the two factors of −90 that add up to +1) gives $4x(x^2 + 10)(x^2 - 9) = 0$. Set each factor equal to 0 and solve for x:

$$4x = 0 \qquad x = 0$$

$$x^2 + 10 = 0 \qquad x^2 = -10 \qquad x = \pm\sqrt{-1 \cdot 10} \qquad x = \pm i\sqrt{10}$$

$$x^2 - 9 = 0 \qquad x^2 = 9 \qquad x = \pm 3$$

The solutions are $x = 0$, $x = \pm i\sqrt{10}$, $x = \pm 3$. (B) is the only choice that contains one of the solutions $\left(-i\sqrt{10}\right)$.

6. A
Difficulty: Medium

Category: Additional Topics in Math / Imaginary Numbers

Strategic Advice: You must be very careful when multiplying square roots of negative numbers. You cannot simply multiply the two negative signs together. Instead, write each factor as an imaginary number, and then multiply the results together.

Getting to the Answer: Write the number under each radical as a product of −1 and convenient factors of the number. Then you can move an i to the outside of the radical (because $\sqrt{-1}$ is defined as i):

$$\begin{aligned}\sqrt{-18} \times \sqrt{-50} &= \sqrt{-1 \times 9 \times 2} \times \sqrt{-1 \times 25 \times 2} \\ &= 3i\sqrt{2} \times 5i\sqrt{2} \\ &= 15i^2 \times 2 \\ &= 30i^2 \\ &= -30\end{aligned}$$

(A) is the correct answer.

7. A
Difficulty: Medium

Category: Additional Topics in Math / Imaginary Numbers

Strategic Advice: You could square $(u + v)$ and then substitute the expressions given into the resulting product, but complex numbers add and subtract very nicely, so it is almost always quicker and easier to combine the complex numbers and then perform any additional operations.

Getting to the Answer: Start by substituting the expressions for u and v, combine them, and then square the result. You'll also need to use the fact that $i^2 = -1$:

$$\begin{aligned}(u+v)^2 &= (3-5i+(-6)+i)^2\\ &= (-3-4i)^2\\ &= (-3-4i)(-3-4i)\\ &= 9+12i+12i+16i^2\\ &= 9+24i+16(-1)\\ &= -7+24i\end{aligned}$$

This means (A) is correct.

8. A
Difficulty: Medium

Category: Additional Topics in Math / Imaginary Numbers

Strategic Advice: Because $i = \sqrt{-1}$, you can't leave an i in the denominator of a complex number (it's not considered proper notation to leave a radical in a denominator). This means you'll need to *rationalize* the denominator by multiplying the numerator and denominator by the conjugate of the denominator. The *conjugate* of $(a + bi)$ is $(a - bi)$. In other words, keep the real part the same and change the sign of the imaginary part.

Getting to the Answer: The conjugate of $10-5i$ is $10+5i$; multiply the numerator and denominator of the given expression by this to remove i from the denominator. Use FOIL to expand the binomial pairs, and then combine like terms to simplify. Separate the real and imaginary parts and reduce as needed. The steps for this sequence are shown here:

$$\frac{4+6i}{10-5i}\times\frac{10+5i}{10+5i}$$

$$=\frac{(4\times 10)+(4\times 5i)+(6i\times 10)+(6i\times 5i)}{100-25i^2}$$

$$=\frac{40+20i+60i+30i^2}{100-25(-1)}$$

$$=\frac{40+80i+30(-1)}{125}$$

$$=\frac{10+80i}{125}$$

$$=\frac{10}{125}+\frac{80}{125}i$$

$$=\frac{2}{25}+\frac{16}{25}i$$

(A) is the correct answer.

9. 1/29
Difficulty: Medium

Category: Additional Topics in Math / Imaginary Numbers

Strategic Advice: Multiply the top and bottom of the fraction by the conjugate of the denominator, and then confirm that you've found the requested value (the value of $-b$) before gridding in your answer.

Getting to the Answer: The conjugate of $5-2i$ is $5+2i$; multiply the numerator and denominator of the given expression by this to remove i from the denominator. Use FOIL to expand the binomial pairs, and then combine like terms to simplify. Separate the real and imaginary parts and reduce as needed. The steps for this sequence are shown here:

$$\frac{2-i}{5-2i}\times\frac{5+2i}{5+2i}$$

$$=\frac{(2\times 5)+(2\times 2i)+(-i\times 5)+(-i\times 2i)}{25-4i^2}$$

$$=\frac{10+4i-5i-2i^2}{25-4(-1)}$$

$$=\frac{10-i-2(-1)}{29}$$

$$=\frac{12-i}{29}$$

$$=\frac{12}{29}-\frac{1}{29}i$$

The question asks for the value of $-b$ (and b is the coefficient of the imaginary part), so the correct answer is $-\left(-\frac{1}{29}\right)=\frac{1}{29}$.

10. D

Difficulty: Hard

Category: Additional Topics in Math / Imaginary Numbers

Strategic Advice: Resist the urge to rationalize the denominator right away. Before doing this, you should remove the negative signs from under the radicals.

Getting to the Answer: Rewrite the number under each radical as a product of -1 and convenient factors of the number. Then simplify if possible:

$$10-\sqrt{-12}=10-\sqrt{-1\cdot4\cdot3}=10-2i\sqrt{3}$$
$$1-\sqrt{-27}=1-\sqrt{-1\cdot9\cdot3}=1-3i\sqrt{3}$$

Next, multiply the numerator and denominator of the resulting expression by the conjugate of $1-3i\sqrt{3}$ and simplify:

$$\frac{10-2i\sqrt{3}}{1-3i\sqrt{3}}\times\frac{1+3i\sqrt{3}}{1+3i\sqrt{3}}$$
$$=\frac{10(1)+10(3i\sqrt{3})-2i\sqrt{3}(1)-2i\sqrt{3}(3i\sqrt{3})}{1(1)-3i\sqrt{3}(3i\sqrt{3})}$$
$$=\frac{10+30i\sqrt{3}-2i\sqrt{3}-6i^2(3)}{1-9i^2(3)}$$
$$=\frac{10+28i\sqrt{3}i-18(-1)}{1-27(-1)}$$
$$=\frac{28+28i\sqrt{3}}{28}$$
$$=\frac{28}{28}+\frac{28i\sqrt{3}}{28}$$
$$=1+i\sqrt{3}$$

(D) is the correct answer.

11. B

Difficulty: Hard

Category: Additional Topics in Math / Imaginary Numbers

Strategic Advice: Even though the wording of this question is a bit different, the task is the same as in the previous question—rationalize the denominator and then write the expression in the form $a+bi$.

Getting to the Answer: Start by writing the expression in the requested form: $\frac{w}{z}=\frac{3-2i}{-2+4i}$.

The conjugate of $-2+4i$ is $-2-4i$, so multiply the numerator and denominator by $-2-4i$. To simplify the expression, FOIL the binomials in the numerator and in the denominator, and use the definition $i=\sqrt{-1}$, which can also be written as $i^2=-1$:

$$\frac{3-2i}{-2+4i}\times\frac{-2-4i}{-2-4i}=\frac{-6-12i+4i+8i^2}{4-16i^2}$$
$$=\frac{-6-8i+8(-1)}{4-16(-1)}$$
$$=\frac{-14-8i}{20}$$

This isn't one of the answer choices, so simplify the expression further by dividing each term by 2 and then splitting the result into its real part and its imaginary part:

$$=\frac{-7-4i}{10}$$
$$=\frac{-7}{10}-\frac{4i}{10}$$
$$=\frac{-7}{10}-\frac{2}{5}i$$

Choice (B) is correct.

12. 12

Difficulty: Hard

Category: Additional Topics in Math / Imaginary Numbers

Strategic Advice: There are multiple letters and multiple equations in this question, so start with the simplest thing you know and build from there.

Getting to the Answer: You know that $AB - 15 = 60$. Solve this for AB by adding 15 to both sides of the equation: $AB = 75$. Next, substitute the expressions given for A and B, multiply them together, and see where that takes you:

$$\begin{aligned} AB &= 75 \\ (3 - 4i)(9 + ki) &= 75 \\ 27 + 3ki - 36i - 4ki^2 &= 75 \\ 27 - 36i + 3ki - 4k(-1) &= 75 \\ 27 - 36i + 3ki + 4k &= 75 \end{aligned}$$

You now have two options: The only way the expression on the left can equal plain 75 is if the imaginary terms add up to 0, so you could set the two imaginary terms equal to 0 and solve for k:

$$\begin{aligned} -36i + 3ki &= 0 \\ 3ki &= 36i \\ 3k &= 36 \\ k &= 12 \end{aligned}$$

The other option is to ignore the imaginary terms (because they add up to 0) and set the real terms equal to 75 and solve for k. The result should be the same:

$$\begin{aligned} 27 + 4k &= 75 \\ 4k &= 48 \\ k &= 12 \end{aligned}$$

Either route leads to the correct answer, which is 12.

PRACTICE SET 12: LINES, ANGLES, AND TRIANGLES

Lines and Angles

Lines and angles are the foundation of SAT geometry. Therefore, reviewing a few basic definitions and rules will make answering these questions, as well as related geometry questions, much easier.

- Acute angle: an angle that measures between 0° and 90°
- Obtuse angle: angle that measures between 90° and 180°
- Right angle (formed by perpendicular lines): an angle that measures exactly 90°
- Complementary angles: angles that sum to 90°
- Supplementary angles: angles that sum to 180°
- Vertical angles: angles opposite to each other when two lines intersect. Vertical angles have equal measures.

Parallel Lines Cut by a Transversal

When two parallel lines are intersected by another line (called a transversal), all acute angles are equal, and all obtuse angles are equal. Additionally, the following pairs of angles have equal measures:

- Alternate interior angles
- Alternate exterior angles
- Corresponding angles

Triangles

Lines and angles form the basis of triangles—some of the most commonly occurring shapes on the SAT. The following facts apply to all triangles:

- The sum of the measures of the interior angles of a triangle is 180°.
- The exterior angle of a triangle is equal to the sum of the two opposite interior angles.
- A side opposite a greater angle is longer than a side opposite a smaller angle.
- Triangle Inequality Theorem: The length of any side of a triangle is less than the sum of the other two sides and greater than the positive difference of the other two sides. This can be represented by $a - b < c < a + b$, where a, b, and c are the side lengths of the triangle.

The Pythagorean Theorem

The Pythagorean theorem is an important triangle topic that you are probably familiar with already. If you know any two lengths of a right triangle, you can use the Pythagorean theorem to find the missing side. The theorem states that $a^2 + b^2 = c^2$, where a and b are the shorter sides of the triangle (called legs) and c is the hypotenuse, which is always across from the right angle.

Pythagorean Triplets and Special Right Triangles

Knowing common Pythagorean triplets can save valuable time on Test Day. The two most common are 3-4-5 and 5-12-13. Multiples of these (e.g., 6-8-10, 10-24-26) can also pop up, so watch out for those, too. If you see any two sides of one of these ratios, you can automatically fill in the third.

Another time-saving strategy is recognizing special right triangles (45-45-90 and 30-60-90). The lengths of the sides of these special triangles are always in the same ratio, so you only need to know one side in order to calculate the other two. The ratios are shown below:

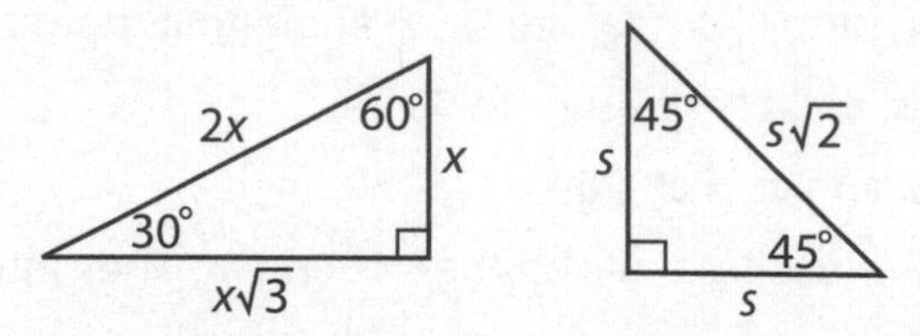

While the Pythagorean theorem can almost always be used to solve right triangle questions, it is often not the most efficient way to proceed. Recognizing Pythagorean triplets and special right triangles allows you to save time, so use them whenever possible!

Finding the Area of a Triangle

On Test Day, you might also be asked to determine the area of a triangle. The formula for finding the area of a triangle is $A = \frac{1}{2}bh$, where b is the length of the base of the triangle and h is the height. When you have a right triangle, you can use the legs as the base and the height. If the triangle isn't a right triangle, you'll need to draw in the height (sometimes called an altitude) from the angle opposite the base. Remember that the height *must* be perpendicular to the base.

Complex Figures

Unusual shapes, comprised of multiple familiar shapes, which can be obvious or cleverly hidden, are also a recurring SAT geometry topic. These figures can always be broken down into squares, rectangles, triangles, and/or circles. No matter how complex the figure, following a few general guidelines will lead you to the correct answer on Test Day.

- Start by transferring information from the question stem to the figure. If a figure isn't provided, draw one!
- Break the figure into familiar shapes.
- Determine how one line segment can play multiple roles in a figure. For example, if a circle and triangle overlap just right, the radius of the circle might also be the hypotenuse of a right triangle.
- Work from the shape with the most information to the shape with the least information.

PRACTICE SET

Easy

1. The angles of a triangle are in the ratio 2:3:4. What is the degree measure of the largest angle?

   A) 40

   B) 80

   C) 90

   D) 120

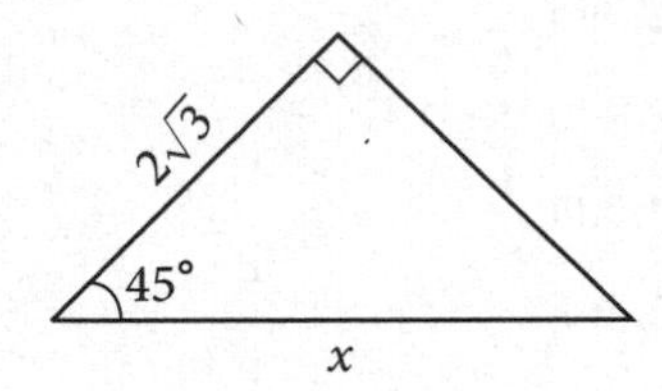

2. In the figure above, what is the value of x?

   A) $2\sqrt{6}$

   B) 6

   C) $6\sqrt{2}$

   D) 8

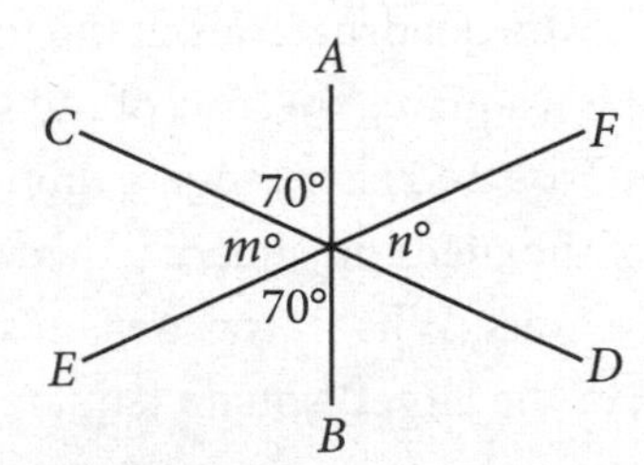

3. In the figure above, $\overline{AB}$, $\overline{CD}$, and $\overline{EF}$ all intersect. What is the value of n?

   A) 30

   B) 40

   C) 50

   D) 60

Medium

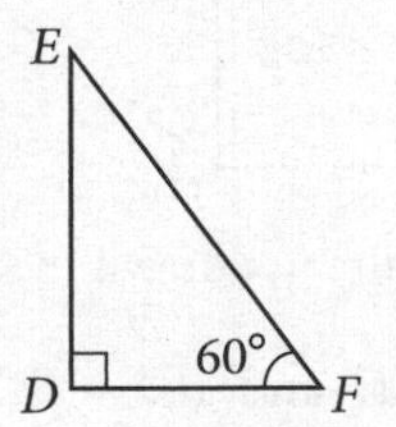

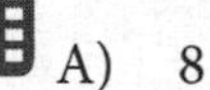

4. The area of triangle DEF is $32\sqrt{3}$ square units. What is the length of $\overline{DF}$?

   A) 8

   B) $8\sqrt{3}$

   C) 16

   D) $16\sqrt{3}$

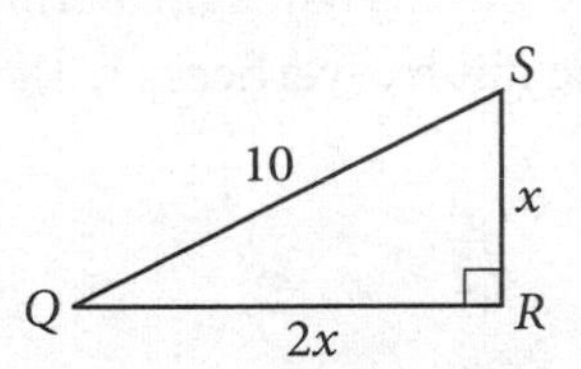

5. In the figure above, what is the area of ΔQRS?

   A) $2\sqrt{5}$

   B) 10

   C) $8\sqrt{5}$

   D) 20

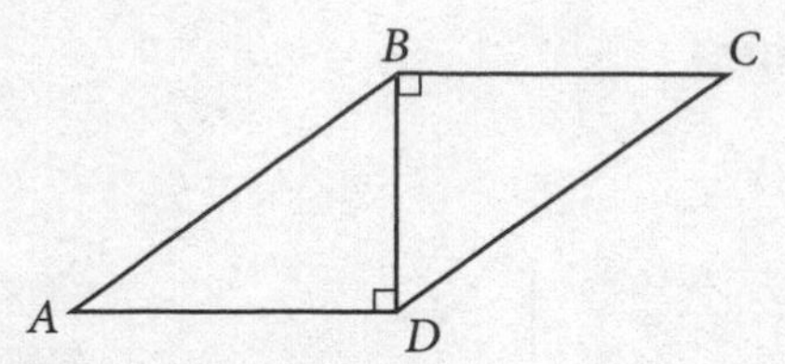

Note: Figure not drawn to scale.

6. If $AD = 2CD$ and $BD = BC = 6$, what is the length of side AB?

   A) $6\sqrt{2}$

   B) 12

   C) $12\sqrt{2}$

   D) 18

7. A triangle is graphed on a standard coordinate plane. What is the perimeter of the triangle if it has vertices (1, 4), (1, 7), and (4, 4)?

   A) $3 + \sqrt{2}$

   B) $3\sqrt{2}$

   C) $6 + 3\sqrt{2}$

   D) $9 + \sqrt{2}$

8. What is the area, in square units, of an isosceles right triangle with a hypotenuse of 2?

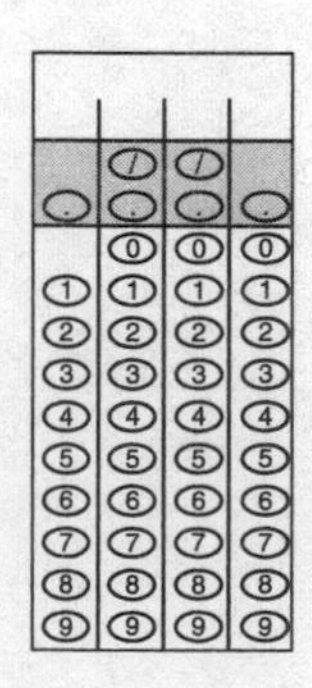

Hard

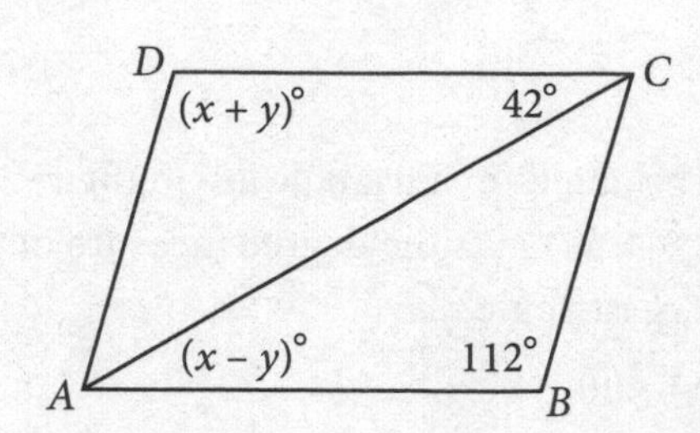

Note: Figure not drawn to scale.

9. Figure $ABCD$ is a parallelogram. What is the product of xy?

   A) 2,695

   B) 2,940

   C) 4,704

   D) 6,468

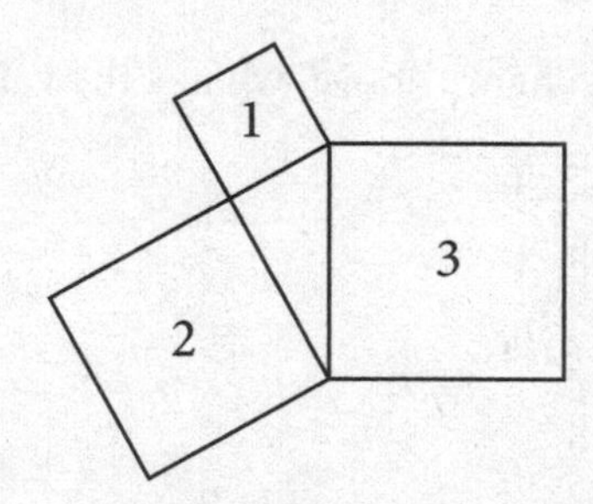

10. Each of the quadrilaterals in the figure above is a square. The area of the smallest square (square 1) is 16 square units, and the area of the medium square (square 2) is 48 square units. What is the area, in square units, of the largest square (square 3)?

    A) 56

    B) 64

    C) 78

    D) 96

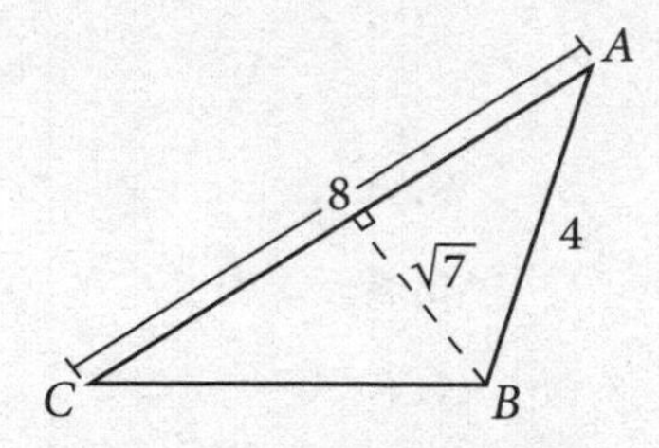

Note: Figure not drawn to scale.

11. In the figure above, what is the length of side BC?

A) $4\sqrt{2}$

B) $4\sqrt{3}$

C) $4\sqrt{5}$

D) $4\sqrt{7}$

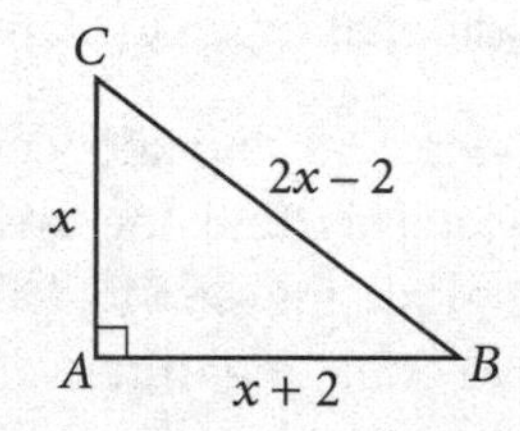

12. In the figure above, ΔABC is a right triangle. What is the value of x?

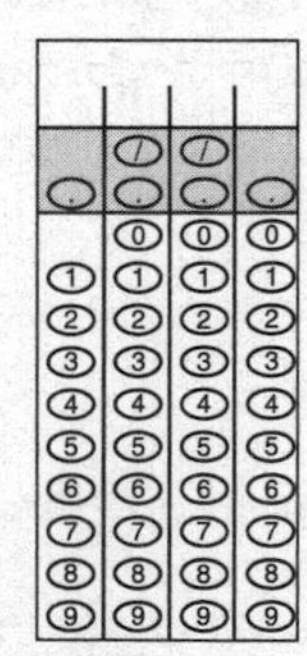

ANSWER KEY

1. B	5. D	9. A
2. A	6. D	10. B
3. B	7. C	11. A
4. A	8. 1	12. 6

ANSWERS & EXPLANATIONS

1. B

Difficulty: Easy

Category: Additional Topics in Math / Geometry

Strategic Advice: Use the fact that the sum of the measures of the interior angles of every triangle is 180°.

Getting to the Answer: Because the angles are in the ratio 2:3:4, use $2x$, $3x$, and $4x$ to represent their measures. This means $2x + 3x + 4x = 180$, or when simplified, $9x = 180$, which yields $x = 20$. The largest angle has measure $4x$, or $4 \times 20 = 80$, making (B) the correct answer.

2. A

Difficulty: Easy

Category: Additional Topics in Math / Geometry

Strategic Advice: Whenever you see a 45 degree angle, look for a special right triangle. On Test Day, don't forget that you can refer to the formula page at the beginning of the math section.

Getting to the Answer: The measure of the third angle is $180° - 90° - 45° = 45°$, so the triangle is a 45-45-90 triangle. The side lengths of a 45-45-90 triangle are always in the ratio $x{:}x{:}x\sqrt{2}$, so the hypotenuse (which happens to be x in this question) is the length of a leg, multiplied by $\sqrt{2}$. This means $x = \sqrt{2} \times 2\sqrt{3} = 2\sqrt{6}$, which is (A).

3. B

Difficulty: Easy

Category: Additional Topics in Math / Geometry

Strategic Advice: Use your knowledge of angle properties to quickly find the answer. Angles that form a straight line sum to 180°, and vertical angles have equal measures.

Getting to the Answer: The three angles on the left side of $\overline{AB}$ form a straight line, so $70° + m° + 70° = 180°$. Solving for m gives 40. The angles marked $m°$ and $n°$ are vertical angles, so $n = 40$ as well. (B) is correct.

4. A

Difficulty: Medium

Category: Additional Topics in Math / Geometry

Strategic Advice: When you see a triangle on Test Day that has a 60-degree angle, check the formula page at the beginning of the math section and jot down the ratio for the sides of a 30-60-90 triangle, $x{:}x\sqrt{3}{:}2$.

Getting to the Answer: You're given the area of the triangle. Because the triangle is a right triangle, the base and the height are the two legs of the triangle. Call the lengths of the legs

x and $x\sqrt{3}$. Substitute these values and the given area into the area formula:

$$A = \frac{1}{2}bh$$

$$32\sqrt{3} = \frac{1}{2}(x)(x\sqrt{3})$$

$$32\sqrt{3} = \frac{1}{2}x^2\sqrt{3}$$

To solve this equation, divide both sides by the radical, and then multiply both sides by 2.

$$32\sqrt{3} = \frac{1}{2}x^2\sqrt{3}$$

$$32 = \frac{1}{2}x^2$$

$$64 = x^2$$

$$x = \pm\sqrt{64} = \pm 8$$

Length can't be negative, so $x = 8$, which is the length of the shortest leg. Side $\overline{DF}$ is across from the 30-degree angle (which is the smallest angle in the triangle), so it is the shortest leg (which is given by x), making (A) correct.

5. D
Difficulty: Medium

Category: Additional Topics in Math / Geometry

Strategic Advice: Treat the variable lengths as you would number lengths; the same rules and theorems still apply.

Getting to the Answer: The area of a triangle is given by the formula $A = \frac{1}{2}bh$. Before you can use the area formula, you'll need to use the Pythagorean theorem to find the lengths of the base and the height, $2x$ and x:

$$x^2 + (2x)^2 = 10^2$$

$$x^2 + 4x^2 = 100$$

$$5x^2 = 100$$

$$x^2 = 20$$

$$x = \sqrt{20} = \sqrt{4 \cdot 5} = 2\sqrt{5}$$

The height (x) is $2\sqrt{5}$, and the base is twice that, or $2 \times 2\sqrt{5} = 4\sqrt{5}$. This makes the area of the triangle $\frac{1}{2}(2\sqrt{5})(4\sqrt{5}) = \frac{8}{2} \times 5 = 20$, which matches (D).

6. D
Difficulty: Medium

Category: Additional Topics in Math / Geometry

Strategic Advice: Whenever you're given a figure, transfer as much information as possible from the question to the figure, and then label the figure with any additional calculations you make.

Getting to the Answer: Because $BD = BC$, ΔBDC is an isosceles right triangle (and therefore a 45-45-90 triangle). You're told that the two legs both measure 6, which means that hypotenuse $\overline{CD}$ measures $6\sqrt{2}$. You also know that $AD = 2CD$, so $AD = 2 \times 6\sqrt{2} = 12\sqrt{2}$. Next, use the Pythagorean theorem to find AB:

$$\begin{aligned} AB^2 &= BD^2 + AD^2 \\ &= 6^2 + (12\sqrt{2})^2 \\ &= 36 + 144(2) \\ &= 36 + 288 \\ AB^2 &= 324 \end{aligned}$$

So $AB = \sqrt{324} = 18$, making (D) the correct answer.

7. C

Difficulty: Medium

Category: Additional Topics in Math / Geometry

Strategic Advice: Draw a quick sketch to help visualize the information. Don't spend a lot of time labeling hash marks—you just need a general idea of what the triangle looks like.

Getting to the Answer: The coordinates given in the question are plotted and connected in the figure below. We've labeled the vertices to aid with the explanation.

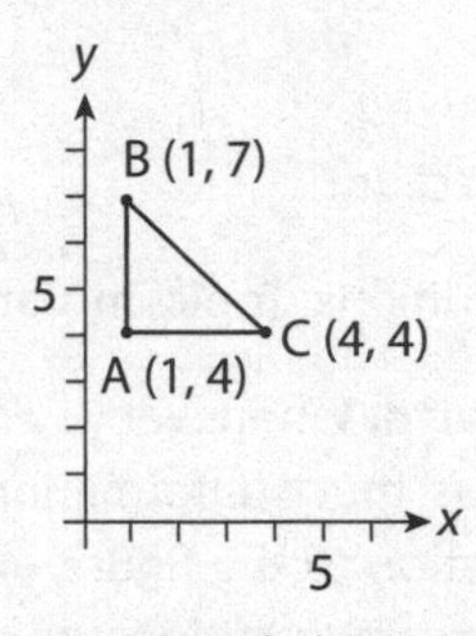

Notice that one side of the triangle is parallel to the x-axis and another is parallel to the y-axis, so you can conclude that the legs are perpendicular, and the triangle formed is a right triangle. The leg lengths are easy to find—just subtract the appropriate coordinates to get lengths of $7 - 4 = 3$ units and $4 - 1 = 3$ units. This means the right triangle is isosceles and therefore a 45-45-90 triangle with legs of length 3. Thus, the hypotenuse measures $3\sqrt{2}$. Add the three sides together to find the perimeter: $3 + 3 + 3\sqrt{2} = 6 + 3\sqrt{2}$, making (C) the correct answer.

8. 1

Difficulty: Medium

Category: Additional Topics in Math / Geometry

Strategic Advice: An isosceles right triangle has two legs of equal length, making it a 45-45-90 triangle.

Getting to the Answer: The side lengths of a 45-45-90 triangle are in the ratio $x{:}x{:}x\sqrt{2}$. The length of the hypotenuse is given as 2, so $x\sqrt{2} = 2$, and consequently $x = \frac{2}{\sqrt{2}}$, which is the length of each leg (or in a right triangle, the base and the height). Substitute this value into the area formula for b and h to find that $A = \frac{1}{2}bh = \frac{1}{2}\left(\frac{2}{\sqrt{2}}\right)\left(\frac{2}{\sqrt{2}}\right) = \frac{4}{4} = 1$.

9. A

Difficulty: Hard

Category: Additional Topics in Math / Geometry

Strategic Advice: Opposite angles of a parallelogram are congruent, and the diagonal acts as a transversal that is cutting two parallel lines (the opposite sides of the parallelogram).

Getting to the Answer: The alternate interior angles formed by the diagonal are congruent. This means $x - y = 42$ and $x + y = 112$. Set up a system of equations and solve it. The system is already arranged nicely to use elimination; combining the equations by adding them eliminates the y terms:

$$\begin{array}{r} x - y = 42 \\ x + y = 112 \\ \hline 2x = 154 \\ x = 77 \end{array}$$

Substitute this value into either of the original equations for x to find y: $77 + y = 112$, so $y = 35$. The question asks for the product of x and y, so the correct answer is $77 \times 35 = 2{,}695$, which is (A).

10. B
Difficulty: Hard

Category: Additional Topics in Math / Geometry

Strategic Advice: Sometimes, working backward is necessary to solve a problem. Here, use the areas of the two smaller squares (squares 1 and 2) to find the side lengths, and then work in the opposite direction to find the area of the larger square (square 3).

Getting to the Answer: The area of a square is given by the formula $A = s^2$. If the area of the smallest square is 16, then each side has length $\sqrt{16} = 4$. If the area of the medium square is 48, then a side has length $\sqrt{48} = \sqrt{16 \times 3} = 4\sqrt{3}$. Use this information to label the diagram:

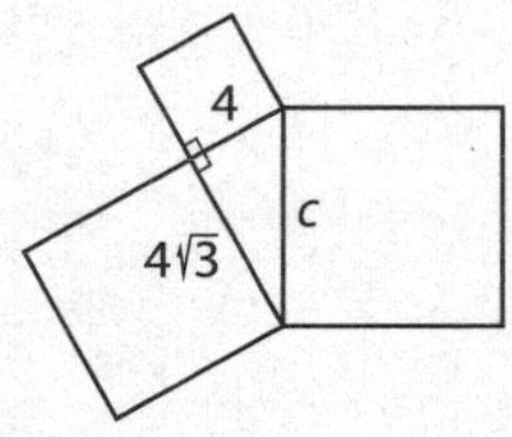

Notice that the figure contained within the three squares is a right triangle with legs of lengths 4 and $4\sqrt{3}$. You might recognize these lengths as two pieces of the ratio of the sides for a 30-60-90 triangle, which is $x{:}x\sqrt{3}{:}2x$. The hypotenuse of the triangle is given by $2x$, so it has length $2(4) = 8$. This means the side of the largest square has length 8, and its area is $8^2 = 64$ square units, which is (B). (Note that you could also use the Pythagorean theorem to find the length of the hypotenuse of the triangle.)

11. A
Difficulty: Hard

Category: Additional Topics in Math / Geometry

Strategic Advice: Work with right triangles, whenever possible, so you can use properties and theorems that allow you to find side lengths.

Getting to the Answer: Add a point (call it *D*) to the diagram where the dashed line meets $\overline{AC}$, as shown below:

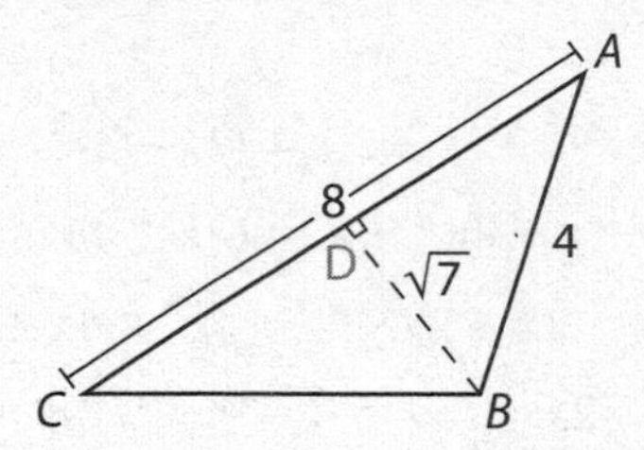

$\overline{BD}$ breaks ΔABC into two smaller right triangles. You know two side lengths for ΔABD, so use the Pythagorean theorem to find the third side (which happens to be a leg of the triangle):

$$(\sqrt{7})^2 + AD^2 = 4^2$$
$$7 + AD^2 = 16$$
$$AD^2 = 9$$
$$AD = 3$$

Now, $CD = AC - AD = 8 - 3 = 5$. Use the Pythagorean theorem again to find *BC*, keeping in mind that *BC* is the hypotenuse this time:

$$(\sqrt{7})^2 + 5^2 = BC^2$$
$$7 + 25 = BC^2$$
$$BC = \sqrt{32}$$
$$BC = \sqrt{16 \times 2} = 4\sqrt{2}$$

Choice (A) is correct.

12. 6
Difficulty: Hard

Category: Additional Topics in Math / Geometry

Strategic Advice: When variable expressions are used to represent side lengths of a right triangle, you treat them the same as you would numerical lengths.

Getting to the Answer: Using the Pythagorean theorem, we know that $AB^2 + AC^2 = BC^2$. Substitute the side lengths and simplify using FOIL:

$$x^2 + (x + 2)^2 = (2x - 2)^2$$
$$x^2 + (x + 2)(x + 2) = (2x - 2)(2x - 2)$$
$$x^2 + x^2 + 2x + 2x + 4 = 4x^2 - 4x - 4x + 4$$
$$2x^2 + 4x + 4 = 4x^2 - 8x + 4$$

Now, solve the quadratic equation by moving everything to one side of the equal sign (so the equation equals 0) and factoring it:

$$0 = 2x^2 - 12x$$
$$0 = 2x(x - 6)$$

Finally, set each factor equal to 0 to find that $x = 0$ and $x = 6$. Because a side length cannot be 0, x (and therefore AC) must be 6.

PRACTICE SET 13: SIMILARITY, CONGRUENCE, AND PROOFS

Similarity and Congruence

Knowing when two figures are similar (or congruent) allows you to use what you know about one figure to derive information about the other figure.

- Two triangles are *similar* if the measures of the angles of one triangle are equal to the measures of the angles of the other triangle or if all pairs of corresponding sides of the two triangles are in the same proportion.
- Two triangles are *congruent* if and only if they have the same angle measures AND the same side lengths.

Notation

The order of the vertices in a similarity (~) or congruence (≅) statement is key to understanding which parts are corresponding parts. For example, if $\Delta CAT \sim \Delta DOG$, then $\angle C$ corresponds to $\angle D$, $\angle A$ corresponds to $\angle O$, and $\angle T$ corresponds to $\angle G$. Similarly, side *CA* corresponds to side *DO*, side *CT* corresponds to side *DG*, and side *AT* corresponds to side *OG*.

Line, Angle, and Triangle Theorems

In the following table, you'll find several properties and theorems that you will likely need to use on Test Day to establish that two triangles are similar (or congruent). Some of these are review from the previous practice set.

Triangle Theorems	**Definition**
Triangle Sum and Exterior Angle Theorems	• The sum of the measures of the interior angles is 180°. • An exterior angle equals the sum of the two opposite interior angles.
Isosceles Triangle Theorems	• Angles opposite congruent sides have equal measures. • Sides opposite congruent angles have equal lengths.
Triangle Inequality Theorem	• The length of any side of a triangle is less than the sum of the other two sides and greater than the positive difference of the other two sides (i.e., $a - b < c < a + b$, where a, b, and c are the side lengths of the triangle).
Side-Angle Relationships	• In a triangle, the longest side is across from the largest angle. • In a triangle, the largest angle is across from the longest side.

Mid-Segment Theorems	• The mid-segment (or midline) of a triangle is parallel to one side of the triangle and joins the midpoints of the other two sides. • The length of the mid-segment is half the length of the side to which it is parallel.

Proving Congruence

There are several theorems that can be used to prove two triangles are congruent; these are summarized in the following table. The good news: You will *not* need to construct a complete proof on Test Day. However, you may need to use these theorems to determine that two triangles are congruent so that you can transfer known lengths or angle measures from one triangle to another.

Theorem/Theorem Notation	Diagram
SSS (Side-side-side): All three sides of one triangle are congruent to corresponding sides of another triangle.	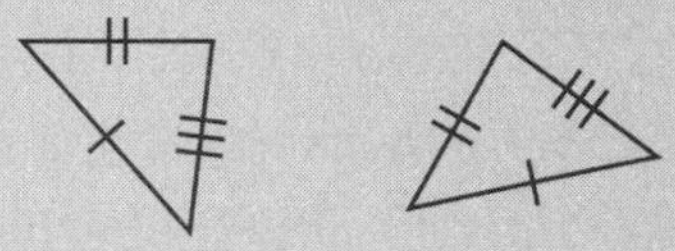
SAS (Side-angle-side): Two sides and the included angle of one triangle are congruent to corresponding parts of another triangle.	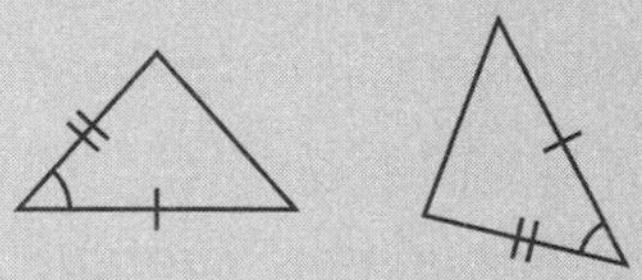
AAS (Angle-angle-side): Two angles and the non-included side are congruent to corresponding parts of another triangle.	
ASA (Angle-side-angle): Two angles and the included side of one triangle are congruent to corresponding parts of another triangle.	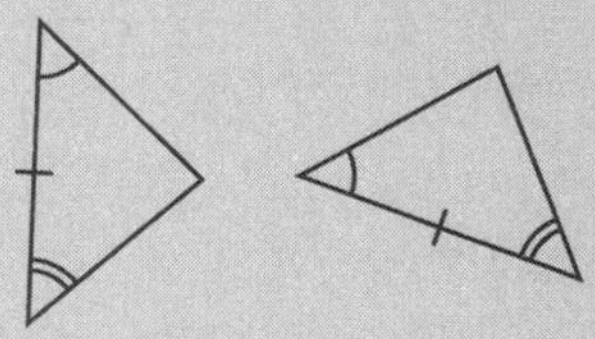
HL (Hypotenuse-leg): The hypotenuse and leg of one right triangle are congruent to corresponding parts of another right triangle.	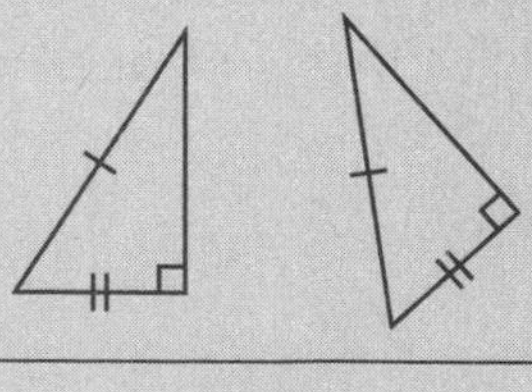
CPCTC: Corresponding parts of congruent triangles are congruent.	N/A

Beware the bogus "theorems" AAA and SSA. Two triangles with identical angles are always similar, but they are not necessarily congruent. SSA is only valid for right triangles, in which case you should use the HL theorem.

PRACTICE SET

Easy

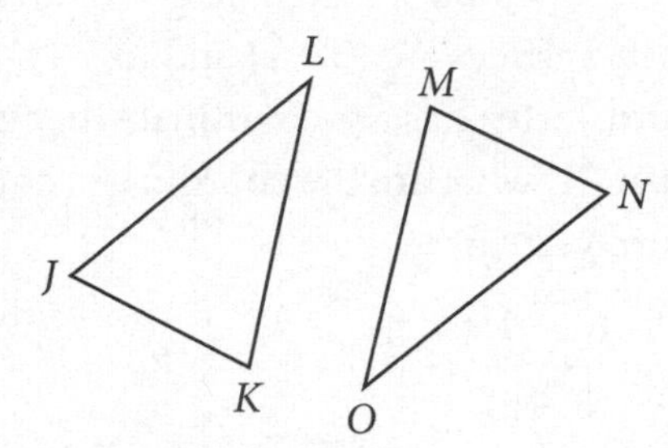

1. In the diagram above, $\overline{JL} \cong \overline{ON}$ and $\overline{KL} \cong \overline{OM}$. Confirmation of which of the following facts would be sufficient to prove that the two triangles are congruent?

   A) $\angle L \cong \angle O$

   B) $\angle K \cong \angle M$

   C) $\angle J \cong \angle M$

   D) $\angle J \cong \angle N$

2. While working on a geometry problem, Raul determines that the angles of one triangle are congruent to the corresponding angles of another triangle. Which of the following is a valid deduction that Raul can make?

   A) The two triangles are congruent but not necessarily similar.

   B) The two triangles are similar but not necessarily congruent.

   C) The two triangles are both similar and congruent.

   D) The two triangles are neither similar nor congruent.

Medium

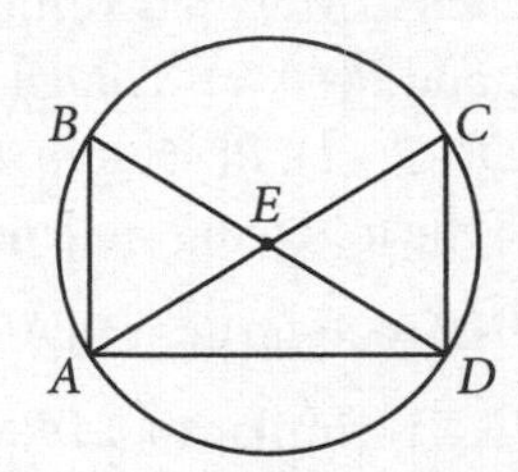

3. $\overline{AC}$ and $\overline{BD}$ are diameters of circle E above. Which triangle congruence theorem can be used to prove that ΔAEB is congruent to ΔDEC?

   A) AAA

   B) HL

   C) SAS

   D) SSA

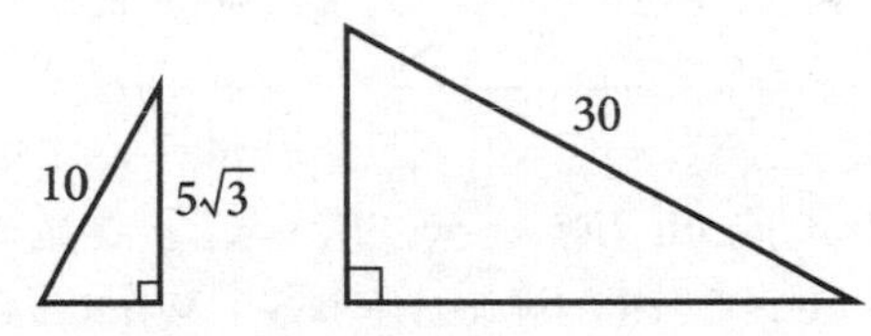

Note: Figure not drawn to scale.

4. If the right triangles in the figure shown are similar triangles, what is the length of the shorter leg of the larger triangle?

   A) 10

   B) 15

   C) $10\sqrt{3}$

   D) $15\sqrt{3}$

5. A triangle with side lengths of 5, 12, and 15 centimeters is similar to another triangle. The longest side of the other triangle has length 24 centimeters. What is the perimeter, in centimeters, of the larger triangle?

   A) 38.4

   B) 44

   C) 51.2

   D) 58

6. Two triangles are graphed on a coordinate plane. Triangle MNP has vertices $M(-4, 2)$, $N(-4, 6)$, and $P(-6, 2)$. Triangle QRS has vertices $Q(-5, -1)$, $R(-5, -5)$, and $S(4, -5)$. Which of the following statements is true?

A) ΔMNP is congruent to ΔQRS.

B) ΔMNP is similar to ΔQRS.

C) ΔMNP is similar to ΔRSQ.

D) The triangles are neither congruent nor similar.

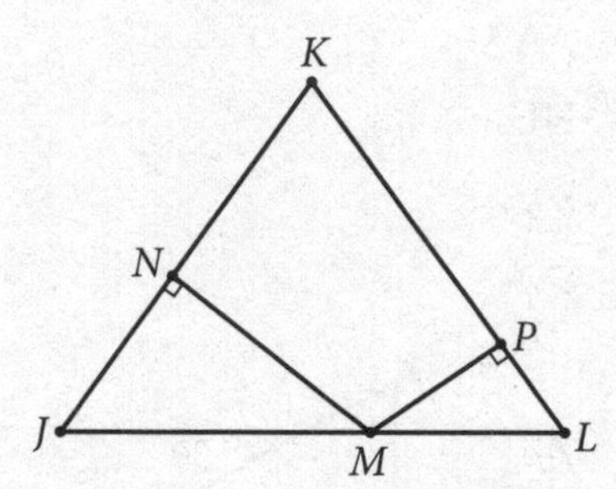

7. In triangle JKL above, $JK = KL$ and $JL = 26$. The ratio of MN to MP is 8:5. What is the length of segment JM?

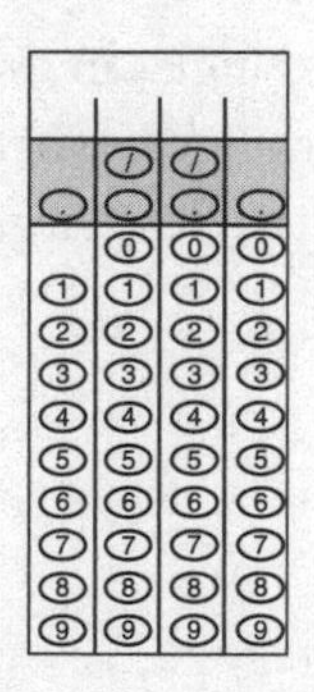

Hard

8. Triangle CAT is an isosceles triangle with vertices (2, 1), (6, 1), and (4, 7). Triangle DOG is similar to triangle CAT, and two of its vertices are (3, −1) and (5, −1). If the third vertex has a y-coordinate that is less than −1, what are the coordinates of the third vertex?

A) (3, −4)

B) (3, −5)

C) (4, −4)

D) (4, −5)

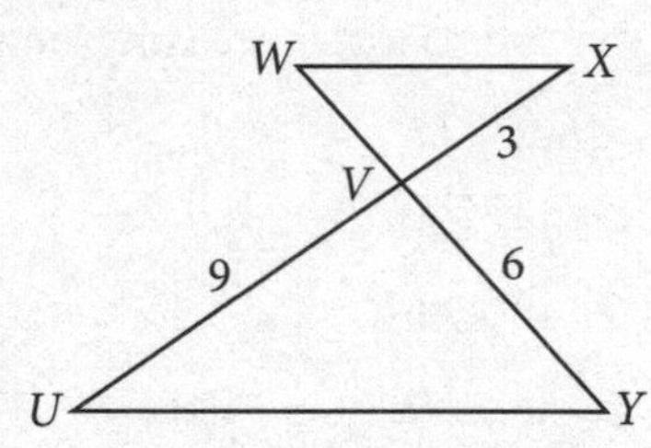

Note: Figure not drawn to scale.

9. In the figure above, $\overline{UY}$ and $\overline{WX}$ are parallel and $\overline{UX}$ intersects $\overline{WY}$ at V. What is the length of $\overline{WY}$?

A) 8

B) 9

C) 10

D) 12

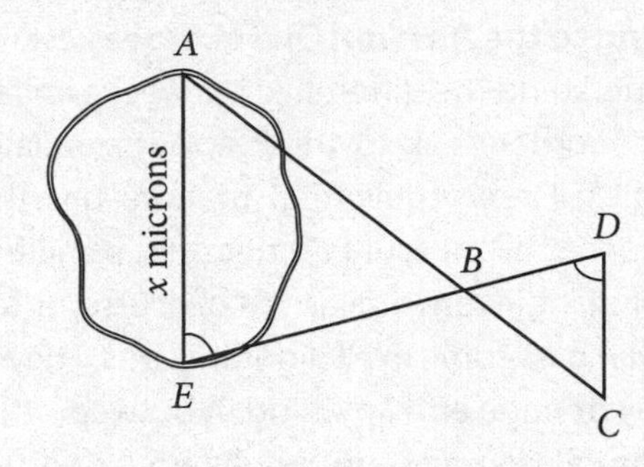

10. A scientist looking at a sample of infected tissue through a microscope wants to find the length *x*, in microns, across a damaged blood cell, as represented in the sketch above. The lengths represented by *AB*, *EB*, *BD*, and *CD* were determined to be 26 microns, 22 microns, 11 microns, and 12 microns, respectively. Given that the measure of $\angle AEB$ is equal to the measure of $\angle CDB$, what is the value of *x*?

ANSWER KEY

1. A	6. D
2. B	7. 16
3. C	8. C
4. B	9. A
5. C	10. 24

ANSWERS & EXPLANATIONS

1. A

Difficulty: Easy

Category: Additional Topics in Math / Geometry

Strategic Advice: Think about the options for proving that two triangles are congruent. Then, find the choice that satisfies one of those options.

Getting to the Answer: You're given two congruent pairs of sides, so your options are SSS and SAS. A quick scan of the choices shows that none of them provides congruence for the missing sides, so SSS is out. In order to have SAS, you need to know that the angles between the two congruent sides are congruent. The middle angles in this case are $\angle L$ and $\angle O$ (as shown below), so the answer must be (A).

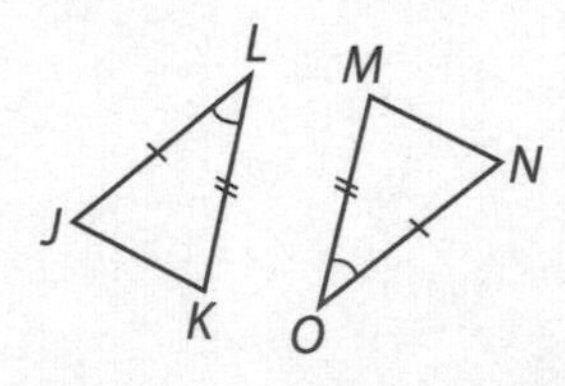

2. B

Difficulty: Easy

Category: Additional Topics in Math / Geometry

Strategic Advice: Before you sift through the choices, take some time to consider the differences between similarity and congruence.

Getting to the Answer: One of the easiest ways to understand the difference between similarity and congruence is to think about equilateral triangles. Every equilateral triangle has three angles that all measure 60 degrees, so all equilateral triangles are similar. But of course, equilateral triangles come in all different sizes: They can be tiny or huge and anything in between. Thus, equilateral triangles are usually not congruent. Raul has two triangles whose corresponding angles are all congruent, so the triangles are similar. They are not, however, necessarily congruent, so the answer is (B).

3. C

Difficulty: Medium

Category: Additional Topics in Math / Geometry

Strategic Advice: Whenever you're asked to show that two triangles within a figure are congruent, mark up the figure to show which pairs of sides and which pairs of angles are congruent.

Getting to the Answer: Angles *BEA* and *CED* are vertical angles, and vertical angles are always congruent. Mark this on the diagram. Also, several sides of the triangles are radii, so they are all equal. Use hash marks to show this relationship. Your diagram should look something like this:

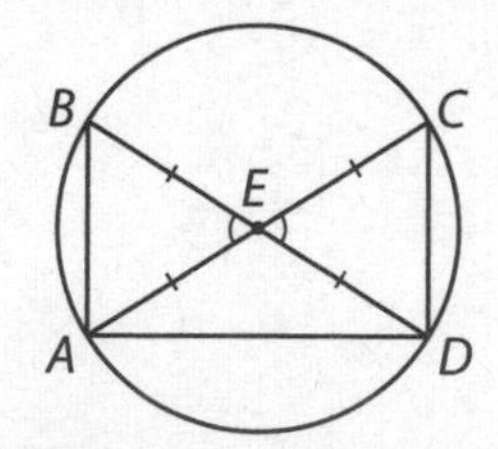

Notice that the angles *between* the two sides with hash marks are congruent. This lets you demonstrate by SAS that $\Delta AEB \cong \Delta DEC$, making (C) the correct answer.

4. B

Difficulty: Medium

Category: Additional Topics in Math / Geometry

Strategic Advice: Whenever you see $\sqrt{3}$ in a triangle problem, think about special right triangles. Don't forget that the ratios of the sides of these triangles are given on the reference page at the beginning of each math section.

Getting to the Answer: Examine the smaller triangle: When the sides of a right triangle are in the ratio $x : x\sqrt{3} : 2x$, the triangle is a 30-60-90 triangle. Knowing that two of the sides of the smaller triangle fit this ratio is enough information to draw the conclusion that it is a 30-60-90 triangle (although you could use the Pythagorean theorem to confirm that the shorter leg has length 5 if you're not convinced). Because the two triangles are similar, you know that their angle measures are equal, and the larger triangle is also a 30-60-90 triangle. In the ratio $x : x\sqrt{3} : 2x$, $2x$ represents the length of the hypotenuse and x represents the length of the shorter leg. According to the figure, the hypotenuse of the larger triangle has length 30, so solve $2x = 30$ to find that the length of the shorter leg of the larger triangle has length $x = 15$, which is (B).

5. C

Difficulty: Medium

Category: Additional Topics in Math / Geometry

Strategic Advice: Corresponding sides of similar triangles are in proportion to each other. Because you know the length of the longest side for both triangles, you can set up and solve a proportion to find the lengths of the other two sides of the larger triangle.

Getting to the Answer: The longest sides are in the ratio 15:24, which simplifies to 5:8. Use this in your proportions:

Shortest side to longest side	Medium side to longest side
$\frac{5}{x} = \frac{5}{8}$	$\frac{12}{x} = \frac{5}{8}$
$5(8) = 5x$	$12(8) = 5x$
$40 = 5x$	$96 = 5x$
$x = 8$	$x = 19.2$

The perimeter of the larger triangle is the sum of the side lengths, 8 + 19.2 + 24 = 51.2 cm, which is (C).

Note: There is a shortcut for this type of question if you happen to remember that the perimeters of similar triangles are also in the same proportion as the sides. The perimeter of the smaller triangle is 5 + 12 + 15 = 32. You could set up and solve the same type of proportion here:

$$\frac{32}{P} = \frac{5}{8}$$

$$32(8) = 5P$$

$$256 = 5P$$

$$P = 51.2$$

6. D

Difficulty: Medium

Category: Additional Topics in Math / Geometry

Strategic Advice: When you're given coordinates for the vertices of two triangles and asked to compare the triangles in some way, drawing a sketch is definitely the best way to go.

Getting to the Answer: Draw a quick sketch and label any lengths that are obvious from the coordinates of the points. Because there are some repeated *x*- and *y*-coordinates, finding side lengths is likely to be easier than you might initially think.

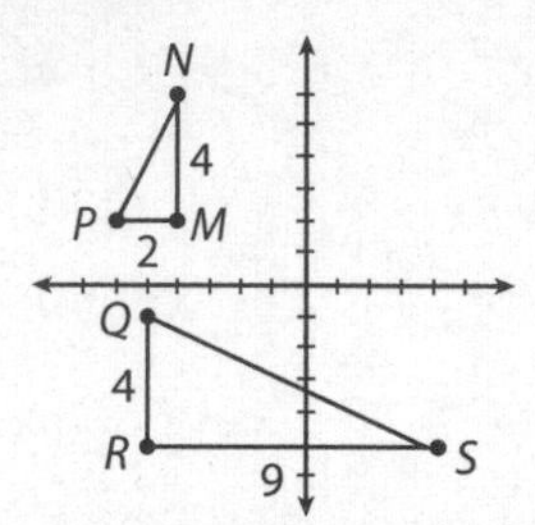

Notice that the triangles are clearly not congruent (the same size), so you can eliminate A. Both triangles are right triangles, so they could be similar. However, the ratios of corresponding pairs of sides are not equal $\left(\frac{2}{4} \neq \frac{4}{9}\right)$, so the triangles are neither similar nor congruent, which is (D).

7. 16

Difficulty: Medium

Category: Additional Topics in Math / Geometry

Strategic Advice: Whenever two sides of a triangle are equal in length, the triangle is isosceles. The base angles of an isosceles triangle are congruent.

Getting to the Answer: The given ratio involves sides of the two small right triangles, and the question asks about a side of one of those triangles, so focus on this. The base angles of ΔJKL are congruent (because it is an isosceles triangle), so $\angle J$ and $\angle L$ are congruent. Additionally, $\angle JNM$ and $\angle LPM$ are both right angles and therefore are congruent. When two pairs of corresponding angles in two triangles are congruent, the third pair must also be congruent, so ΔJNM and ΔLPM are similar by AAA. This means the corresponding sides of these triangles are proportional. Thus, $\frac{JM}{ML} = \frac{MN}{MP}$. Because $\frac{MN}{MP} = \frac{8}{5}$, it follows that $\frac{JM}{ML} = \frac{8}{5}$. If we let $JM = 8x$, then $ML = 5x$. We also know that JL (which is the sum of the lengths of segments JM and ML) equals 26, so $8x + 5x = 26$. Solving this equation yields $13x = 26$, or $x = 2$. Finally, $JM = 8(2) = 16$.

8. C

Difficulty: Hard

Category: Additional Topics in Math / Geometry

Strategic Advice: Drawing a sketch is the only way to approach a question like this.

Getting to the Answer: Even though you don't have graph paper, draw a quick sketch to get an idea of how the triangles look. Don't forget that the question tells you that the *y*-coordinate of the third vertex is less than −1, so you know the third vertex is below the other two vertices. Also, label any distances that are obvious from the coordinates of the points. Your sketch might look like the following:

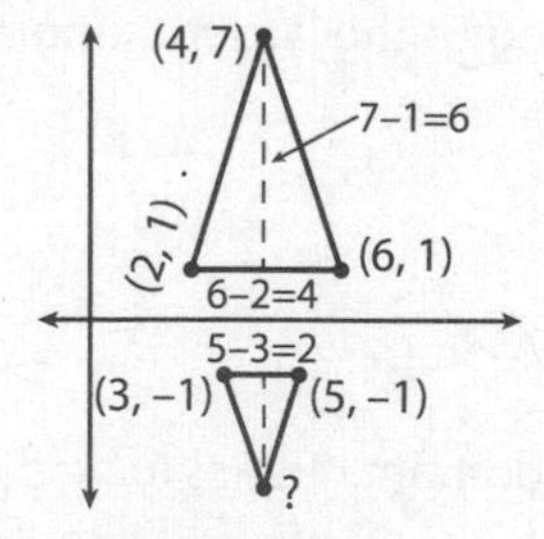

Now, examine the figure more closely. The question states that the triangles are similar, so look for the ratio of the corresponding sides. The ratio of the base of the smaller triangle to the base of the larger triangle is $\frac{2}{4} = \frac{1}{2}$, so the ratio of the other pairs of sides, and therefore the height, must be the same. This means the height of triangle *DOG* is half the height of triangle *CAT*, or half of 6, which is 3 (which means the vertex is 3 units below –1). The vertex that creates the height is exactly halfway between the two base vertices, so the coordinates of the third vertex is $\left(\frac{3+5}{2}, -1-3\right) = (4, -4)$, which is (C).

9. A

Difficulty: Hard

Category: Additional Topics in Math / Geometry

Strategic Advice: When comparing similar (or congruent) triangles that are oriented differently, be sure to mark any corresponding sides and angles with hash marks so that you keep them in the correct order.

Getting to the Answer: Start by filling in as much information as you can on the diagram. $\angle WVX$ is congruent to $\angle YVU$ because they are vertical angles, so mark these angles with a single hash mark. Also note that because *UY* is parallel to *WX*, $\angle W$ is congruent to $\angle Y$, and $\angle X$ is congruent to $\angle U$ (both pairs are alternate interior angles). Mark these pairs with double and triple hash marks. Your diagram might look like the following:

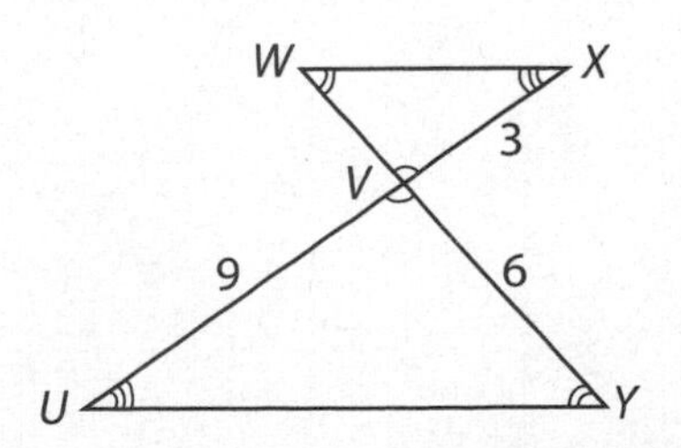

Use the hash marks to write a similarity statement—start with the angles marked with one hash mark, then two hash marks, then three: $\Delta VWX \sim \Delta VYU$. The order of the letters in the similarity statement makes it clear that *VX* (first and last letter) corresponds to *VU* (first and last letter) and is triple the length (based on the figure). Segment *VW* (first and second letter) is therefore one-third the length of *VY* (first and second letter), which is $6 \div 3 = 2$. The question asks for the length of *WY*, which is $2 + 6 = 8$. (A) is the correct answer.

10. 24

Difficulty: Hard

Category: Additional Topics in Math / Geometry

Strategic Advice: Mark up the diagram, and then write a similarity statement to help you answer the question.

Getting to the Answer: Start by filling in as much information as you can on the diagram. The given angles are already marked with a single hash mark. You also know that $m\angle ABE = m\angle CBD$ because they are vertical angles, so mark this pair with double hash marks. You can reason that when two angles of a triangle are congruent to two angles of another triangle, all three pairs of angles must be congruent (because the angles of a triangle always add up to 180 degrees). Thus, $m\angle A = m\angle C$, which can be marked with triple hash marks. Then add all the known lengths. Your diagram should look like the following:

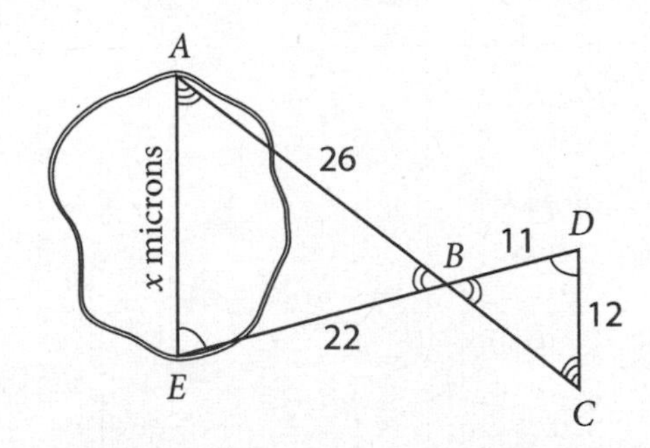

Use the hash marks to write a similarity statement—start with the angles marked with one hash mark, then two hash marks, then three: $\Delta EBA \sim \Delta DBC$. Now you can see from the order of the letters that *EB* and *DB* are corresponding sides, and one is twice the other. Thus, *AE*, which the question asks about, corresponds to *CD*, and it is twice as long. Twice 12 is 24, which is the correct answer.

PRACTICE SET 14: CIRCLES

Terminology and Basic Formulas

The SAT Math Test includes questions that test your knowledge of circles, including circles drawn on a coordinate plane. Here are some key features that you should review before Test Day:

- Radius (r): the distance from the center of a circle to its edge
- Chord: a line segment that connects two points on a circle
- Diameter (d): a chord that passes through the center of a circle. The diameter is always the longest chord a circle can have and is twice the length of the radius.
- Circumference (C): the distance around a circle given by the formula $C = 2\pi r = \pi d$
- Area (A): the space a circle takes up given by the formula $A = \pi r^2$
- Total number of degrees in a circle: 360°

Tip: Finding the radius of a circle is often the key to unlocking several other components of the circle. Therefore, your first step for many circle questions will be to find the radius.

Angles, Arcs, and Sectors

The SAT can also ask you about parts of a circle. These parts include angles, arcs, and sectors.

- When radii cut a circle into multiple (but not necessarily equal) pieces, the angle at the center of the circle contained by the radii is the central angle. The measure of a central angle cannot be greater than the number of degrees in a circle (360°).
- When two chords share a common endpoint on the circumference of a circle, the angle between the chords is an inscribed angle.
- An arc is part of a circle's circumference. Thus, an arc length can never be greater than the circumference. Both chords and radii can cut a circle into arcs. The number of arcs present depends on how many chords and/or radii are present. If only two arcs are present, the smaller arc is called the minor arc, and the larger one is the major arc. On the SAT, if an arc is named with only its two endpoints (such as $\overparen{PQ}$), then you can assume it is a minor arc, which always has a measure that is less than 180°. If a diameter cuts a circle in half, the two arcs formed are called semicircles and measure exactly 180°.
- A sector is a pie-shaped piece of a circle enclosed by two radii and an arc. The area of a sector cannot be greater than the total area of the circle.

The following ratios can be used to find the measure or size of part of a circle based on what you know about the whole circle:

$$\frac{\text{arc length}}{\text{circumference}} = \frac{\text{central angle}}{360^\circ} = \frac{\text{area of sector}}{\text{area of circle}}$$

Radian Measure

Radian measure is simply another way to describe an angle. The size of a radian is defined by the requirement that there are 2π radians in a circle, which means $2\pi = 360°$, or $\pi = 180°$. To convert between radians and degrees, use the conversion factor $\frac{\pi}{180°}$ or $\frac{180°}{\pi}$, whichever is needed to cancel the unit that you are trying to eliminate. For example, to change 60° to radians, use the conversion factor with degrees in the denominator so that the degrees cancel:

$$60° \times \frac{\pi}{180°} = \frac{60\pi}{180} = \frac{\pi}{3}$$

Knowing how to work with radians also allows you to use two additional properties of circles:

- The length of an arc is equal to the radian measure of the central angle that subtends (forms) the arc.
- The length of an arc is equal to twice the radian measure of the inscribed angle that subtends the arc.

Tangent Lines

A tangent line touches a circle at exactly one point and is perpendicular to the radius of the circle at the point of contact. The following diagram demonstrates what this looks like:

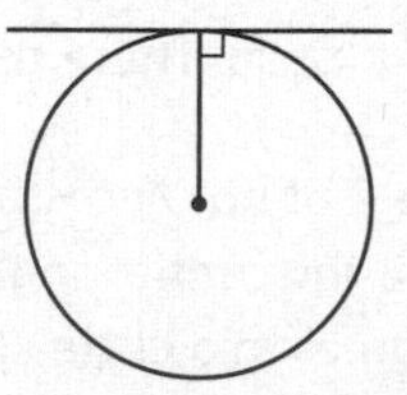

Tip: The presence of a right angle opens up the opportunity to draw otherwise hidden shapes, so pay special attention to tangent lines when they're mentioned in a question.

Circles on the Coordinate Plane and Their Equations

When a circle is drawn on a coordinate plane, its equation is given by $(x - h)^2 + (y - k)^2 = r^2$. This is called standard form. The variables h and k represent the x- and y-coordinates of the center of the circle, and r is the length of the radius. You might also be given the general form of a circle, which is $x^2 + y^2 + Cx + Dy + E = 0$, where C, D, and E are constants. Here are some tips for working with equations of circles:

- To convert from standard form to general form, square the two binomials, $(x - h)^2$ and $(y - k)^2$, square r, move everything to the left side of the equal sign, and simplify as much as possible.
- To convert from general form to standard form, complete the square for the x terms, and then repeat for the y terms.
- To find the center or radius of a circle, write the equation in standard form.

PRACTICE SET

Easy

1. If the area of a circle is 64π, what is the circumference of the circle?

   A) 8π

   B) 16π

   C) 32π

   D) 64π

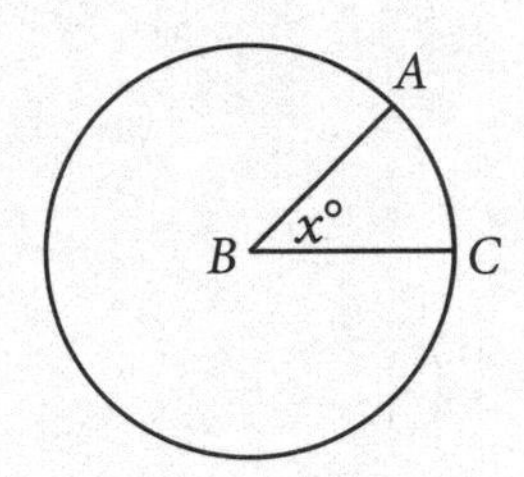

2. In the figure above, the ratio of the circumference of circle B to the length of minor arc AC is 8:1. What is the value of x?

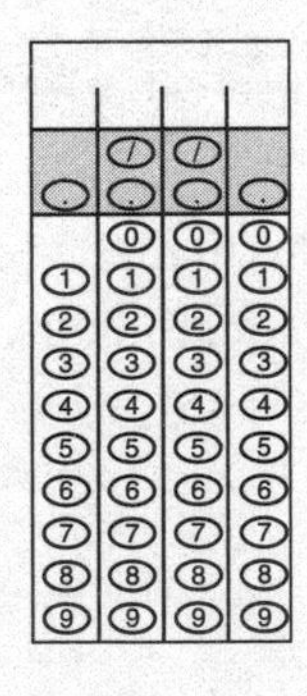

Medium

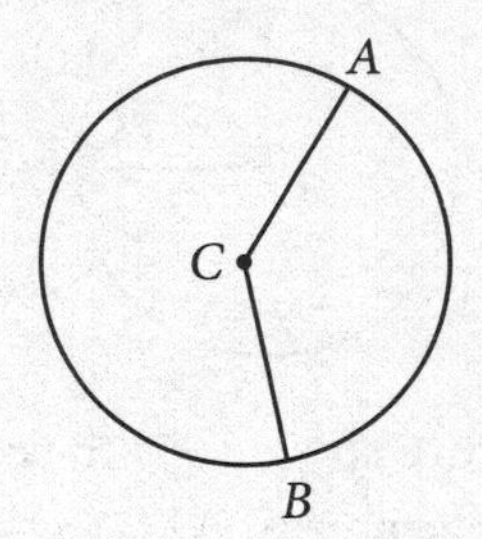

3. Points A and B lie on circle C as shown. The measure of angle ACB is 120°. If the area of circle C is 81π square units, what is the length of minor arc AB?

   A) 6π

   B) 9π

   C) 18π

   D) 27π

4. Circle C (not shown) is drawn on a coordinate plane, centered at the origin. If the point (a, b) lies on the circumference of the circle, what is the radius of the circle in terms of a and b?

   A) $a - b$

   B) $a + b$

   C) $\sqrt{a^2 + b^2}$

   D) $a^2 + b^2$

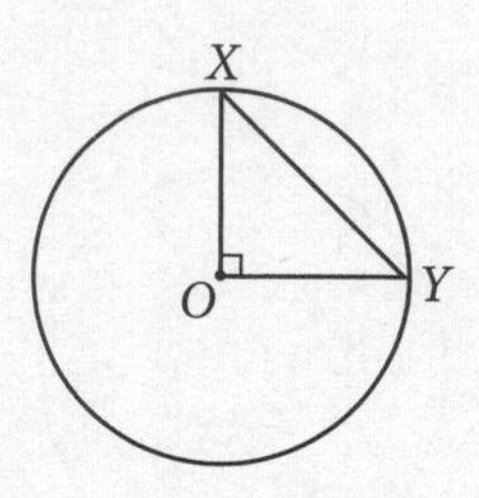

5. In the figure above, O is the center of the circle. If the area of ΔXOY is 25, what is the area of the circle?

A) 25π

B) $25\pi\sqrt{2}$

C) 50π

D) 625π

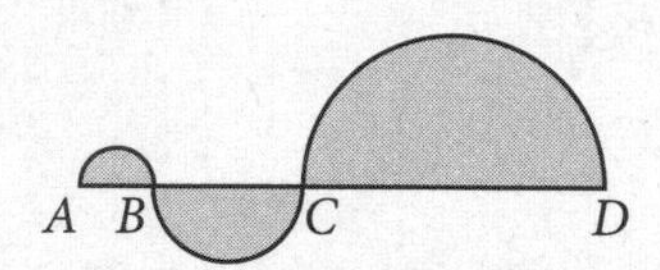

6. Each of the three shaded regions above is a semicircle. If $AB = 4$, $BC = 2AB$, and $CD = 2BC$, what is the area of the entire shaded region?

A) 28π

B) 42π

C) 84π

D) 96π

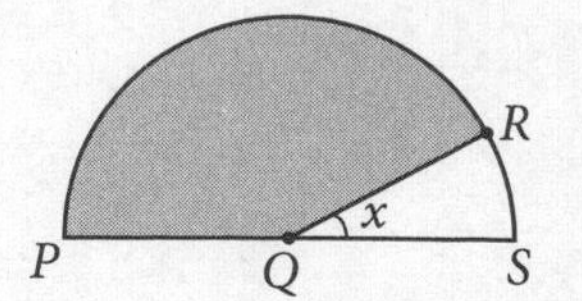

7. The semicircle shown has its center at point Q. If the measure of the central angle of the shaded sector is 160 degrees, what is the value of x in radians?

A) $\frac{\pi}{20}$

B) $\frac{\pi}{12}$

C) $\frac{\pi}{9}$

D) $\frac{\pi}{6}$

Hard

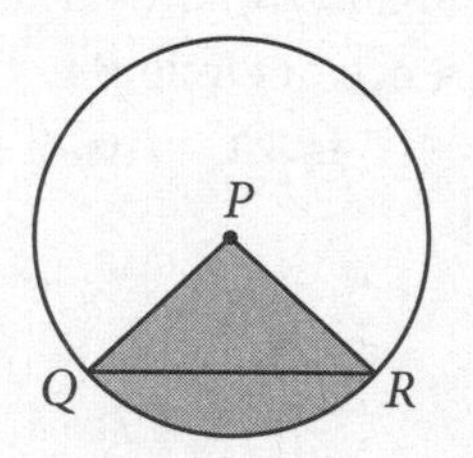

8. The area of the shaded sector in circle P above is 18π square units. If the measure of angle PQR is 45°, what is the length of chord QR?

A) 6

B) 9

C) $9\sqrt{2}$

D) 12

9. The center of circle O (not shown) falls on the point where the line $y = \frac{4}{3}x + 4$ intersects the x-axis on the coordinate plane. The point (3, 8) lies on the circumference of the circle. Which of the following could be the equation for circle O?

A) $x^2 + y^2 = 25$

B) $(x + 3)^2 + y^2 = 25$

C) $(x + 3)^2 + y^2 = 100$

D) $(x + 3)^2 + (y - 8)^2 = 100$

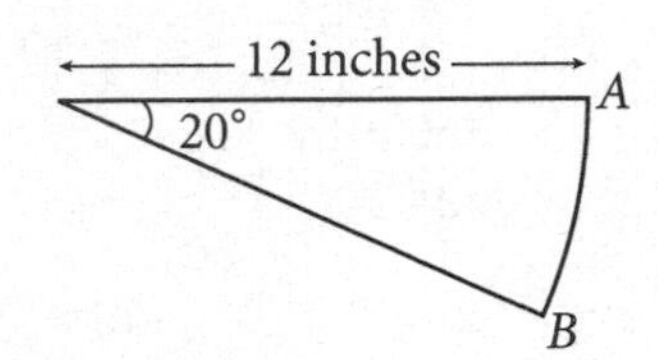

10. The figure above shows the path traced by the hand of a scale as it moves from A to B. What is the area, in square inches, of the region passed over by the scale's hand?

A) 2π

B) 8π

C) 12π

D) 16π

11. If arc AB has a length of 12π and represents three-fourths of the circumference of circle O (not shown), what is the shortest distance between the endpoints of the arc?

A) 4

B) $4\sqrt{2}$

C) 8

D) $8\sqrt{2}$

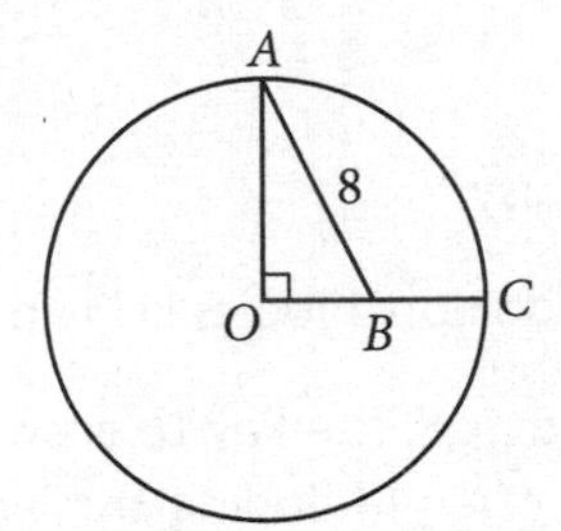

12. In the figure above, circle O has a circumference of 12π. If $AB = 8$, what is BC?

A) $2\sqrt{7}$

B) $2(3 - \sqrt{7})$

C) $2(6 - \sqrt{7})$

D) $4\sqrt{5}$

ANSWER KEY

1. B	**5. C**	**9. C**
2. 45	**6. B**	**10. B**
3. A	**7. C**	**11. D**
4. C	**8. D**	**12. B**

ANSWERS & EXPLANATIONS

1. B

Difficulty: Easy

Category: Additional Topics in Math / Geometry

Strategic Advice: The key to answering circle questions is often in finding the length of the radius. Make that the first thing you look for.

Getting to the Answer: The area of a circle is given by $A = \pi r^2$. Substitute the given area (64π) into the formula to get $64\pi = \pi r^2$, which simplifies to $r = 8$. The circumference of a circle is equal to $2\pi r$, so the circumference of this particular circle is $2\pi \times 8 = 16\pi$. Therefore, (B) is correct.

2. 45

Difficulty: Easy

Category: Additional Topics in Math / Geometry

Strategic Advice: Circle ratios used in proportions usually compare a part to a whole. In this case, you're comparing an arc (part) to the full circumference (whole), and an internal angle measure (part) to the number of degrees in a full circle (whole).

Getting to the Answer: Use the ratio of the circumference to the arc length and your knowledge of degree measures in a circle to write a proportion, and then solve for the missing length:

$$\frac{1}{8} = \frac{x}{360}$$
$$360 = 8x$$
$$45 = x$$

Grid in 45.

3. A

Difficulty: Medium

Category: Additional Topics in Math / Geometry

Strategic Advice: The length of an arc is found by multiplying the circumference of the circle by the portion of the circumference that the arc represents.

Getting to the Answer: You know that the area of the circle is 81π. The area of a circle is given by $A = \pi r^2$, so $r = 9$. This means the circumference of the circle, which is given by $C = 2\pi r$, is $C = 2\pi(9) = 18\pi$. Now, find the portion of the circumference that minor arc AB represents by writing the measure of the central angle that subtends the arc over 360° (the number of degrees in a full circle): $\frac{120°}{360°} = \frac{1}{3}$. The arc represents one-third of the circumference, or $18\pi \div 3 = 6\pi$, which is (A).

4. C

Difficulty: Medium

Category: Additional Topics in Math / Geometry

Strategic Advice: You could use algebra to answer a question like this, or you could draw a quick sketch and use geometry. Use whichever method gets you to the answer quicker.

Getting to the Answer: *Algebraic solution*: The equation of a circle centered at the origin is given by the equation $x^2 + y^2 = r^2$, where x and y represent any ordered pair (point) that lies on the circumference of the circle, and r is the length of the radius. Substitute the point (a, b) for x and y, and then solve for r:

$$x^2 + y^2 = r^2$$
$$a^2 + b^2 = r^2$$
$$\sqrt{a^2 + b^2} = \sqrt{r^2}$$
$$\sqrt{a^2 + b^2} = r$$

This matches (C).

Geometric solution: Draw a sketch of a circle with center (0, 0). Add point (a, b) to the circle and draw in a radius to that point. Use the coordinates of the point to draw and label a right triangle. Your sketch should look like the following:

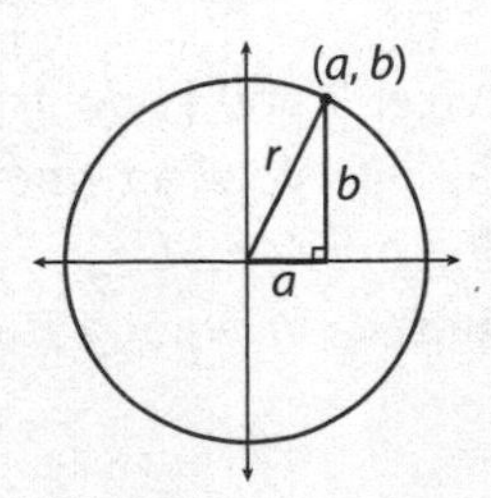

Now use the Pythagorean theorem to find that $r = \sqrt{a^2 + b^2}$, which agrees with the algebraic solution.

5. C

Difficulty: Medium

Category: Additional Topics in Math / Geometry

Strategic Advice: The two legs of ΔXOY are both radii of the circle and therefore have the same length. Use this fact to find a link between the formulas used to find the area of a triangle and the area of a circle.

Getting to the Answer: Let r represent the length of each leg of ΔXOY. Because the area of a triangle is given by $A_\Delta = \frac{1}{2}bh$ and the area of ΔXOY is 25, it follows that $25 = \frac{1}{2} \times r \times r = \frac{1}{2}r^2$. Therefore, $r^2 = 50$. Substitute this value for r^2 into the formula for finding the area of a circle: $A_\circ = \pi r^2 = \pi \times 50 = 50\pi$. Choice (C) is therefore the correct answer.

(Note that you did not have to find the length of the circle's radius to answer this question; doing so would have cost you extra time.)

6. B

Difficulty: Medium

Category: Additional Topics in Math / Geometry

Strategic Advice: Use the only diameter given ($AB = 4$) to determine the other two diameters. Once there, use $A = \pi r^2$ to find the area of each. Remember to multiply each area by $\frac{1}{2}$ to account for the fact that you have semicircles, not whole circles.

Getting to the Answer: The diameter of semicircle AB is given as 4, so its radius is 2. Therefore, the area of semicircle AB is $\frac{1}{2}\left(\pi \times 2^2\right) = 2\pi$. If $BC = 2AB$, then $BC = 8$, and the radius of semicircle BC is 4. Its area, therefore, is $\frac{1}{2}(\pi \times 4^2) = 8\pi$. Similarly, because $CD = 2BC$, CD must equal 16, and the radius of semicircle CD is 8, giving it an area of $\frac{1}{2}(\pi \times 8^2) = 32\pi$. Add the three areas together to get $2\pi + 8\pi + 32\pi = 42\pi$, which matches (B).

7. C

Difficulty: Medium

Category: Additional Topics in Math / Geometry

Strategic Advice: In most cases, it is easier to carry out calculations in degrees, and then convert your answer to radians.

Getting to the Answer: There are 180° in a semicircle (half of 360°). If the central angle of the shaded sector is 160°, then the value of x, in degrees, is 180° − 160° = 20°. To convert this to radians, multiply by the conversion factor that has degrees in the denominator: $20° \times \frac{\pi}{180°} = \frac{20\pi}{180} = \frac{\pi}{9}$, which is (C).

8. D

Difficulty: Hard

Category: Additional Topics in Math / Geometry

Strategic Advice: The area of a sector is found by multiplying the area of the whole circle by the portion of the circle that the sector represents. To find the portion of the circle's area that the shaded sector represents, write the measure of the central angle of the sector over 360°.

Getting to the Answer: This question requires thinking backward to come up with a plan. You need to find the length of chord QR. To find this length, you'll need to know the radius of the circle. To find the radius, you need to find the area of the whole circle by finding the portion of the circle that the shaded sector represents. To do this, you need to know the measure of the central angle of the sector. This is your starting point.

Because PQ and PR are radii, triangle PQR is isosceles. You're given that the measure of angle PQR is 45°, so the measure of angle PRQ is also 45°, making triangle PQR a 45-45-90 triangle. Therefore, the central angle of the shaded sector is 90°, which means the sector represents $\frac{90°}{360°} = \frac{1}{4}$ of the circle. If the area of this sector is 18π, then the area of the whole circle is $18\pi \times 4 = 72\pi$. Now, use the formula for finding the area of a circle to find the radius:

$$A = \pi r^2$$
$$72\pi = \pi r^2$$
$$72 = r^2$$
$$r = \sqrt{72}$$

Simplify the radical to make the numbers more manageable: $\sqrt{72} = \sqrt{36 \times 2} = 6\sqrt{2}$. This means each leg of triangle PQR has a length of $6\sqrt{2}$. The sides of a 45-45-90 triangle are in the ratio $x : x : x\sqrt{2}$, so the hypotenuse of triangle PQR (which is chord QR) has a length of $6\sqrt{2} \times \sqrt{2} = 6 \times 2 = 12$. Choice (D) is correct.

9. C

Difficulty: Hard

Category: Additional Topics in Math / Geometry

Strategic Advice: If you know the center and the radius of a circle, you can write its equation in the form $(x - h)^2 + (y - k)^2 = r^2$, where (h, k) is the center of the circle and r is the length of its radius.

Getting to the Answer: Start by finding the center of the circle. You're told that the center falls on the point where the given line intersects the x-axis, which is the same as the x-intercept of the line. An x-intercept is always of the form $(x, 0)$, so set $y = 0$ and solve for x:

$$y = \frac{4}{3}x + 4$$

$$0 = \frac{4}{3}x + 4$$

$$-\frac{4}{3}x = 4$$

$$-4x = 12$$

$$x = -3$$

You now know that the center of circle O is (–3, 0), so its equation should look like $(x - (-3))^2 + (y - 0)^2 = r^2$ or, when simplified, $(x + 3)^2 + y^2 = r^2$. This means you can eliminate A and D.

To find the radius of the circle, draw a quick sketch that includes the center of the circle and the point given in the question, (3, 8).

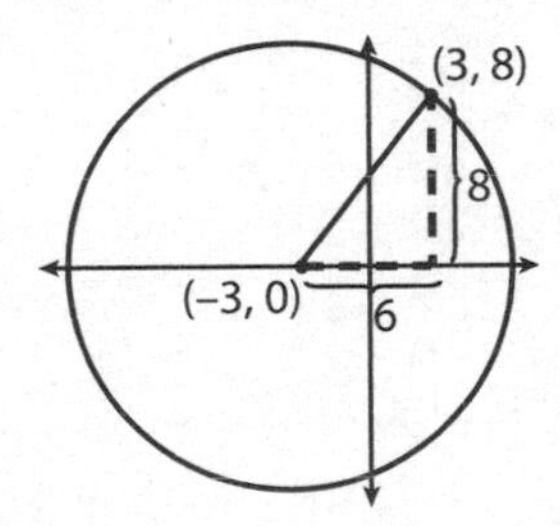

Notice that drawing in a right triangle from the center of the circle to the point on the circumference gives you a nice 6-8-10 Pythagorean triplet, making the radius of the circle 10 (or you could use the Pythagorean theorem to find the length of the radius). Thus, the equation of the circle is $(x + 3)^2 + y^2 = 10^2$, or $(x + 3)^2 + y^2 = 100$, making (C) correct.

10. B

Difficulty: Hard

Category: Additional Topics in Math / Geometry

Strategic Advice: Sometimes, the first step to answering a question is thinking logically: If the hand of the scale were to make a full rotation, its path would be a circle. Use this fact and the information provided in the figure to write and solve a proportion to answer the question.

Getting to the Answer: The length of the scale hand is also the radius of a circle created from one full rotation of the hand around the entire scale. Therefore, the full area passed over by the hand of the scale is $A = \pi r^2 = \pi \times 12^2 = 144\pi$. You can use this in conjunction with the degree measures of the sector in the diagram (20°) and the full circle (360°) to find the area of the sector (the area passed over by the scale hand):

$$\frac{20}{360} = \frac{x}{144\pi}$$

$$\frac{1}{18} = \frac{x}{144\pi}$$

$$x = \frac{144\pi}{18} = 8\pi$$

(B) is the correct answer.

11. D

Difficulty: Hard

Category: Additional Topics in Math / Geometry

Strategic Advice: Draw a sketch to visualize the situation and to reveal helpful hidden shapes.

Getting to the Answer: You're told that arc AB measures 12π and represents $\frac{3}{4}$ of a full circumference, which means it is the major arc and sits across from an internal angle that has a measure of $\frac{3}{4} \times 360° = 270°$. This, in turn, means minor arc AB is across from a 90° (right) angle, as shown here:

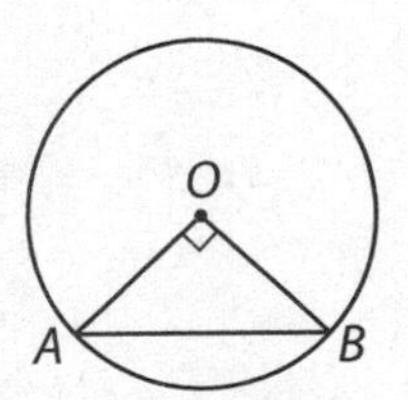

Because $\frac{3}{4}$ of the full circumference is 12π, the full circumference itself must be 16π. A circle's circumference is equal to $2\pi r$, so the radius of the circle is 8. $\overline{AO}$ and $\overline{BO}$ are both radii and therefore equal, so ΔABO is an isosceles right (and 45-45-90) triangle with sides in the ratio $x:x:x\sqrt{2}$. The legs (radii) have length 8, so the length of the hypotenuse (which represents the shortest distance between the endpoints of the major arc) is $8\sqrt{2}$. (D) is therefore correct.

12. B

Difficulty: Hard

Category: Additional Topics in Math / Geometry

Strategic Advice: Identify pieces of the figure that play roles in multiple shapes present (for example, a radius of the circle that is also a leg of a triangle), and transfer information as you work toward the answer.

Getting to the Answer: The circumference of circle O is 12π, so its radius must be $6 (C = 2\pi r)$. You now know the lengths of two sides of ΔAOB (the given hypotenuse and the leg represented by the radius), and because it's a right triangle, you can use the Pythagorean theorem to find the length of the missing side (OB):

$$AO^2 + OB^2 = AB^2$$
$$6^2 + OB^2 = 8^2$$
$$OB^2 = 64 - 36$$
$$OB = \sqrt{28} = 2\sqrt{7}$$

Don't stop yet; the question asks for BC, not OB. BC is equal to $OC - OB$, which is $6 - 2\sqrt{7}$ (because $\overline{OC}$ is also a radius). Factor out a 2 to get $2(3 - \sqrt{7})$, which matches (B).

PRACTICE SET 15: 3-D SHAPES

Introduction to 3-D Shapes

In addition to expertise in two-dimensional geometry, you'll also want to have a good understanding of solids, which are often referred to as three-dimensional shapes. You'll want to be familiar with both the terminology and formulas related to these shapes.

Terminology

A face (or surface) is a 2-D shape (plane) that acts as one of the sides of the solid. Two faces meet at a line segment called an edge, and three faces meet at a single point called a vertex.

The following diagram illustrates these terms:

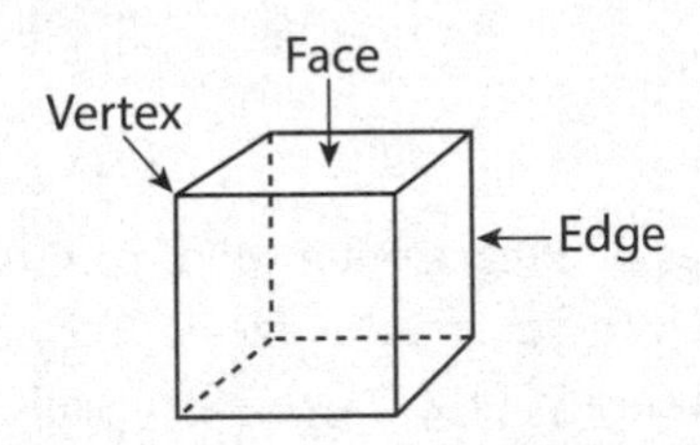

Volume

Volume is the amount of 3-D space occupied by a solid. This is analogous to the area of a 2-D shape like a triangle or circle. You can find the volume of many 3-D shapes by finding the area of the base and multiplying it by the height. We've highlighted the base area components of the formulas in the following table with parentheses. You'll be given the volume formulas on Test Day, so you don't absolutely have to memorize them all; however, knowing the formulas in advance will save valuable time.

Rectangular Solid	**Cube**	**Right Cylinder**
h, w, l	s, s, s	r, h
$V = (l \times w) \times h$	$V = (s \times s) \times s = s^3$	$V = (\pi r^2) \times h$

The shapes in the table above are all prisms. Almost all prisms on the SAT are right prisms; that is, all faces are perpendicular to those with which they share edges. Less commonly seen prisms (e.g., triangular, hexagonal, octagonal, etc.) have the same general volume formula as these: $V = A_{base} \times h$.

More complicated 3-D shapes include the right pyramid, right cone, and sphere. The vertex of a right pyramid or right cone will always be centered above the middle of the base. Their volume formulas are similar to those of prisms, with an added coefficient.

Right (Rectangular) Pyramid	Right Cone	Sphere
h, w, l	h, r	r
$V = \frac{1}{3} \times (l \times w) \times h$	$V = \frac{1}{3} \times (\pi r^2) \times h$	$V = \frac{4}{3} \times \pi r^3$

Surface Area

Surface area is the sum of the areas of all faces of a solid. You might liken this to determining the amount of wrapping paper needed to cover all its faces. To calculate the surface area of a solid, simply find the area of each face using your 2-D geometry skills, and then add them all together. Formulas for calculating surface area are NOT provided on the reference page.

You might think that finding the surface area of a solid with many sides, such as a 10-sided right octagonal prism, is a tall order. However, you can save time by noticing a vital trait: The prism has two identical octagonal faces and eight identical rectangular faces. Don't waste time finding the area of each of the 10 sides; find the area of one octagonal face and one rectangular face instead. Once complete, multiply the area of the octagonal face by 2 and the area of the rectangular face by 8, add the products together, and you're done! The same is true for other 3-D shapes, such as rectangular solids (including cubes), other right prisms, and certain pyramids.

PRACTICE SET

You may use your calculator for all questions in this practice set.

Easy

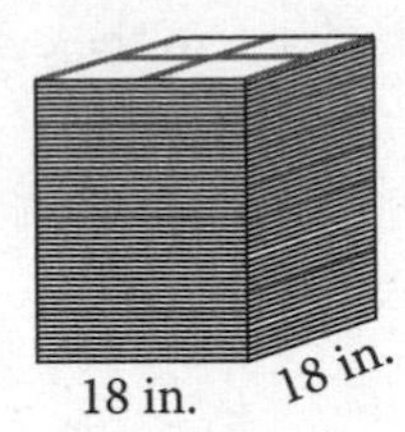

1. A flooring company stores its marble tiles in vertical stacks as shown above. Each tile measures $18" \times 18" \times \frac{1}{2}"$. How many cubic feet of tile are there in one stack of 48 of these tiles?

   A) 4.5
   B) 54
   C) 162
   D) 648

2. If a right cylinder with a radius of 2 cm has a volume of 100π cm^3, what is the height, in centimeters, of the cylinder?

   A) 20
   B) 25
   C) 40
   D) 50

3. A cube and a rectangular solid are equal in volume. If the lengths of the edges of the rectangular solid are 4, 8, and 16, what is the length of an edge of the cube?

4. What is the maximum number of boxes with dimensions 2 inches by 3 inches by 4 inches that could fit in a cube-shaped container that has a volume of 1 cubic foot?

Medium

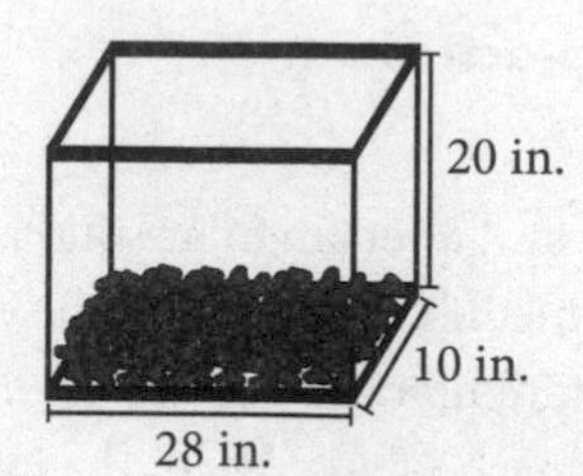

5. The bottom of the fish tank shown is filled with rocks. The tank is then filled with water to a height of 18 inches. When the rocks are removed, the height of the water drops to 16.5 inches. How many cubic inches of water do the rocks displace?

   A) 280

   B) 420

   C) 560

   D) 980

6. What is the radius of the largest sphere that can be placed inside a cube that has a volume of 64 cubic units?

   A) 2

   B) $2\sqrt{2}$

   C) 4

   D) 8

7. A cylinder has a volume of 72π cubic inches and a height of 8 inches. If the height is increased by 4 inches, what will be the new volume of the cylinder in cubic inches?

   A) 76π

   B) 108π

   C) 328π

   D) 576π

8. Milk is poured from a full rectangular container with dimensions 4 inches by 9 inches by 10 inches into a cylindrical container with a diameter of 6 inches. Assuming all the milk is transferred without spillage, how many inches high will the milk reach in the cylindrical container?

   A) $\frac{40}{\pi}$

   B) $\frac{60}{\pi}$

   C) 24

   D) 30

9. A solid, cone-shaped lead crystal paperweight has a height of 5 centimeters and a base diameter that is 20% larger than the height. If the density of lead crystal is 3.1 g/cm^3, what is the approximate mass of the paperweight? Round your answer to the nearest gram.

Hard

10. A jeweler makes beads for a bracelet by melting silver into spherical molds. Each mold has a diameter of 1 centimeter. After the silver hardens, the jeweler drills a small cylindrical hole through the center of the bead for the chain as shown. The drill bit has a diameter of 4 millimeters. If the jeweler strings a total of 15 beads on an elastic band, what is the approximate volume of silver, in cubic centimeters, on the bracelet? (Note: There are 10 millimeters in 1 centimeter.)

    Note: The volume is *approximate* because the top and bottom of the cylindrical piece that is drilled out is slightly curved.

    A) $\frac{19}{150}\pi$

    B) $\frac{1}{6}\pi$

    C) $\frac{19}{10}\pi$

    D) $\frac{5}{2}\pi$

11. A rectangular block with a volume of 250 cubic inches is sliced into two cubes of equal volume. How much greater, in square inches, is the combined surface area of the two cubes than the surface area of the original rectangular block?

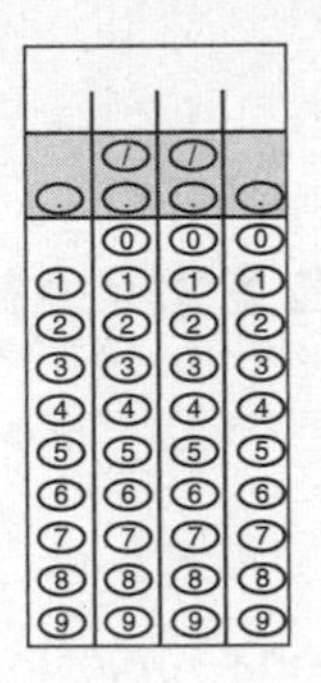

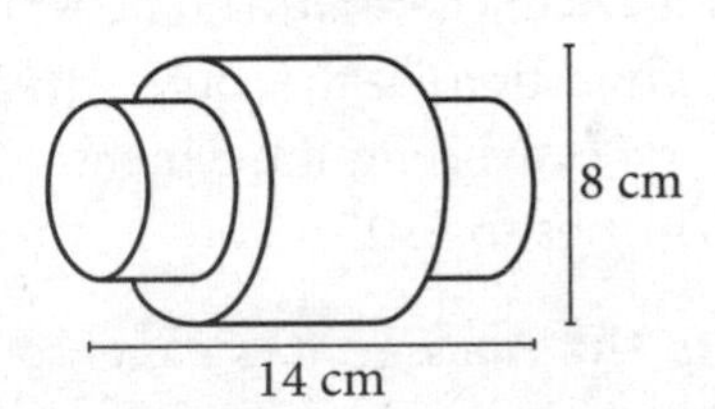

12. A locking pin is often made using a cylinder-cylinder pair in which a narrow cylinder fits tightly inside a wider cylinder. The inner cylinder protrudes from the outer cylinder, usually by equal amounts on both ends. In the diagram above, the radius of the inner cylinder is half the radius of the outer cylinder, and it protrudes from the outer cylinder by 4 centimeters on each end. What is the volume of the locking pin? Round your answer to the nearest cubic centimeter.

ANSWER KEY

1. A	5. B	9. 146
2. B	6. A	10. C
3. 8	7. B	11. 50
4. 72	8. A	12. 402

ANSWERS & EXPLANATIONS

1. A
Difficulty: Easy

Category: Additional Topics in Math / Geometry

Strategic Advice: There are a number of ways to approach a question like this, but each approach must include converting the dimensions of the tiles from inches to feet.

Getting to the Answer: It is easier to convert inches to feet than cubic inches to cubic feet, so start by converting each of the dimensions of the tile to feet. There are 12 inches in 1 foot, so the dimensions of a single piece of tile are $(18 \div 12) \times (18 \div 12) \times (0.5 \div 12)$ or, when simplified, $1.5 \times 1.5 \times \frac{1}{24}$. The stack of tiles takes on the shape of a right rectangular prism, so you can find the volume using the formula $V = lwh$. You already know the length (1.5 ft) and the width (1.5 ft). To find the height, multiply the height of one tile by the number of tiles in the stack: $\frac{1}{24} \times 48 = \frac{48}{24} = 2$ feet. This means the number of cubic feet in the stack of tiles is $1.5 \times 1.5 \times 2 = 4.5$, making (A) correct.

2. B
Difficulty: Easy

Category: Additional Topics in Math / Geometry

Strategic Advice: Use the formula for finding the volume of a cylinder to determine the missing dimension. On Test Day, don't forget that the formulas are provided for you on the reference page.

Getting to the Answer: The volume of a right cylinder is given by the formula $V = \pi r^2 h$. Plug the radius (2) and the volume (100π) into the formula, and then solve for the height:

$$100\pi = \pi \times 2^2 \times h$$
$$100 = 4h$$
$$h = 25$$

The height of the cylinder is 25 cm, which is (B).

3. 8
Difficulty: Easy

Category: Additional Topics in Math / Geometry

Strategic Advice: Use the dimensions of the rectangular solid to find its volume, and then apply the properties of a cube to find the length of one of the cube's edges.

Getting to the Answer: The volume of the rectangular solid is $4 \times 8 \times 16 = 512$ cubic units. The cube has the same volume, and its edges are all equal in length, so express this as $s^3 = 512$ and take the cube root of 512 to solve for s. The result is 8, which gives the length of one edge of the cube.

4. 72
Difficulty: Easy

Category: Additional Topics in Math / Geometry

Strategic Advice: Whenever a geometry question involves *filling* one shape with another, you are looking for volume.

Getting to the Answer: Before finding the volume of the cube-shaped container, convert its dimensions to inches so all the units match. The cube-shaped container has a volume of $12 \times 12 \times 12$ cubic inches. Divide this quantity by the volume of one smaller box to get the answer: $\frac{12 \times 12 \times 12}{2 \times 3 \times 4} = 72$.

5. B
Difficulty: Medium

Category: Additional Topics in Math / Geometry

Strategic Advice: Sometimes, it is necessary to think about a question logically before making any calculations. Here, the amount of water displaced is equal to the difference between the volume of water in the tank with the rocks and the volume of water in the tank without the rocks.

Getting to the Answer: You could find each of the volumes described. However, because the length and width of the tank do not change (only the height of the water changes), it would be quicker to find the difference in the heights of the water and find the volume of only that part of the water. The height of the water with the rocks is 18 inches and the height without the rocks is 16.5 inches, so 1.5 inches of water is displaced. Now, use the formula $V = lwh$ to find that the amount of water displaced by the rocks is $28 \times 10 \times 1.5 = 420$ cubic inches. This means (B) is correct.

6. A
Difficulty: Medium

Category: Additional Topics in Math / Geometry

Strategic Advice: Often, a verbal description of a geometric scenario is simply not enough to answer a question. Instead, you need to draw a quick sketch to visualize the situation.

Getting to the Answer: The figure below shows the sphere inside the cube.

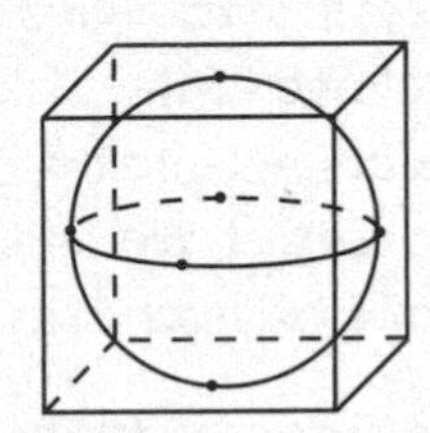

The sphere will touch the cube at six points. Each point will be an endpoint of a diameter and will be at the center of one of the cube's faces. Therefore, the diameter of the sphere extends directly from one face of the cube to the opposite face and is perpendicular to both faces that it touches. This means that the diameter must have the same length as an edge of the cube. The volume of the cube is given (64), so express this as $s^3 = 64$, and take the cube root of 64 to find that the length of each edge is 4. The diameter of the sphere is therefore also 4, so its radius must be 2, making (A) the correct answer.

7. B
Difficulty: Medium

Category: Additional Topics in Math / Geometry

Strategic Advice: To find the volume of the new cylinder, you'll need to determine the area of its base first. Remember, the volume of any prism is equal to the area of its base times its height.

Getting to the Answer: Plug the volume and height of the first cylinder into the correct volume formula to find the area of its base: $V = \pi r^2 \times h \rightarrow 72\pi = \pi r^2 \times 8 \rightarrow 9\pi = \pi r^2$.

Notice that you don't need to solve for the radius, as it will remain constant. Plug the area of the base back into the volume formula with the new height, which is $8 + 4 = 12$ inches: $V = 9\pi \times 12 \rightarrow V = 108\pi$. (B) is correct.

8. A
Difficulty: Medium

Category: Additional Topics in Math / Geometry

Strategic Advice: In a real-world scenario, think logically: The volume of the milk itself will not change after it's poured into the cylindrical container, so you can use that volume to determine the height that the milk will reach in its new container.

Getting to the Answer: The volume of milk present is $4 \times 9 \times 10 = 360$ cubic inches. The diameter of the base of the cylindrical container measures 6 inches, so its radius is 3 inches. Substitute this value and the volume of the milk that you found earlier into the formula for finding the volume of a cylinder, and then solve for h: $360 = \pi \times 3^2 \times h \rightarrow h = \frac{360}{9\pi} = \frac{40}{\pi}$. This means (A) is the correct answer.

9. 146
Difficulty: Medium

Category: Additional Topics in Math / Geometry

Strategic Advice: If you don't recall how to find density, let the units (g/cm^3) in this question guide you. The unit g (grams) indicates mass, and the unit cm^3 indicates volume. So density can be found by dividing mass by volume. To find an unknown mass from its density, do the inverse: Multiplying the density times the volume will give you the mass.

Getting to the Answer: To calculate the volume of the paperweight, start by finding the radius. The question states that the diameter of the paperweight's base is 20% larger than (120% of) its height, which is $1.2 \times 5 = 6$ centimeters, so the radius is half that, or 3 centimeters.

The volume of a cone (remember to check the formula page) is given by $V_{cone} = \frac{1}{3}\pi r^2 h$, or here $\frac{1}{3}\pi \times 3^2 \times 5 = 15\pi$ cubic centimeters. Now, multiply this volume by the given density (3.1 g/cm^3) to find that the mass of the paperweight is $15\pi \text{ cm}^3 \times \frac{3.1 \text{ g}}{1 \text{ cm}^3} = 46.5\pi$ g. Rounded to the nearest gram, the correct answer is $46.5 \times 3.14 \approx 146$.

10. C
Difficulty: Hard

Category: Additional Topics in Math / Geometry

Strategic Advice: Staying organized is the key to answering a question like this. Write down what you need to find in words, and then apply the math. Take note that the answers are given as fractions of π, so don't spend time converting numbers to decimals.

Getting to the Answer: You need to find the volume of one bead (after the hole has been drilled) and then multiply this volume by the number of beads (15). Write down something like this: Volume of bead minus volume of drilled-out part = volume of sphere minus volume of cylinder. Check the formula page and be sure to label dimensions as you go. Both formulas involve a radius, but the radii are not the same for the bead and the drilled-out part.

Volume of bead = volume of sphere – volume of cylinder translates to:

$$V_{bead} = \frac{4}{3}\pi\left(r_{bead}\right)^3 - \pi\left(r_{hole}\right)^2 \times h_{hole}$$

At this point, you should jot down the value of each variable in the equation based on the information provided in the question:

r_{bead} = half the diameter of the mold = $\frac{1}{2}$ cm

r_{hole} = half the diameter of the drill bit = $\frac{4}{2}$ mm = 2 mm = $\frac{2}{10} = \frac{1}{5}$ cm

h_{hole} = diameter of the whole bead = 1 cm

Now, carefully substitute each value into the equation and simplify:

$$\begin{aligned} V_{\text{bead}} &= \frac{4}{3}\pi\left(r_{\text{bead}}\right)^3 - \pi\left(r_{\text{hole}}\right)^2 \times h_{\text{hole}} \\ &= \frac{4}{3}\pi\left(\frac{1}{2}\right)^3 - \pi\left(\frac{1}{5}\right)^2 \times 1 \\ &= \frac{4}{3}\pi\left(\frac{1}{2}\times\frac{1}{2}\times\frac{1}{2}\right) - \pi\left(\frac{1}{5}\times\frac{1}{5}\right) \\ &= \frac{1}{6}\pi - \frac{1}{25}\pi \\ &= \frac{25}{150}\pi - \frac{6}{150}\pi \\ &= \frac{19}{150}\pi \end{aligned}$$

Don't forget to multiply this by 15 (because there are 15 beads). The final volume is $15\times\frac{19}{150}\pi = \frac{19}{10}\pi$, which matches (C).

11. 50

Difficulty: Hard

Category: Additional Topics in Math / Geometry

Strategic Advice: Think about this question conceptually before you start calculating anything. When a rectangular solid is cut into two identical cubes, two new faces are formed (where the slice was made). The difference in surface area between the two cubes combined and the original rectangular solid is the area of the two new faces, so plan a route to finding this quantity.

Getting to the Answer: If the two newly formed cubes are identical in volume, that means they must each have a volume of $\frac{250}{2} = 125$ cubic inches. The length of one of the cube's edges is therefore $\sqrt[3]{125}$ inches, which is equal to 5 inches. You can thus conclude that each face of the cubes is 5^2, or 25 square inches. Recall that two new faces were formed, so the two cubes combined will have $25 \times 2 = 50$ square inches more surface area than the original solid.

12. 402

Difficulty: Hard

Category: Additional Topics in Math / Geometry

Strategic Advice: Breaking apart a complex figure usually makes finding its volume easier. The locking pin in this question is comprised of three separate cylinders, so look for a way to find the volume of each one. Once there, all you need to do is add them together.

Getting to the Answer: The three cylinders of the locking pin with their dimensions are shown below:

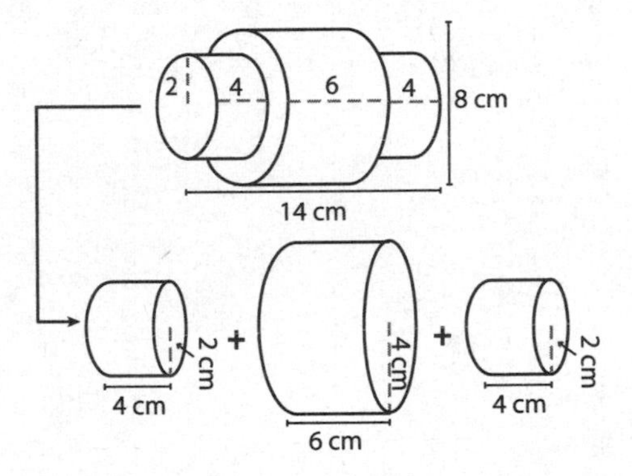

The volume of a cylinder is equal to $\pi r^2 h$, so each of the two smaller end cylinders have a volume of $\pi \times 2^2 \times 4 = 16\pi$ cubic centimeters; double this to get 32π cubic centimeters for the ends. Repeat for the middle cylinder: $V = \pi \times 4^2 \times 6 = 96\pi$ cubic centimeters. Adding the volumes together yields $32\pi + 96\pi = 128\pi$ cubic centimeters. Rounded to the nearest cubic centimeter, this value becomes $128 \times 3.14 \approx 402$ cubic centimeters.

PRACTICE SET 16: TRIGONOMETRY

Introduction to Trigonometry

The word *trigonometry* comes from Greek words meaning "triangle" and "measure," so the one or two trig questions that you may encounter on Test Day will have to do with a triangle. In fact, these questions almost always involve a right triangle, which you already know quite a bit about.

Trigonometric Ratios

You probably remember learning the acronym SOH CAH TOA, a mnemonic device for remembering the definitions of the sine, cosine, and tangent ratios. These ratios are summarized in the following table:

Sine (sin)	Cosine (cos)	Tangent (tan)
$\frac{\text{opposite}}{\text{hypotenuse}}$	$\frac{\text{adjacent}}{\text{hypotenuse}}$	$\frac{\text{opposite}}{\text{adjacent}}$

The following example shows how to set up the trig ratios for a specific angle:

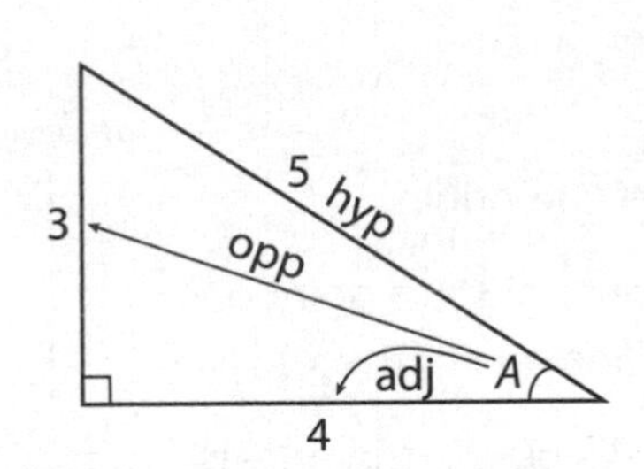

$$\sin A = \frac{\text{opp}}{\text{hyp}} = \frac{3}{5}$$

$$\cos A = \frac{\text{adj}}{\text{hyp}} = \frac{4}{5}$$

$$\tan A = \frac{\text{opp}}{\text{adj}} = \frac{3}{4}$$

Complementary Angle Relationship

One particularly useful property to know about trig functions is that the sine of an acute angle is equal to the cosine of the complement of the angle. For example, $\sin(20°) = \cos(70°)$ because 20° and 70° are complementary angles. The same is true in reverse: $\cos(20°) = \sin(70°)$. This can be stated mathematically as $\sin(x°) = \cos(90° - x°)$ or $\cos(x°) = \sin(90° - x°)$.

Radians

Most geometry questions present angle measures in degrees. In trigonometry (and in some circle questions), you may encounter a different unit: the radian. In the practice set on circles, you learned that $180° = \pi$ radians and that you can use the conversion factor $\frac{\pi}{180°}$ or $\frac{180°}{\pi}$ to convert between units, whichever is needed to cancel the unit that you are trying to eliminate.

Benchmark Angles

Knowing the trig functions for the most commonly tested "benchmark" angles will save time on Test Day. You will not be asked to evaluate trig functions for angles that require a calculator. The table below gives the most commonly tested angles in degrees, their radian equivalents, and their trig values. (Note: "Und" means undefined.)

x	$0°\ (0\pi)$	$30°\left(\frac{\pi}{6}\right)$	$45°\left(\frac{\pi}{4}\right)$	$60°\left(\frac{\pi}{3}\right)$	$90°\left(\frac{\pi}{2}\right)$	$180°\ (\pi)$	$270°\left(\frac{3\pi}{2}\right)$	$360°\ (2\pi)$
$\sin x$	0	$\frac{1}{2}$	$\frac{\sqrt{2}}{2}$	$\frac{\sqrt{3}}{2}$	1	0	–1	0
$\cos x$	1	$\frac{\sqrt{3}}{2}$	$\frac{\sqrt{2}}{2}$	$\frac{1}{2}$	0	–1	0	1
$\tan x$	0	$\frac{\sqrt{3}}{3}$	1	$\sqrt{3}$	Und	0	Und	0

The Unit Circle

You can also use a unit circle (a circle of radius 1) to derive trig values. Here's an example:

For a 60° angle:

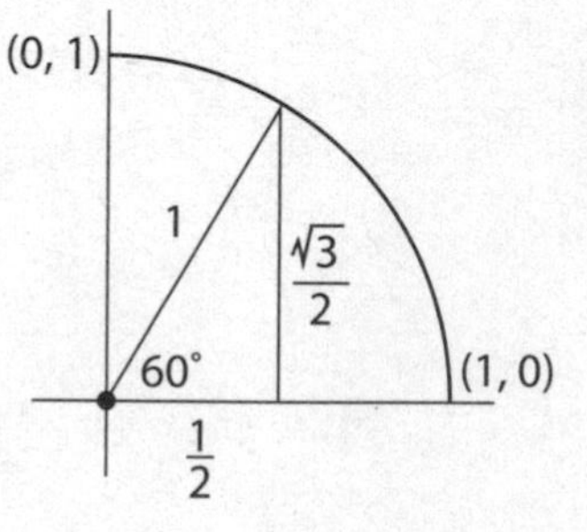

- Draw a circle with a radius of 1 centered at the origin.
- In the first quadrant, draw a radius from the origin to a point on the circle in such a way that it forms a 60° angle with the x-axis.
- Recall that the ratio for the sides of a 30-60-90 triangle is $x{:}x\sqrt{3}{:}2x$. Use the length of the hypotenuse (which is 1 because it is the radius of the circle) to find the lengths of the legs. If $2x = 1$, then $x = \frac{1}{2}$ and $x\sqrt{3} = \frac{1}{2} \cdot \sqrt{3} = \frac{\sqrt{3}}{2}$.
- Use SOH CAH TOA: $\sin(60°) = \frac{\frac{\sqrt{3}}{2}}{1} = \frac{\sqrt{3}}{2}$, $\cos(60°) = \frac{\frac{1}{2}}{1} = \frac{1}{2}$, $\tan(60°) = \frac{\frac{\sqrt{3}}{\cancel{2}}}{\frac{1}{\cancel{2}}} = \sqrt{3}$.
- You can use this same process for any of the benchmark angles (or multiples of these angles) using either a 30-60-90 triangle or a 45-45-90 triangle.

The legs of the triangle also give the coordinates of the point where the radius intersects the circle: $(x, y) = (\cos A, \sin A)$. Knowing the signs of the x- and y-values in each of the quadrants of the coordinate plane gives the signs of the trig values. For example, for any angle that lands in quadrant II, the cosine will be negative (because the x-coordinate of any point in QII is negative) and the sine will be positive (because the y-coordinate of any point in QII is positive).

PRACTICE SET

Easy

1. Davis drew a unit circle and labeled the cosine and sine of 45° as $\left(\frac{\sqrt{2}}{2}, \frac{\sqrt{2}}{2}\right)$. Assuming that Davis is correct, which of the following statements must be true?

A) $\cos \frac{\pi}{4} = \frac{\sqrt{2}}{2}$

B) $\cos \frac{\pi}{3} = \frac{\sqrt{2}}{2}$

C) $\cos \frac{\pi}{2} = \frac{\sqrt{2}}{2}$

D) $\cos \pi = \frac{\sqrt{2}}{2}$

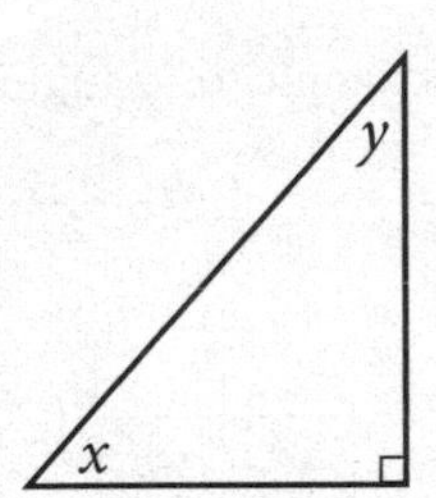

Note: Figure not drawn to scale.

2. In the right triangle above, $\cos x = 0.8$. What is the value of $\cos y$?

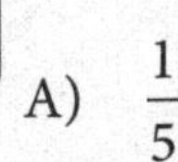

A) $\frac{1}{5}$

B) $\frac{3}{5}$

C) $\frac{3}{4}$

D) $\frac{4}{5}$

3. Right triangle ABC has side lengths 7, 24, and 25. If $\angle B$ is the second-largest interior angle of the triangle, what is the cosine of $\angle B$?

A) $\frac{7}{25}$

B) $\frac{7}{24}$

C) $\frac{24}{25}$

D) $\frac{25}{24}$

Medium

4. If the longer leg of a right triangle has length 32 centimeters, and the measure of the angle that is adjacent to that leg is 30°, which of the following represents the length, in centimeters, of the hypotenuse of the triangle?

A) $32 \times \sin 30°$

B) $32 \times \cos 30°$

C) $\frac{32}{\sin 30°}$

D) $\frac{32}{\cos 30°}$

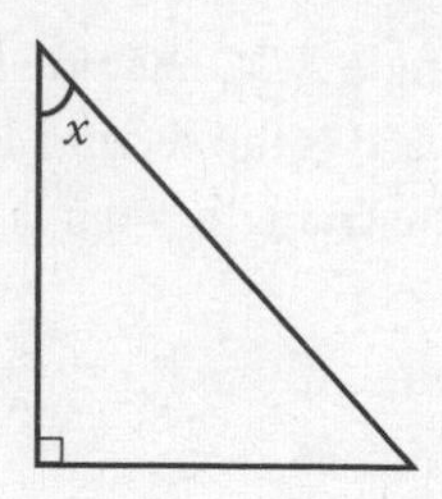

5. In the triangle above, $\tan x = \frac{\sqrt{7}}{3}$. What is $\cos x$?

A) $\frac{\sqrt{7}}{4}$

B) $\frac{\sqrt{2}}{3}$

C) $\frac{3\sqrt{2}}{2}$

D) $\frac{3}{4}$

6. If $\frac{\cos t}{\sin t} = \frac{1}{3}$, then what is $\tan t$?

A) $\frac{1}{3}$

B) $\frac{\sqrt{3}}{3}$

C) $\sqrt{3}$

D) 3

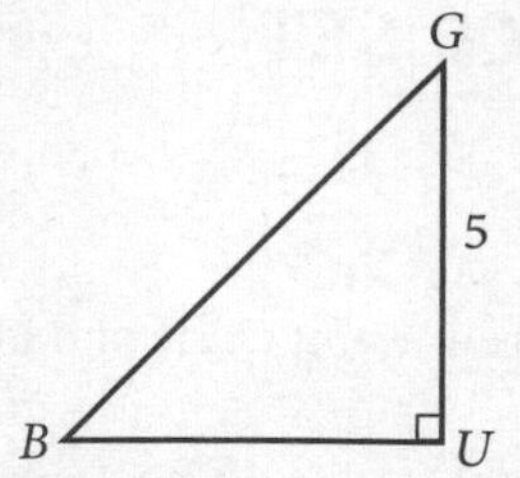

Note: Figure not drawn to scale.

7. Triangle BUG shown is an isosceles right triangle. If the length of side UG is 5 units, what is the sine of $\angle G$?

A) $\frac{\sqrt{5}}{5}$

B) $\frac{\sqrt{2}}{2}$

C) $\frac{\sqrt{3}}{2}$

D) 1

8. In a right triangle, one angle measures $x°$, where $\sin x° = \frac{5}{13}$. What is $\cos(90° - x°)$?

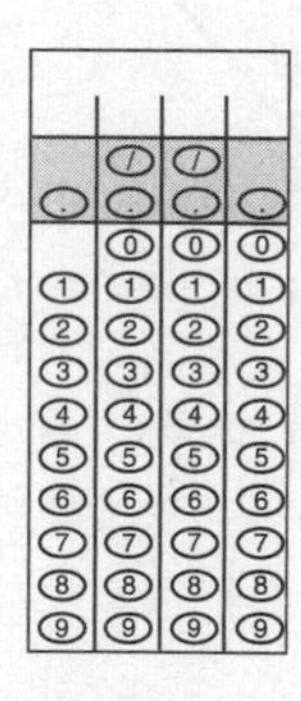

9. The angles of a triangle are in the ratio 1:2:3. What is the sine of the smallest angle?

Hard

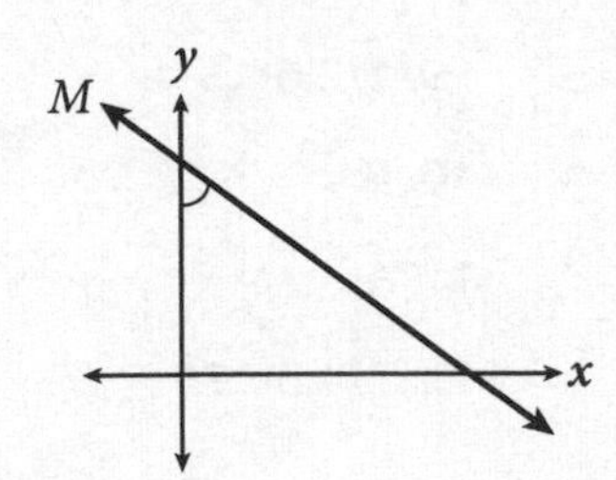

Note: Figure not drawn to scale.

10. The equation of line M shown above is $y = -\frac{3}{4}x + 5$. Given that angle A is the acute angle formed by the intersection of line M and the y-axis, which expression could be used to find the measure of angle A?

A) $\cos A = \frac{3}{4}$

B) $\sin A = \frac{4}{3}$

C) $\tan A = \frac{4}{3}$

D) $\cos A = \frac{4}{5}$

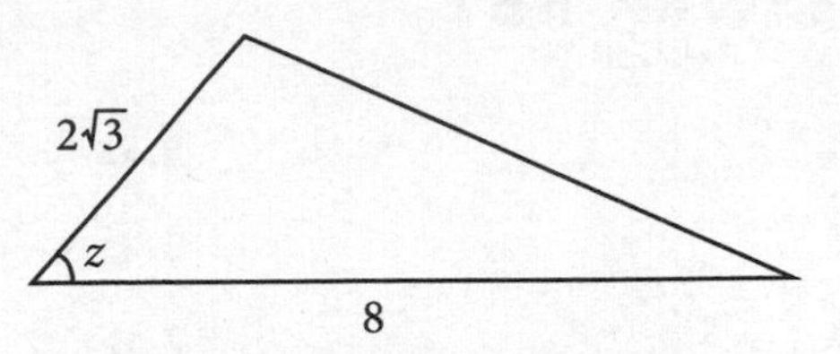

Note: Figure not drawn to scale.

11. If the area of the triangle shown above is 12 square inches, what is the value of $\cos z$?

A) $\frac{1}{2}$

B) $\frac{\sqrt{2}}{2}$

C) $\frac{\sqrt{3}}{2}$

D) 1

12. In triangle XYZ (not shown), the measure of $\angle Y$ is 90°, $YZ = 12$, and $XZ = 15$. Triangle HJK is similar to triangle XYZ, where vertices H, J, and K correspond to vertices X, Y, and Z, respectively, and each side of triangle HJK is $\frac{1}{5}$ the length of the corresponding side of triangle XYZ. What is the value of $\tan K$?

ANSWER KEY

1. A	**5. D**	**9. 1/2** or **.5**
2. B	**6. D**	**10. C**
3. A	**7. B**	**11. A**
4. D	**8. 5/13**	**12. 3/4** or **.75**

ANSWERS & EXPLANATIONS

1. A
Difficulty: Easy

Category: Additional Topics in Math / Trigonometry

Strategic Advice: Skim the answer choices to see what you're dealing with. All the values on the right-hand side of the equal sign are the same, so all you need to do is convert 45° to radians.

Getting to the Answer: Because $\pi = 180°$, converting between radians and degrees always involves using one of the two ratios, $\frac{180°}{\pi}$ or $\frac{\pi}{180°}$. Use the ratio that will cancel the unit you want to get rid of (degrees):

$45° = 45° \times \frac{\pi}{180°} = \frac{45\pi}{180} = \frac{\pi}{4}$. Therefore:

$\cos 45° = \cos \frac{\pi}{4} = \frac{\sqrt{2}}{2}$, which is (A).

You may not have recognized that $\frac{45\pi}{180} = \frac{\pi}{4}$. If not, divide easy numbers (such as 5 or 9) out of the numerator and denominator until you reach one of the answer choices: $\frac{45\pi \div 9}{180 \div 9} = \frac{5\pi}{20} = \frac{\pi}{4}$.

2. B
Difficulty: Easy

Category: Additional Topics in Math / Trigonometry

Strategic Advice: Whenever trig values are given as decimal numbers, convert them to fractions so you can use the ratios given by SOH CAH TOA.

Getting to the Answer: The decimal 0.8 equals $\frac{8}{10} = \frac{4}{5}$, which should make you think of a 3:4:5 triangle. Indeed, that's exactly what this triangle is. Draw a quick sketch and use SOH CAH TOA to place the side lengths in the correct places, ensuring that $\cos x = \frac{\text{adj}}{\text{hyp}} = \frac{4}{5}$. That is, label the side adjacent to angle x as 4 and the hypotenuse 5. The remaining side must have length 3.

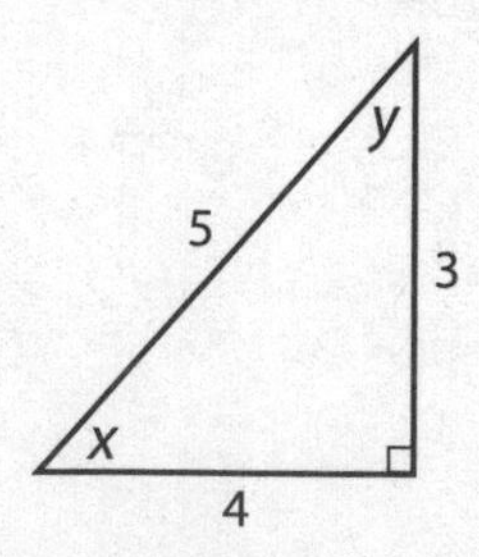

Again by SOH CAH TOA (and looking at angle y this time), $\cos y = \frac{\text{adj}}{\text{hyp}} = \frac{3}{5}$, choice (B).

3. A
Difficulty: Easy

Category: Additional Topics in Math / Trigonometry

Strategic Advice: The second-largest angle of a triangle must be opposite the second-longest side. Knowing this allows you to draw a sketch and use SOH CAH TOA to find the cosine of the indicated angle.

Getting to the Answer: Draw a quick sketch of a right triangle with one leg that is considerably longer than the other. (The hypotenuse of a right triangle is always the longest side.) The sketch should help you see that $\angle B$ is across from the side labeled 24:

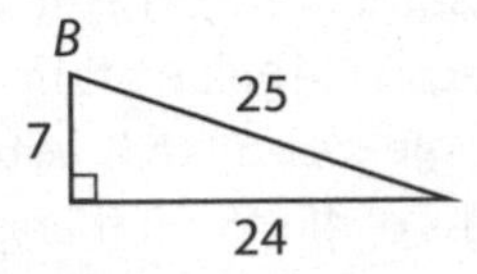

By SOH CAH TOA, $\cos \angle B = \frac{\text{adj}}{\text{hyp}} = \frac{7}{25}$, which is (A).

4. D
Difficulty: Medium

Category: Additional Topics in Math / Trigonometry

Strategic Advice: Take a peek at the answer choices—they involve trig functions, so set up a proportion using SOH CAH TOA.

Getting to the Answer: Draw a quick sketch of a right triangle. Label the longer leg 32 and the angle adjacent to it 30°:

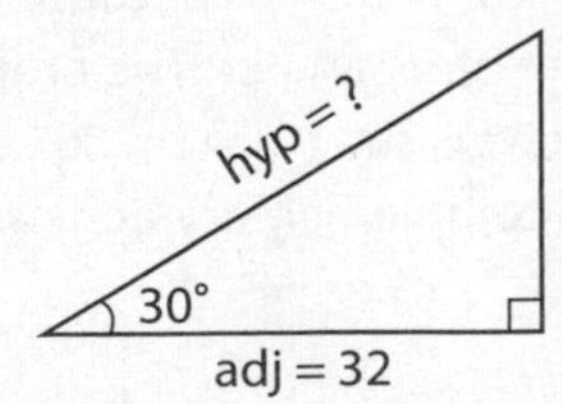

Notice that you have the adjacent side (32) and you're looking for the hypotenuse, so use cosine:

$$\cos x = \frac{\text{adj}}{\text{hyp}}$$

$$\cos 30° = \frac{32}{\text{hyp}}$$

$$\text{hyp} \times \cos 30° = 32$$

$$\text{hyp} = \frac{32}{\cos 30°}$$

This matches (D).

5. D
Difficulty: Medium

Category: Additional Topics in Math / Trigonometry

Strategic Advice: Mark up the triangle with what you know. Then use SOH CAH TOA: $\tan x = \frac{\text{opp}}{\text{adj}}$ and $\cos x = \frac{\text{adj}}{\text{hyp}}$.

Getting to the Answer: You're given that $\tan x = \frac{\sqrt{7}}{3}$, so label the side that is opposite angle x as $\sqrt{7}$, and label the side that is adjacent to angle x (the leg that touches it) 3:

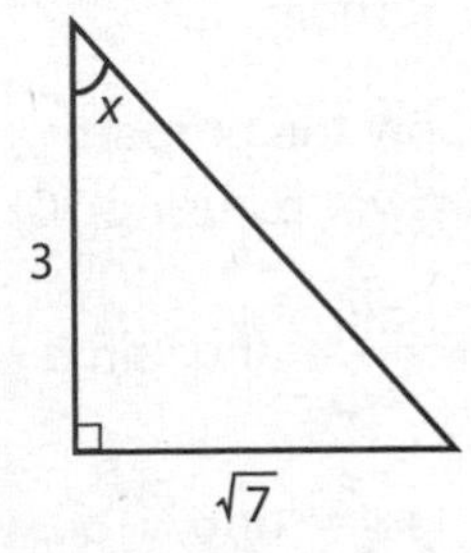

Next, use the Pythagorean theorem to find the length of the hypotenuse:

$$c^2 = a^2 + b^2$$
$$c^2 = (3)^2 + (\sqrt{7})^2$$
$$c^2 = 9 + 7$$
$$c^2 = 16$$
$$c = 4$$

Finally, use the definition of cosine: $\cos x = \frac{\text{adj}}{\text{hyp}} = \frac{3}{4}$, which is (D).

6. D

Difficulty: Medium

Category: Additional Topics in Math / Trigonometry

Strategic Advice: A useful trig property to know is that $\tan a = \frac{\sin a}{\cos a}$.

Getting to the Answer: Given the ratio $\tan a = \frac{\sin a}{\cos a}$, it follows that taking the reciprocal results in $\frac{\cos a}{\sin a} = \frac{1}{\tan a}$. Thus, if $\frac{\cos a}{\sin a} = \frac{1}{3}$, then $\frac{1}{\tan a} = \frac{1}{3}$. Again, taking the reciprocal yields $\tan a = 3$, choice (D).

If you didn't know the property, you could also arrive at the answer by using SOH CAH TOA:

$$\cos a = \frac{\text{adj}}{\text{hyp}} \text{ and } \sin a = \frac{\text{opp}}{\text{hyp}}$$

$$\frac{\cos a}{\sin a} = \frac{\frac{\text{adj}}{\text{hyp}}}{\frac{\text{adj}}{\text{hyp}}} = \frac{\text{adj}}{\text{hyp}} \div \frac{\text{opp}}{\text{hyp}} = \frac{\text{adj}}{\text{hyp}} \times \frac{\text{hyp}}{\text{opp}} = \frac{\text{adj}}{\text{opp}}$$

Because $\tan a = \frac{\text{opp}}{\text{adj}}$, it follows from the above that $\tan a$ is the reciprocal of $\frac{\cos a}{\sin a}$.

7. B

Difficulty: Medium

Category: Additional Topics in Math / Trigonometry

Strategic Advice: When a question doesn't appear to have enough information to answer it, look for clues in definitions of math terms.

Getting to the Answer: An isosceles right triangle is a right triangle that has two legs of equal length, making it a 45-45-90 triangle. This means $\angle G$ is a 45° angle, and the sine of a 45° angle is $\frac{\sqrt{2}}{2}$, which is (B). You should consider memorizing the trig values for common angles (such as 30°, 45°, 60°, 90°, 180°, and 360°) to save yourself some time on Test Day. However, if you didn't recall the value of sin 45°, you could label the triangle using what you know about the side lengths of 45-45-90 triangles (that they are in the ratio $x:x:x\sqrt{2}$ or here, $5:5:5\sqrt{2}$), and then use SOH CAH TOA to find that:

$$\sin \angle G = \frac{\text{opp}}{\text{hyp}} = \frac{5}{5\sqrt{2}} = \frac{5 \times \sqrt{2}}{5\sqrt{2} \times \sqrt{2}} = \frac{\sqrt{2}}{2}$$

8. 5/13

Difficulty: Medium

Category: Additional Topics in Math / Trigonometry

Strategic Advice: This question is quick if you know the complementary angle relationship: $\sin x° = \cos(90° - x°)$. More simply stated, this means that the sine of an acute angle is equal to the cosine of the angle's complement. For example, sin 20° = cos 70° because 20° and 70° are complementary angles. The property works in reverse as well: cos 10° = sin 80° because 10° and 80° are complementary angles.

Getting to the Answer: By the relationship given above, because $\sin x° = \frac{5}{13}$, $\cos(90° - x°)$ must equal the same thing, so grid in 5/13.

Note: If you didn't know the property, you could also sketch a triangle and fill it in using SOH CAH TOA. Sine equals $\frac{\text{opp}}{\text{hyp}}$, so label one angle *x*, the side opposite that angle 5, and the hypotenuse 13. The third angle in the triangle must be $(90° - x°)$ because the angles must sum to 180°.

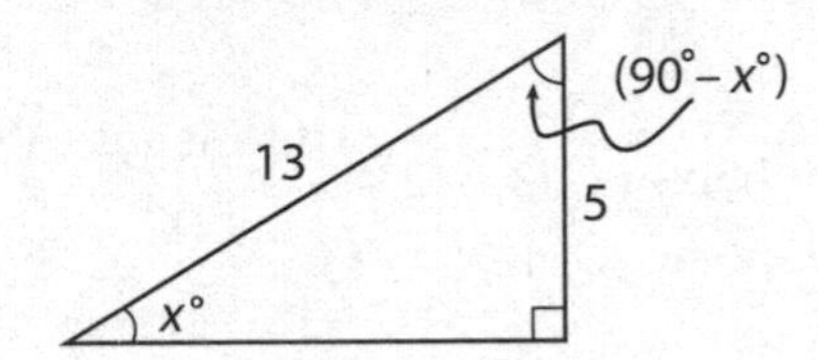

Using SOH CAH TOA again:

$$\cos(90° - x°) = \frac{\text{adj}}{\text{hyp}}$$

$$\cos(90° - x°) = \frac{5}{13}$$

9. 1/2 or **.5**
Difficulty: Medium

Category: Additional Topics in Math / Trigonometry

Strategic Advice: Use your knowledge of ratios to determine what kind of triangle you have. Once you know the measure of the angle being asked about, you can easily find the sine.

Getting to the Answer: You're told that the angles of the triangle are in the ratio 1:2:3. Call the smallest angle $x°$, the next smallest $(2x)°$, and the largest $(3x)°$. Because the angles of a triangle add to 180°:

$$x° + (2x)° + (3x)° = 180°$$
$$(6x)° = 180°$$
$$x° = 30°$$

The question asks about the smallest angle, which is $x° = 30°$. The sine of 30° is $\frac{1}{2}$ or 0.5.

Note: If you didn't already know that, you could draw a diagram of a 30:60:90 triangle and use SOH CAH TOA to deduce the sine:

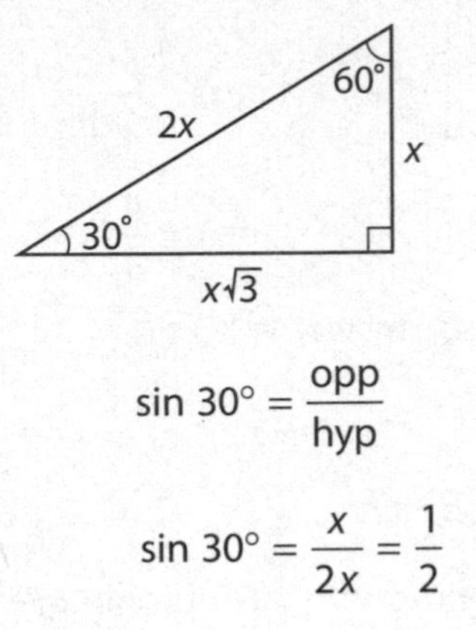

$$\sin 30° = \frac{\text{opp}}{\text{hyp}}$$

$$\sin 30° = \frac{x}{2x} = \frac{1}{2}$$

10. C
Difficulty: Hard

Category: Additional Topics in Math / Trigonometry

Strategic Advice: The answer choices are all trig expressions, so use SOH CAH TOA and your equation solving skills.

Getting to the Answer: To find an angle in a triangle using trig, you need to know the measure of another angle and a side length, or two side lengths. It doesn't appear that you have any of these here. However, you do know the slope of the line $(-\frac{3}{4})$, which you can use to sketch in a triangle that contains angle *A*. Your sketch might look like the following:

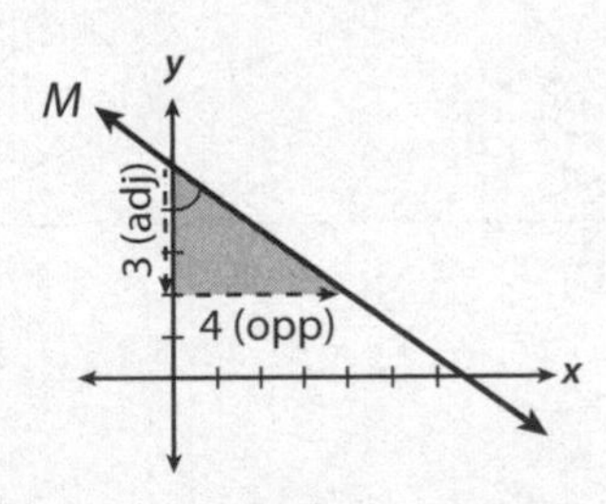

You now know the lengths of the sides that are opposite and adjacent to angle A. This means you can use tangent $\left(\frac{\text{opp}}{\text{adj}}\right)$ to represent the measure of the angle. (Don't worry about whether 3 is positive or negative—you are using the length of the side, which is positive.)

$$\tan A = \frac{\text{opp}}{\text{adj}}$$

$$\tan A = \frac{4}{3}$$

The correct answer is (C).

11. A
Difficulty: Hard

Category: Additional Topics in Math / Trigonometry

Strategic Advice: Sometimes, it is necessary to draw in the height of a triangle to get an idea of which side lengths you need. Use the information given about the area of the triangle to find its height.

Getting to the Answer: Find the height of the triangle and add it to the figure:

$$A = \frac{1}{2}bh$$

$$12 = \frac{1}{2}(8)h$$

$$12 = 4h$$

$$3 = h$$

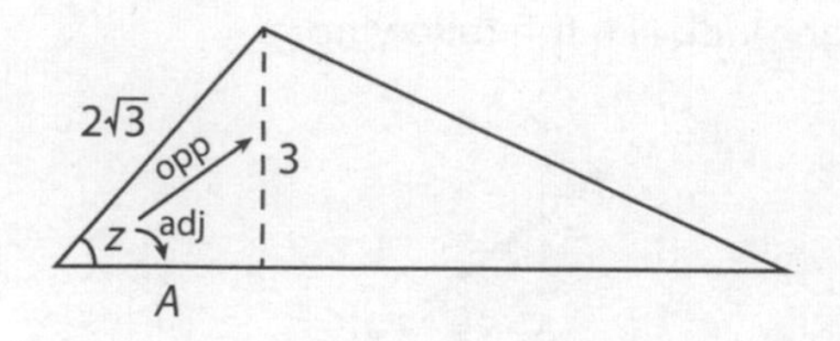

Next, use the Pythagorean theorem to find the missing side of the right triangle that contains z:

$$3^2 + A^2 = (2\sqrt{3})^2$$
$$9 + A^2 = (2\sqrt{3})(2\sqrt{3})$$
$$9 + A^2 = 4 \times 3$$
$$9 + A^2 = 12$$
$$A^2 = 3$$
$$A = \sqrt{3}$$

Finally, use SOH CAH TOA:

$\cos z = \frac{\text{adj}}{\text{hyp}} = \frac{\sqrt{3}}{2\sqrt{3}} = \frac{1}{2}$. This means (A) is correct.

12. 3/4 or **.75**
Difficulty: Hard

Category: Additional Topics in Math / Trigonometry

Strategic Advice: Some questions are very conceptual. Focus on what you know about trig functions, and the answer will come with a little reasoning and the use of SOH CAH TOA.

Getting to the Answer: Trigonometric functions are ratios. This means that it doesn't matter whether x is an angle in a huge triangle or in a small one; tan x will always be the same. In this question, triangles XYZ and HJK are similar, which means that their corresponding angles are equal. The question asks for the tangent of K, which is the same as the tangent of Z (since Z and K are corresponding and thus congruent). Don't bother drawing HJK and finding $\frac{1}{5}$ of all the side lengths; this is unnecessary. Just draw XYZ and find tan Z:

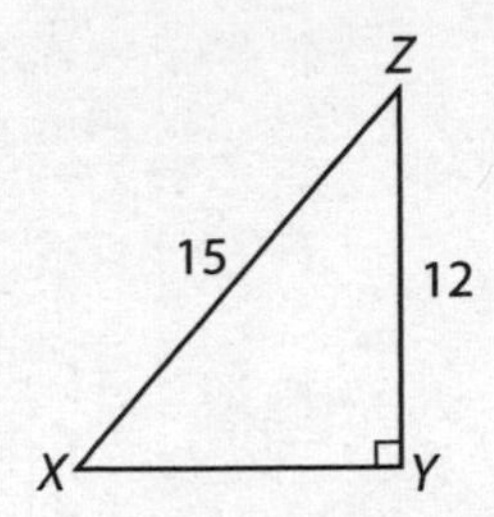

This is a multiple of a 3:4:5 triangle, specifically a 9:12:15 triangle (which makes the missing side 9), so $\tan Z = \frac{\text{opp}}{\text{adj}} = \frac{9}{12} = \frac{3}{4} = 0.75$.

SECTION THREE

Practice Test

HOW TO SCORE YOUR PRACTICE TEST

For this Practice Test, convert your raw score, or the number of questions you answered correctly, to a scaled score using the table below. To get your raw score for Math, add the number of questions you answered correctly for the Math—No Calculator and Math—Calculator sections.

Math			
Raw Score	Scaled Score	Raw Score	Scaled Score
0	200	30	530
1	200	31	540
2	210	32	550
3	230	33	560
4	240	34	560
5	260	35	570
6	280	36	580
7	290	37	590
8	310	38	600
9	320	39	600
10	330	40	610
11	340	41	620
12	360	42	630
13	370	43	640
14	380	44	650
15	390	45	660
16	410	46	670
17	420	47	670
18	430	48	680
19	440	49	690
20	450	50	700
21	460	51	710
22	470	52	730
23	480	53	740
24	480	54	750
25	490	55	760
26	500	56	780
27	510	57	790
28	520	58	800
29	520		

SAT PRACTICE TEST ANSWER SHEET (MATH—SECTIONS 3 AND 4 ONLY)

Remove (or photocopy) this answer sheet and use it to complete the test. See the answer key following the test when finished.

Start with number 1 for each section. If a section has fewer questions than answer spaces, leave the extra spaces blank.

SECTION

3

1. Ⓐ Ⓑ Ⓒ Ⓓ
2. Ⓐ Ⓑ Ⓒ Ⓓ
3. Ⓐ Ⓑ Ⓒ Ⓓ
4. Ⓐ Ⓑ Ⓒ Ⓓ
5. Ⓐ Ⓑ Ⓒ Ⓓ
6. Ⓐ Ⓑ Ⓒ Ⓓ
7. Ⓐ Ⓑ Ⓒ Ⓓ
8. Ⓐ Ⓑ Ⓒ Ⓓ
9. Ⓐ Ⓑ Ⓒ Ⓓ
10. Ⓐ Ⓑ Ⓒ Ⓓ
11. Ⓐ Ⓑ Ⓒ Ⓓ
12. Ⓐ Ⓑ Ⓒ Ⓓ
13. Ⓐ Ⓑ Ⓒ Ⓓ
14. Ⓐ Ⓑ Ⓒ Ⓓ
15. Ⓐ Ⓑ Ⓒ Ⓓ

16.

	/	/	
.	.	.	.
	0	0	0
1	1	1	1
2	2	2	2
3	3	3	3
4	4	4	4
5	5	5	5
6	6	6	6
7	7	7	7
8	8	8	8
9	9	9	9

17.

	/	/	
.	.	.	.
	0	0	0
1	1	1	1
2	2	2	2
3	3	3	3
4	4	4	4
5	5	5	5
6	6	6	6
7	7	7	7
8	8	8	8
9	9	9	9

18.

	/	/	
.	.	.	.
	0	0	0
1	1	1	1
2	2	2	2
3	3	3	3
4	4	4	4
5	5	5	5
6	6	6	6
7	7	7	7
8	8	8	8
9	9	9	9

19.

	/	/	
.	.	.	.
	0	0	0
1	1	1	1
2	2	2	2
3	3	3	3
4	4	4	4
5	5	5	5
6	6	6	6
7	7	7	7
8	8	8	8
9	9	9	9

20.

	/	/	
.	.	.	.
	0	0	0
1	1	1	1
2	2	2	2
3	3	3	3
4	4	4	4
5	5	5	5
6	6	6	6
7	7	7	7
8	8	8	8
9	9	9	9

right in Section 3

wrong in Section 3

SECTION 4

1. Ⓐ Ⓑ Ⓒ Ⓓ
2. Ⓐ Ⓑ Ⓒ Ⓓ
3. Ⓐ Ⓑ Ⓒ Ⓓ
4. Ⓐ Ⓑ Ⓒ Ⓓ
5. Ⓐ Ⓑ Ⓒ Ⓓ
6. Ⓐ Ⓑ Ⓒ Ⓓ
7. Ⓐ Ⓑ Ⓒ Ⓓ
8. Ⓐ Ⓑ Ⓒ Ⓓ
9. Ⓐ Ⓑ Ⓒ Ⓓ
10. Ⓐ Ⓑ Ⓒ Ⓓ
11. Ⓐ Ⓑ Ⓒ Ⓓ
12. Ⓐ Ⓑ Ⓒ Ⓓ
13. Ⓐ Ⓑ Ⓒ Ⓓ
14. Ⓐ Ⓑ Ⓒ Ⓓ
15. Ⓐ Ⓑ Ⓒ Ⓓ
16. Ⓐ Ⓑ Ⓒ Ⓓ
17. Ⓐ Ⓑ Ⓒ Ⓓ
18. Ⓐ Ⓑ Ⓒ Ⓓ
19. Ⓐ Ⓑ Ⓒ Ⓓ
20. Ⓐ Ⓑ Ⓒ Ⓓ
21. Ⓐ Ⓑ Ⓒ Ⓓ
22. Ⓐ Ⓑ Ⓒ Ⓓ
23. Ⓐ Ⓑ Ⓒ Ⓓ
24. Ⓐ Ⓑ Ⓒ Ⓓ
25. Ⓐ Ⓑ Ⓒ Ⓓ
26. Ⓐ Ⓑ Ⓒ Ⓓ
27. Ⓐ Ⓑ Ⓒ Ⓓ
28. Ⓐ Ⓑ Ⓒ Ⓓ
29. Ⓐ Ⓑ Ⓒ Ⓓ
30. Ⓐ Ⓑ Ⓒ Ⓓ

right in Section 4

wrong in Section 4

31.

32.

33.

34.

35.

36.

37.

38.

MATH TEST

25 Minutes—20 Questions

NO-CALCULATOR SECTION

Turn to Section 3 of your answer sheet to answer the questions in this section.

Directions: For this section, solve each problem and decide which is the best of the choices given. Fill in the corresponding oval on the answer sheet. You may use any available space for scratch work.

Notes:

1. Calculator use is NOT permitted.
2. All numbers used are real numbers.
3. All figures used are necessary to solving the problems that they accompany. All figures are drawn to scale EXCEPT when it is stated that a specific figure is not drawn to scale.
4. Unless stated otherwise, the domain of any function f is assumed to be the set of all real numbers x, for which $f(x)$ is a real number.

Information:

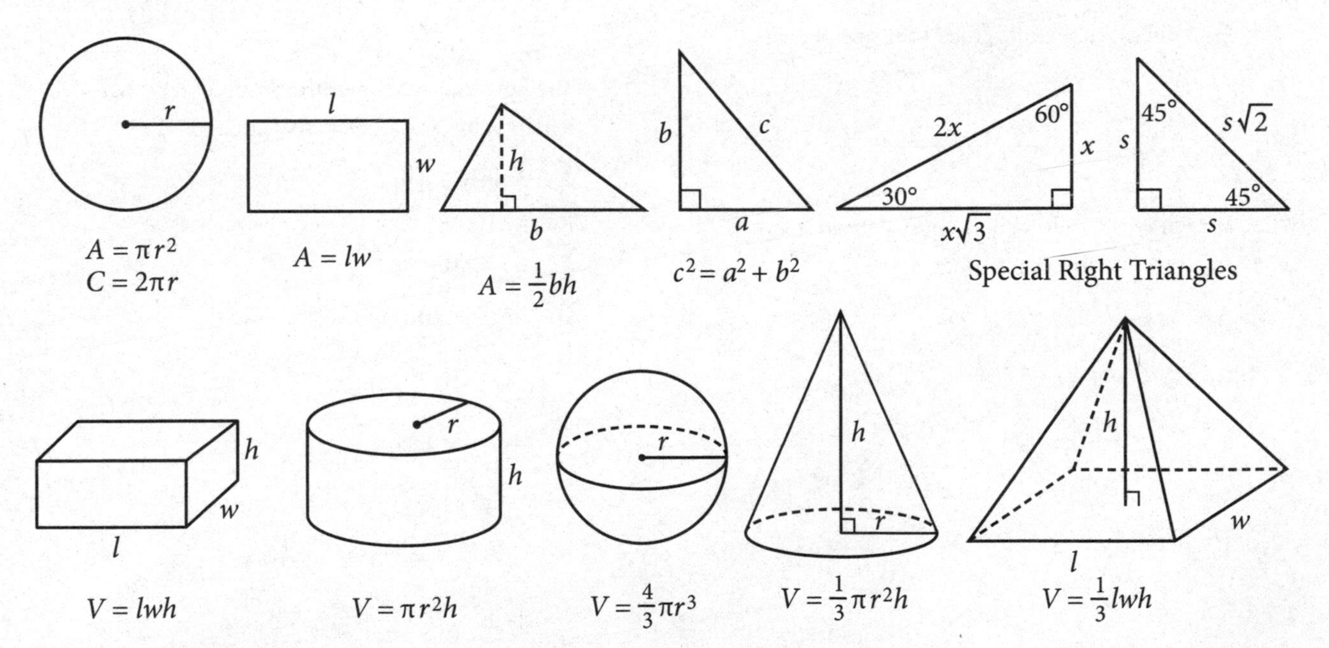

The sum of the degree measures of the angles in a triangle is 180.

The number of degrees of arc in a circle is 360.

The number of radians of arc in a circle is 2π.

GO ON TO THE NEXT PAGE

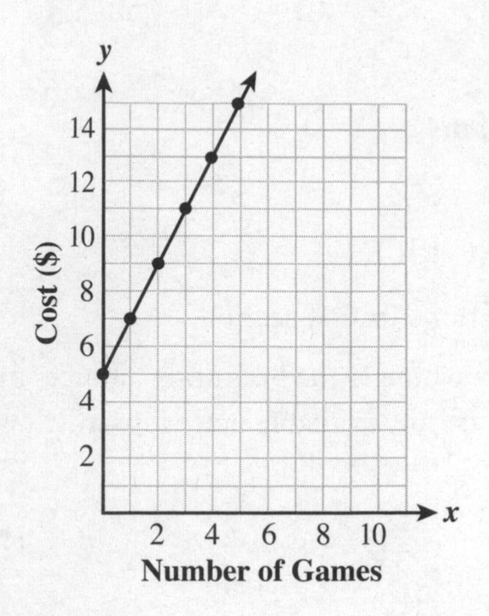

1. The graph above shows the amount that a new, high-tech video arcade charges its customers. What could the *y*-intercept of this graph represent?

A) The cost of playing 5 games

B) The cost per game, which is $5

C) The entrance fee to enter the arcade

D) The number of games that are played

$$\frac{3x}{x+5} \div \frac{6}{4x+20}$$

2. Which of the following is equivalent to the expression above, given that $x \neq -5$?

A) $2x$

B) $\frac{x}{2}$

C) $\frac{9x}{2}$

D) $2x + 4$

$$(x+3)^2 + (y+1)^2 = 25$$

3. The graph of the equation above is a circle. What is the area, in square units, of the circle?

A) 4π

B) 5π

C) 16π

D) 25π

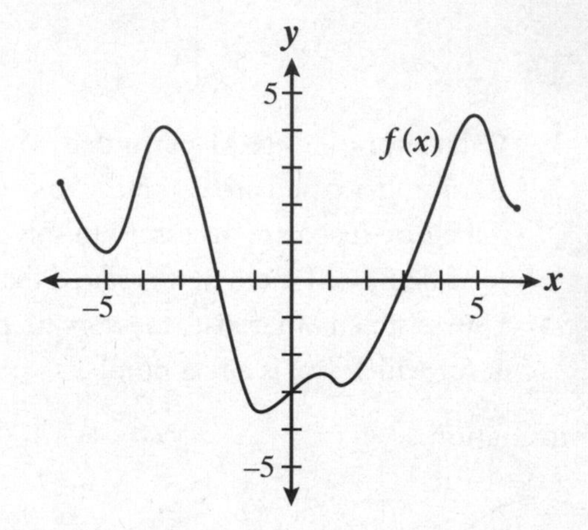

4. The figure above shows the graph of $f(x)$. For which value(s) of x does $f(x)$ equal 0?

A) 3 only

B) –3 only

C) –2 and 3

D) –3, –2, and 3

GO ON TO THE NEXT PAGE

$$\frac{4(d+3)-9}{8}=\frac{10-(2-d)}{6}$$

5. In the equation above, what is the value of d?

A) $\frac{23}{16}$

B) $\frac{23}{8}$

C) $\frac{25}{8}$

D) $\frac{25}{4}$

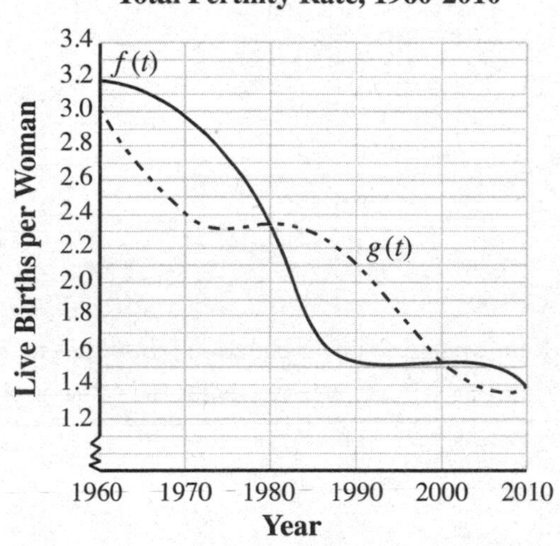

Source: Data from Eurostat.

6. One indicator of a declining economy is a continued decline in birth rates. In 2010, birth rates in Europe were at an all-time low, with the average number of children that a woman has in her lifetime at well below two. In the figure above, $f(t)$ represents birth rates for Portugal between 1960 and 2010, and $g(t)$ represents birth rates in Slovakia for the same time period. For which value(s) of t is $f(t) > g(t)$?

A) $1960 < t < 1980$ only

B) $1980 < t < 2000$ only

C) $1960 < t < 1980$ and $1990 < t < 2000$

D) $1960 < t < 1980$ and $2000 < t < 2010$

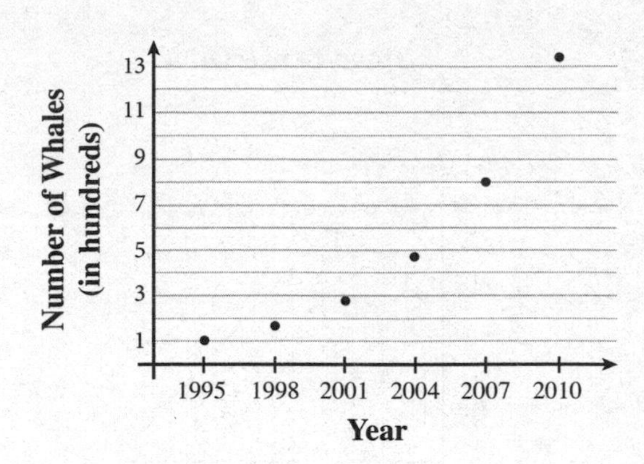

7. The blue whale is the largest creature in the world and has been found in every ocean in the world. A marine biologist surveyed the blue whale population in Monterey Bay, off the coast of California, every three years between 1995 and 2010. The figure above shows her results. If w is the number of blue whales present in Monterey Bay and t is the number of years since the study began in 1995, which of the following equations best represents the blue whale population of Monterey Bay?

A) $w = 100 + 2t$

B) $w=100+\frac{t^2}{4}$

C) $w=100\times 2^t$

D) $w=100\times 2^{\frac{t}{4}}$

GO ON TO THE NEXT PAGE

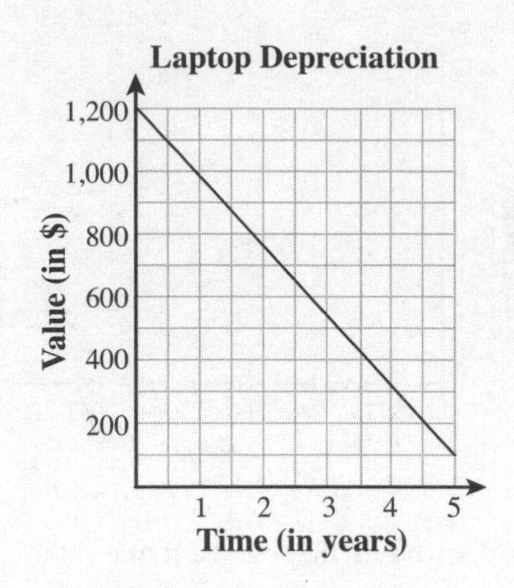

8. The figure above shows the straight-line depreciation of a laptop computer over the first five years of its use. According to the figure, what is the average rate of change in dollars per year of the value of the computer over the five-year period?

A) −1,100

B) −220

C) −100

D) 100

9. What is the coefficient of x^2 when $6x^2 - \frac{2}{5}x + 1$ is multiplied by $10x + \frac{1}{3}$?

A) −4

B) −2

C) 2

D) 4

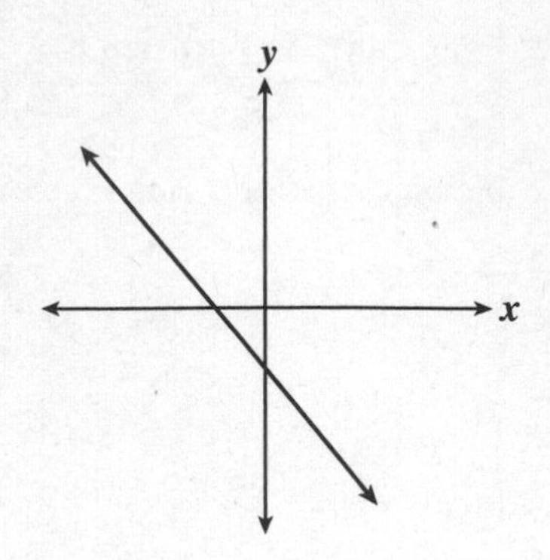

10. The graph above could represent which of the following equations?

A) $-6x - 4y = 5$

B) $-6x - 4y = -5$

C) $-6x + 4y = 5$

D) $-6x + 4y = -5$

$$\begin{cases} \frac{3}{4}x - \frac{1}{2}y = 12 \\ kx - 2y = 22 \end{cases}$$

11. If the system of linear equations above has no solution, and k is a constant, what is the value of k?

A) $-\frac{4}{3}$

B) $-\frac{3}{4}$

C) 3

D) 4

GO ON TO THE NEXT PAGE

12. In Delray Beach, Florida, you can take a luxury golf cart ride around downtown. The driver charges \$4 for the first $\frac{1}{4}$ mile, plus \$1.50 for each additional $\frac{1}{2}$ mile. Which inequality represents the number of miles, m, that you could ride and pay no more than \$10?

A) $3.25 + 1.5m \leq 10$

B) $3.25 + 3m \leq 10$

C) $4 + 1.5m \leq 10$

D) $4 + 3m \leq 10$

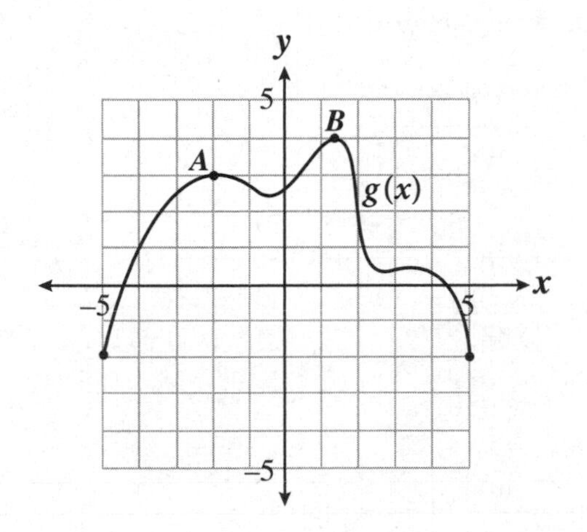

13. The graph of $g(x)$ is shown in the figure above. If $h(x) = -g(x) + 1$, which of the following statements is true?

A) The range of $h(x)$ is $-3 \leq y \leq 3$.

B) The minimum value of $h(x)$ is -4.

C) The coordinates of point A on the function $h(x)$ are $(2, 4)$.

D) The graph of $h(x)$ is increasing between $x = -5$ and $x = -2$.

14. If $a + bi$ represents the complex number that results from multiplying $3 + 2i$ times $5 - i$, what is the value of a?

A) 2

B) 13

C) 15

D) 17

$$\frac{1}{x} + \frac{4}{x} = \frac{1}{72}$$

15. In order to create safe drinking water, cities and towns use water treatment facilities to remove contaminants from surface water and groundwater. Suppose a town has a treatment plant but decides to build a second, more efficient facility. The new treatment plant can filter the water in the reservoir four times as quickly as the older facility. Working together, the two facilities can filter all the water in the reservoir in 72 hours. The equation above represents the scenario. Which of the following describes what the term $\frac{1}{x}$ represents?

A) The portion of the water the older treatment plant can filter in 1 hour

B) The time it takes the older treatment plant to filter the water in the reservoir

C) The time it takes the older treatment plant to filter $\frac{1}{72}$ of the water in the reservoir

D) The portion of the water the new treatment plant can filter in 4 hours

GO ON TO THE NEXT PAGE

Directions: For questions 16-20, solve the problem and enter your answer in the grid, as described below, on the answer sheet.

1. Although not required, it is suggested that you write your answer in the boxes at the top of the columns to help you fill in the circles accurately. You will receive credit only if the circles are filled in correctly.
2. Mark no more than one circle in any column.
3. No question has a negative answer.
4. Some problems may have more than one correct answer. In such cases, grid only one answer.
5. **Mixed numbers** such as $3\frac{1}{2}$ must be gridded as 3.5 or $\frac{7}{2}$.

   (If $3\frac{1}{2}$ is entered into the grid as 3 1 / 2, it will be interpreted as $\frac{31}{2}$, not $3\frac{1}{2}$.)
6. **Decimal answers:** If you obtain a decimal answer with more digits than the grid can accommodate, it may be either rounded or truncated, but it must fill the entire grid.

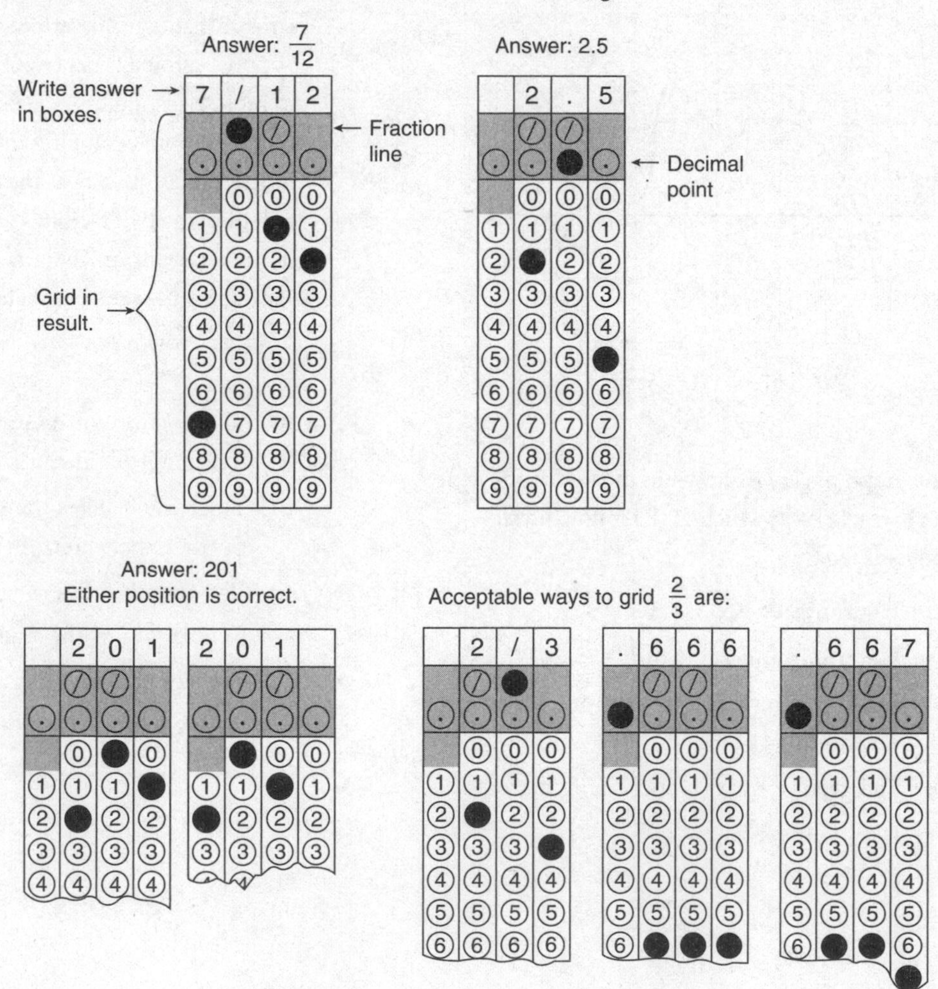

16. If $\frac{1}{4}x = 5 - \frac{1}{2}y$, what is the value of $x + 2y$?

$$\begin{cases} x + 3y \leq 18 \\ 2x - 3y \leq 9 \end{cases}$$

17. If (a, b) is a point in the solution region for the system of inequalities shown above and $a = 6$, what is the minimum possible value for b?

$$\frac{\sqrt{x} \cdot x^{\frac{5}{6}} \cdot x}{\sqrt[3]{x}}$$

18. If x^n is the simplified form of the expression above, what is the value of n?

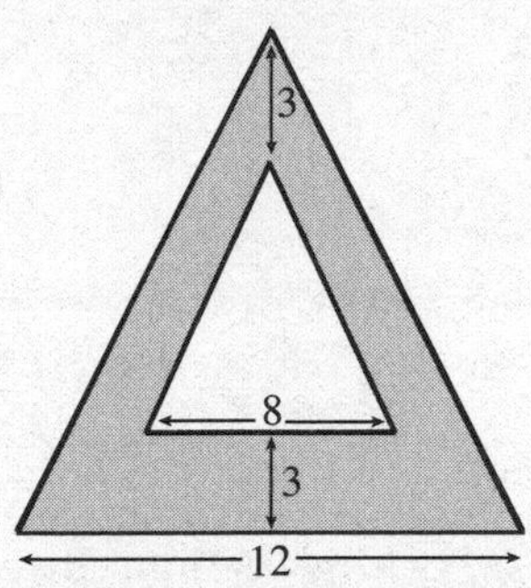

Note: Figure not drawn to scale.

19. In the figure above, the area of the shaded region is 52 square units. What is the height of the larger triangle?

20. If $y = ax^2 + bx + c$ passes through the points $(-3, 10)$, $(0, 1)$, and $(2, 15)$, what is the value of $a + b + c$?

IF YOU FINISH BEFORE TIME IS CALLED, YOU MAY CHECK YOUR WORK ON THIS SECTION ONLY. DO NOT TURN TO ANY OTHER SECTION IN THE TEST. **STOP**

MATH TEST

55 Minutes—38 Questions

CALCULATOR SECTION

Turn to Section 4 of your answer sheet to answer the questions in this section.

Directions: For this section, solve each problem and decide which is the best of the choices given. Fill in the corresponding oval on the answer sheet. You may use any available space for scratch work.

Notes:

1. Calculator use is permitted.
2. All numbers used are real numbers.
3. All figures used are necessary to solving the problems that they accompany. All figures are drawn to scale EXCEPT when it is stated that a specific figure is not drawn to scale.
4. Unless stated otherwise, the domain of any function *f* is assumed to be the set of all real numbers *x*, for which *f*(*x*) is a real number.

Information:

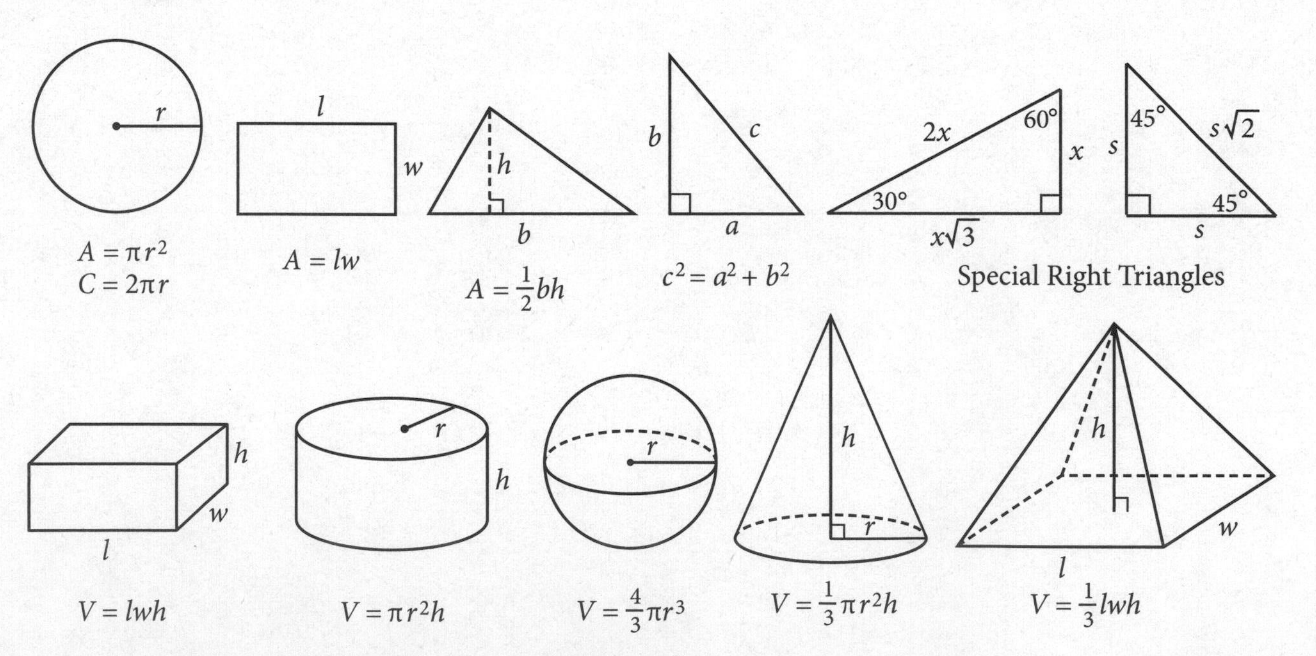

The sum of the degree measures of the angles in a triangle is 180.

The number of degrees of arc in a circle is 360.

The number of radians of arc in a circle is 2π.

GO ON TO THE NEXT PAGE

1. Oceans, seas, and bays represent about 96.5% of Earth's water, including the water found in our atmosphere. If the volume of the water contained in oceans, seas, and bays is about 321,000,000 cubic miles, which of the following best represents the approximate volume, in cubic miles, of all the world's water?

A) 308,160,000

B) 309,765,000

C) 332,642,000

D) 334,375,000

2. An electrician charges a one-time site visit fee to evaluate a potential job. If the electrician accepts the job, he charges an hourly rate plus the cost of any materials needed to complete the job. The electrician also charges for tax, but only on the cost of the materials. If the total cost of completing a job that takes h hours is given by the function $C(h) = 45h + 1.06(82.5) + 75$, then the term 1.06(82.5) represents

A) the hourly rate.

B) the site visit fee.

C) the cost of the materials, including tax.

D) the cost of the materials, not including tax.

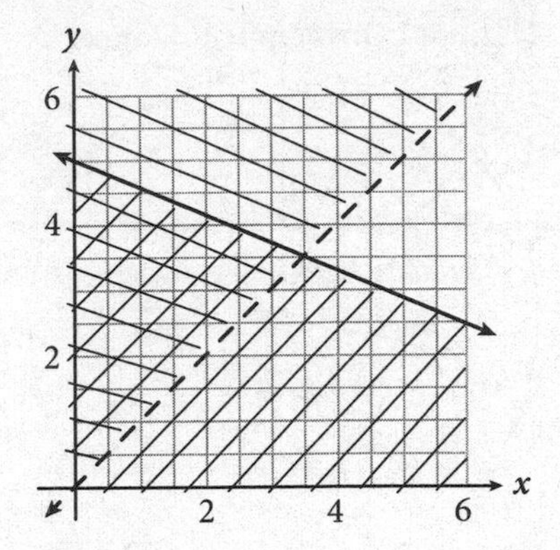

3. The figure above shows the solution set for the system $\begin{cases} y > x \\ y \leq -\frac{3}{7}x + 5 \end{cases}$. Which of the following is not a solution to the system?

A) (0, 3)

B) (1, 2)

C) (2, 4)

D) (3, 3)

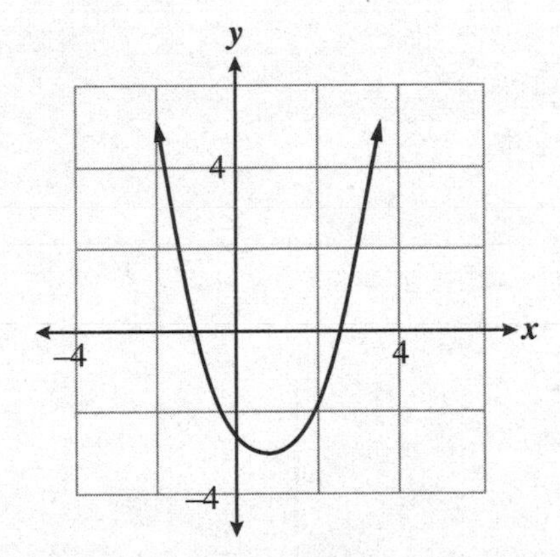

4. Each of the following quadratic equations represents the graph shown above. Which equation reveals the exact values of the x-intercepts of the graph?

A) $y = \frac{1}{2}(2x - 5)(x + 1)$

B) $y = x^2 - \frac{3}{2}x - \frac{5}{2}$

C) $y + \frac{49}{16} = \left(x - \frac{3}{4}\right)^2$

D) $y = \left(x - \frac{3}{4}\right)^2 - \frac{49}{16}$

GO ON TO THE NEXT PAGE

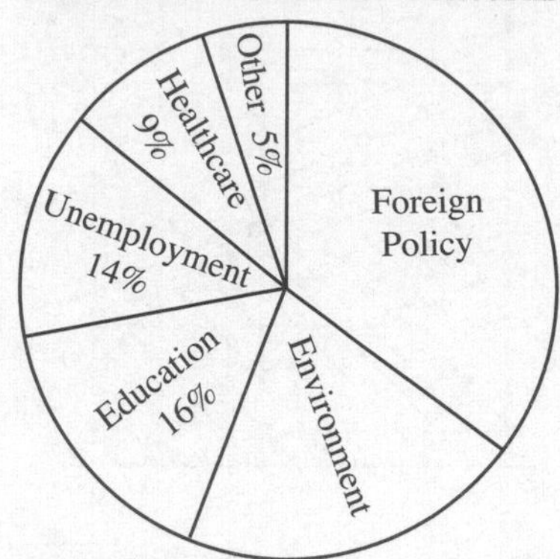

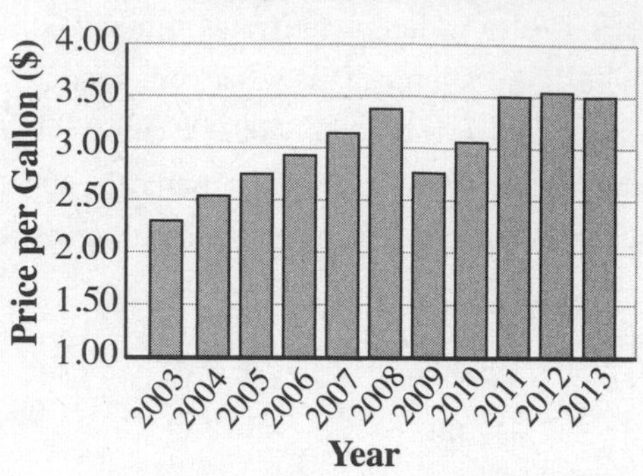

Data from U.S. Energy Information Administration.

5. Margo surveyed all the students in the government classes at her school to see what they thought should be the most important concern of a national government. The results of the survey are shown in the figure above. If the ratio of students who answered "Foreign Policy" to those who answered "Environment" was 5:3, what percentage of the students answered "Environment"?

A) 16%

B) 21%

C) 24%

D) 35%

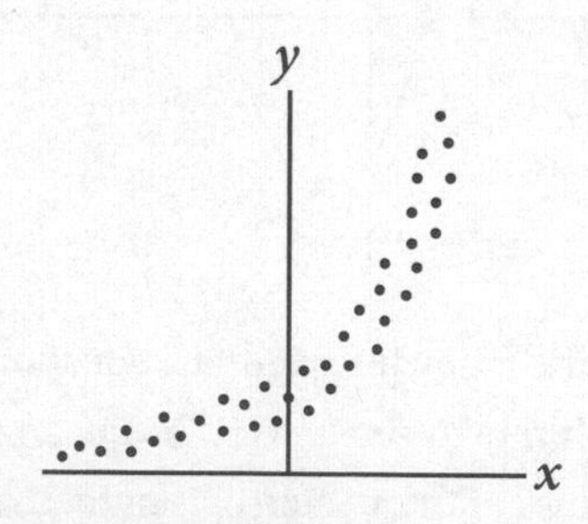

6. Which of the following best describes the type of association shown in the scatterplot above?

A) linear, positive

B) linear, negative

C) exponential, positive

D) exponential, negative

7. The figure above shows the average annual gas prices in the United States from 2003 to 2013. Based on the information shown, which of the following conclusions is valid?

A) A gallon of gas cost more in 2008 than in 2013.

B) The price more than doubled between 2003 and 2013.

C) The drop in price from 2008 to 2009 was more than $1.00 per gallon.

D) The overall change in price was greater between 2003 and 2008 than it was between 2008 and 2013.

$$\begin{cases} -2x+5y=1 \\ 7x-10y=-11 \end{cases}$$

8. If (x, y) is a solution to the system of equations above, what is the sum of x and y?

A) $-\frac{137}{30}$

B) -4

C) $-\frac{10}{3}$

D) -3

GO ON TO THE NEXT PAGE

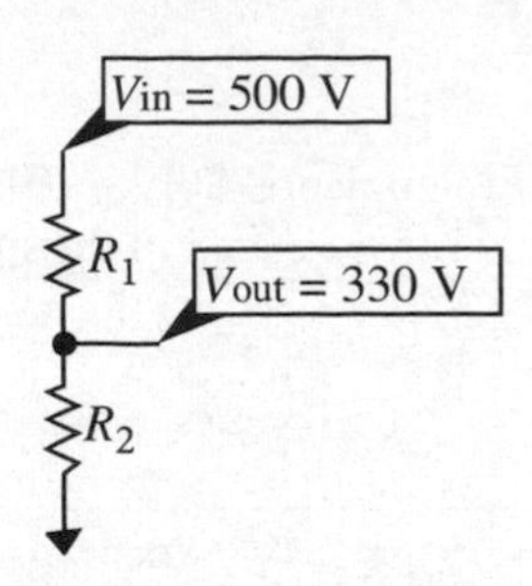

9. A voltage divider is a simple circuit that converts a large voltage into a smaller one. The figure above shows a voltage divider that consists of two resistors that together have a total resistance of 294 ohms. To produce the desired voltage of 330 volts, R_2 must be 6 ohms less than twice R_1. Solving which of the following systems of equations gives the individual resistances for R_1 and R_2?

A) $\begin{cases} R_2 = 2R_1 - 6 \\ R_1 + R_2 = 294 \end{cases}$

B) $\begin{cases} R_1 = 2R_2 + 6 \\ R_1 + R_2 = 294 \end{cases}$

C) $\begin{cases} R_2 = 2R_1 - 6 \\ R_1 + R_2 = \frac{294}{330} \end{cases}$

D) $\begin{cases} R_1 = 2R_2 + 6 \\ R_1 + R_2 = 330(294) \end{cases}$

10. If $\frac{2}{5}(5x)+2(x-1)=4(x+1)-2$, what is the value of x?

A) $x = -2$

B) $x = 2$

C) There is no value of x for which the equation is true.

D) There are infinitely many values of x for which the equation is true.

11. Crude oil is being transferred from a full rectangular storage container with dimensions 4 meters by 9 meters by 10 meters into a cylindrical transportation container that has a diameter of 6 meters. What is the minimum possible length for a transportation container that will hold all of the oil?

A) 40π

B) $\frac{40}{\pi}$

C) 60π

D) $\frac{120}{\pi}$

12. The percent increase from 5 to 12 is equal to the percent increase from 12 to what number?

A) 16.8

B) 19.0

C) 26.6

D) 28.8

$$b = \frac{L}{4\pi d^2}$$

13. The brightness of a celestial body, like a star, decreases as you move away from it. In contrast, the luminosity of a celestial body is a constant number that represents its intrinsic brightness. The inverse square law, shown above, is used to find the brightness, b, of a celestial body when you know its luminosity, L, and the distance, d, in meters to the body. Which equation shows the distance to a celestial body, given its brightness and luminosity?

A) $d = \frac{1}{2}\sqrt{\frac{L}{\pi b}}$

B) $d = \sqrt{\frac{L}{2\pi b}}$

C) $d = \frac{\sqrt{L}}{2\pi b}$

D) $d = \frac{L}{2\sqrt{\pi b}}$

GO ON TO THE NEXT PAGE

Questions 14 and 15 refer to the following information.

Each month, the Bureau of Labor Statistics conducts a survey called the Current Population Survey (CPS) to measure unemployment in the United States. Across the country, about 60,000 households are included in the survey sample. These households are grouped by geographic region. A summary of the January 2014 survey results for male respondents in one geographic region is shown in the table below.

Age Group	Employed	Unemployed	Not in the Labor Force	Total
16 to 19	8	5	10	23
20 to 24	26	7	23	56
25 to 34	142	11	28	157
35 to 44	144	8	32	164
45 to 54	66	6	26	98
Over 54	65	7	36	152
Total	451	44	155	650

14. According to the data in the table, for which age group did the smallest percentage of men report that they were unemployed in January 2014?

A) 20 to 24 years

B) 35 to 44 years

C) 45 to 54 years

D) Over 54 years

15. If one unemployed man from this sample is chosen at random for a follow-up survey, what is the probability that he will be between the ages of 45 and 54?

A) 6.0%

B) 13.6%

C) 15.1%

D) 44.9%

GO ON TO THE NEXT PAGE

16. Which of the following are solutions to the quadratic equation $(x-1)^2 = \frac{4}{9}$?

A) $x = -\frac{5}{3}, x = \frac{5}{3}$

B) $x = \frac{1}{3}, x = \frac{5}{3}$

C) $x = \frac{5}{9}, x = \frac{13}{9}$

D) $x = 1 \pm \sqrt{\frac{2}{3}}$

17. Damien is throwing darts. He has a total of 6 darts to throw. He gets 5 points for each dart that lands in a blue ring and 10 points for each dart that lands in a red ring. If x of his darts land in a blue ring and the rest land in a red ring, which expression represents his total score?

A) $10x$

B) $10x + 5$

C) $5x + 30$

D) $60 - 5x$

18. Red tide is a form of harmful algae that releases toxins as it breaks down in the environment. A marine biologist is testing a new spray, composed of clay and water, hoping to kill the red tide that almost completely covers a beach in southern Florida. He applies the spray to a representative sample of 200 square feet of the beach. By the end of the week, 184 square feet of the beach is free of the red tide. Based on these results, and assuming the same general conditions, how much of the 10,000-square-foot beach would still be covered by red tide if the spray had been used on the entire area?

A) 800 sq ft

B) 920 sq ft

C) 8,000 sq ft

D) 9,200 sq ft

$$\begin{cases} y = \frac{1}{2}x - 2 \\ y = -x^2 + 1 \end{cases}$$

19. If (a, b) is a solution to the system of equations above, which of the following could be the value of b?

A) −3

B) −2

C) 1

D) 2

20. Given the function $g(x) = \frac{2}{3}x + 7$, what domain value corresponds to a range value of 3?

A) −6

B) −2

C) 6

D) 9

21. A landscaper buys a new commercial-grade lawn mower that costs $2,800. Based on past experience, he expects it to last about 8 years, and then he can sell it for scrap metal with a salvage value of about $240. Assuming the value of the lawn mower depreciates at a constant rate, which equation could be used to find its approximate value after x years, given that $x < 8$?

A) $y = -8x + 2{,}560$

B) $y = -240x + 2{,}800$

C) $y = -320x + 2{,}800$

D) $y = 240x - 2{,}560$

GO ON TO THE NEXT PAGE

22. A microbiologist is studying the effects of a new antibiotic on a culture of 20,000 bacteria. When the antibiotic is added to the culture, the number of bacteria is reduced by half every hour. What kind of function best models the number of bacteria remaining in the culture after the antibiotic is added?

A) A linear function

B) A quadratic function

C) A polynomial function

D) An exponential function

23. An airline company purchased two new airplanes. One can travel at speeds of up to 600 miles per hour and the other at speeds of up to 720 miles per hour. How many more miles can the faster airplane travel in 12 seconds than the slower airplane?

A) $\frac{1}{30}$

B) $\frac{2}{5}$

C) 2

D) 30

State	Minimum Wage per Hour
Idaho	\$7.25
Montana	\$7.90
Oregon	\$9.10
Washington	\$9.32

24. The table above shows the 2014 minimum wages for several states that share a border. Assuming an average workweek of between 35 and 40 hours, which inequality represents how much more a worker who earns minimum wage can earn per week in Oregon than in Idaho?

A) $x \geq 1.85$

B) $7.25 \leq x \leq 9.10$

C) $64.75 \leq x \leq 74$

D) $253.75 \leq x \leq 364$

25. In the United States, the maintenance and construction of airports, transit systems, and major roads are largely funded through a federal excise tax on gasoline. Based on the 2011 statistics given below, how much did the average household pay per year in federal gasoline taxes?

- The federal gasoline tax rate was 18.4 cents per gallon.
- The average motor vehicle was driven approximately 11,340 miles per year.
- The national average fuel economy for noncommercial vehicles was 21.4 miles per gallon.
- The average American household owned 1.75 vehicles.

A) \$55.73

B) \$68.91

C) \$97.52

D) \$170.63

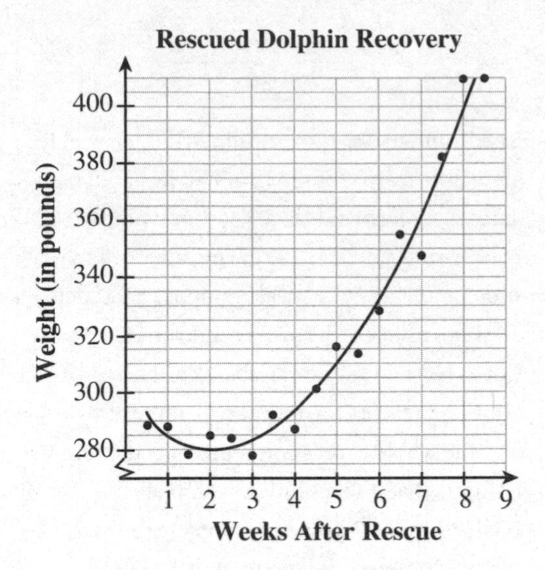

26. Following the catastrophic oil spill in the Gulf of Mexico in April of 2010, more than 900 bottlenose dolphins were found dead or stranded in the oil spill area. The figure above shows the weight of a rescued dolphin during its recovery. Based on the quadratic model fit to the data shown, which of the following is the closest to the average rate of change in the dolphin's weight between week 2 and week 8 of its recovery?

A) 4 pounds per week

B) 16 pounds per week

C) 20 pounds per week

D) 40 pounds per week

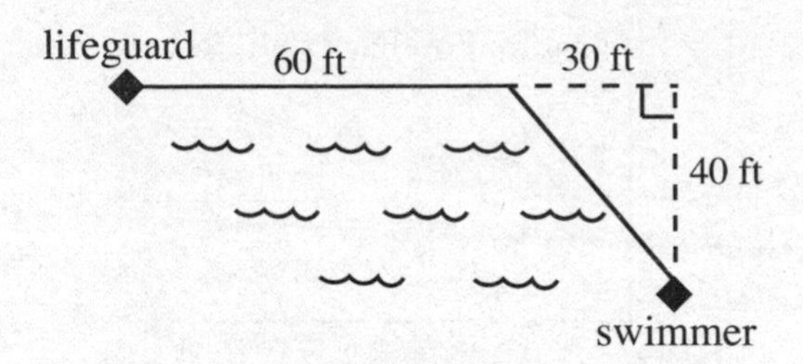

27. As shown in the figure above, a lifeguard sees a struggling swimmer who is 40 feet from the beach. The lifeguard runs 60 feet along the edge of the water at a speed of 12 feet per second. He pauses for 1 second to locate the swimmer again, and then dives into the water and swims along a diagonal path to the swimmer at a speed of 5 feet per second. How many seconds go by between the time the lifeguard sees the struggling swimmer and the time he reaches the swimmer?

A) 16

B) 22

C) 50

D) 56

28. What was the initial amount of gasoline in a fuel trailer, in gallons, if there are now x gallons, y gallons were pumped into a storage tank, and then 50 gallons were added to the trailer?

A) $x + y + 50$

B) $x + y - 50$

C) $y - x + 50$

D) $x - y - 50$

GO ON TO THE NEXT PAGE

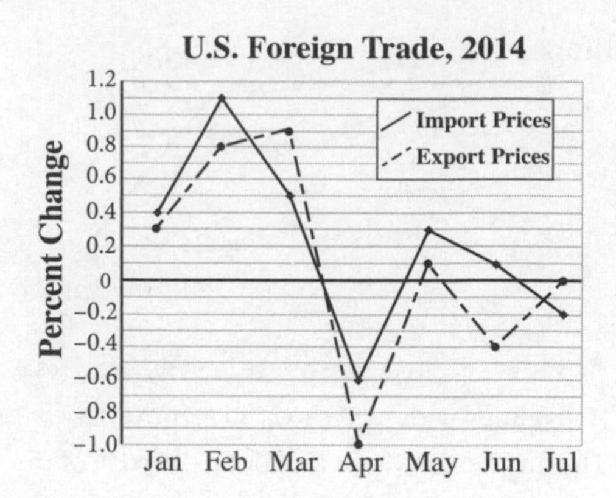

29. The figure above shows the net change, as a percentage, for U.S. import and export prices from January to July 2014 as reported by the Bureau of Labor Statistics. For example, U.S. import prices declined 0.2 percent in July while export prices remained unchanged for that month. Based on this information, which of the following statements is true for the time period shown in the figure?

A) On average, export prices increased more than import prices.

B) Import prices showed an increase more often than export prices.

C) Import prices showed the greatest change between two consecutive months.

D) From January to July, import prices showed a greater overall decrease than export prices.

$$\frac{3.86}{x}+\frac{180.2}{10x}+\frac{42.2}{5x}$$

30. The Ironman Triathlon originated in Hawaii in 1978. The format of the Ironman has not changed since then: It consists of a 3.86-km swim, a 180.2-km bicycle ride, and a 42.2-km run, all raced in that order and without a break. Suppose an athlete bikes 10 times as fast as he swims and runs 5 times as fast as he swims. The variable x in the expression above represents the rate at which the athlete swims, and the whole expression represents the number of hours that it takes him to complete the race. If it takes him 16.2 hours to complete the race, how many kilometers did he swim in 1 hour?

A) 0.85

B) 1.01

C) 1.17

D) 1.87

GO ON TO THE NEXT PAGE

Directions: For questions 31-38, solve the problem and enter your answer in the grid, as described below, on the answer sheet.

1. Although not required, it is suggested that you write your answer in the boxes at the top of the columns to help you fill in the circles accurately. You will receive credit only if the circles are filled in correctly.
2. Mark no more than one circle in any column.
3. No question has a negative answer.
4. Some problems may have more than one correct answer. In such cases, grid only one answer.
5. **Mixed numbers** such as $3\frac{1}{2}$ must be gridded as 3.5 or $\frac{7}{2}$.

   (If $3\frac{1}{2}$ is entered into the grid as 3 1 / 2, it will be interpreted as $\frac{31}{2}$, not $3\frac{1}{2}$.)
6. **Decimal answers:** If you obtain a decimal answer with more digits than the grid can accommodate, it may be either rounded or truncated, but it must fill the entire grid.

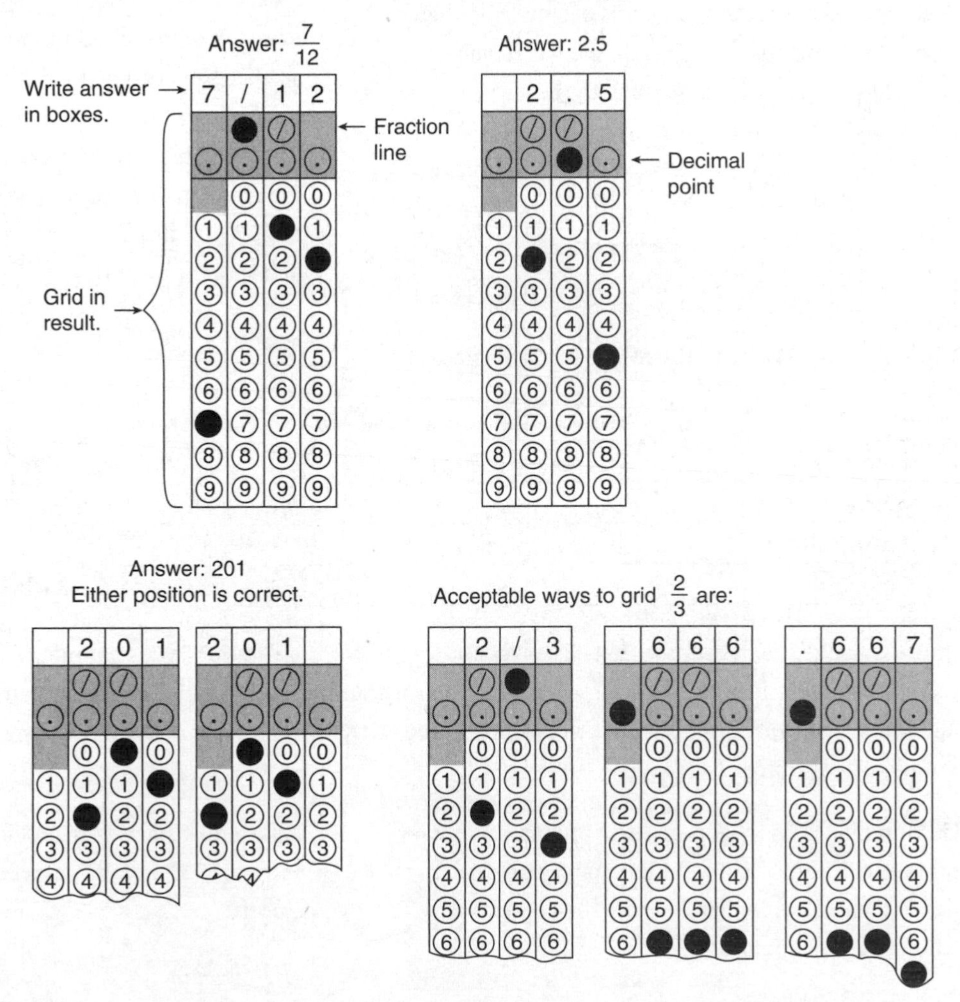

GO ON TO THE NEXT PAGE

31. What value of x satisfies the equation $\frac{2}{3}(5x+7)=8x$?

32. Some doctors base the dosage of a drug to be given to a patient on the patient's body surface area (BSA). The most commonly used formula for calculating BSA is $BSA=\sqrt{\frac{wh}{3,600}}$, where w is the patient's weight (in kg), h is the patient's height (in cm), and BSA is measured in square meters. How tall (in cm) is a patient who weighs 150 kg and has a BSA of $2\sqrt{2}$ m^2?

33. A college math professor informs her students that rather than curving final grades, she will replace each student's lowest test score with the next to lowest test score, and then re-average the test grades. If Leeza has test scores of 86, 92, 81, 64, and 83, by how many points does her final test average change based on the professor's policy?

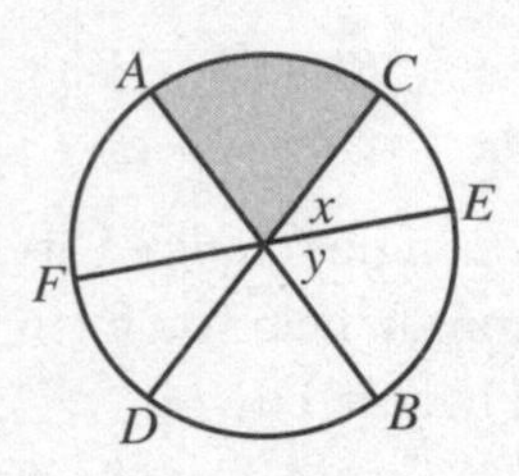

34. In the figure above, $\overline{AB}$, $\overline{CD}$, and $\overline{EF}$ are diameters of the circle. If $y = 2x - 12$, and the shaded area is $\frac{1}{5}$ of the circle, what is the value of x?

35. If the slope of a line is $-\frac{7}{4}$ and a point on the line is (4, 7), what is the y-intercept of the line?

36. Rory left home and drove straight to the airport at an average speed of 45 miles per hour. He returned home along the same route, but traffic slowed him down and he only averaged 30 miles per hour on the return trip. If his total travel time was 2 hours and 30 minutes, how far is it, in miles, from Rory's house to the airport?

Questions 37 and 38 refer to the following information.

Chemical Makeup of One Mole of Chloroform

Element	Number of Moles	Mass per Mole (grams)
Carbon	1	12.011
Hydrogen	1	1.008
Chlorine	3	35.453

A chemical solvent is a substance that dissolves another to form a solution. For example, water is a solvent for sugar. Unfortunately, many chemical solvents are hazardous to the environment. One eco-friendly chemical solvent is chloroform, also known as trichloromethane ($CHCl_3$). The table above shows the chemical makeup of one mole of chloroform.

37. Carbon makes up what percent of the mass of one mole of chloroform? Round your answer to the nearest whole percent and ignore the percent sign when entering your answer.

38. If a chemist starts with 1,000 grams of chloroform and uses 522.5 grams, how many moles of chlorine are left?

IF YOU FINISH BEFORE TIME IS CALLED, YOU MAY CHECK YOUR WORK ON THIS SECTION ONLY. DO NOT TURN TO ANY OTHER SECTION IN THE TEST. **STOP**

ANSWER KEY

MATH—NO CALCULATOR

1. C
2. A
3. D
4. C
5. B
6. D
7. D
8. B
9. B
10. A
11. C
12. B
13. A
14. D
15. A
16. 20
17. 1
18. 2
19. 14
20. 6

MATH—CALCULATOR

1. C
2. C
3. D
4. A
5. B
6. C
7. D
8. B
9. A
10. C
11. B
12. D
13. A
14. D
15. B
16. B
17. D
18. A
19. A
20. A
21. C
22. D
23. B
24. C
25. D
26. C
27. A
28. B
29. B
30. D
31. 1
32. 192
33. 3.4
34. 40
35. 14
36. 45
37. 10
38. 12

ANSWERS AND EXPLANATIONS

MATH TEST: NO-CALCULATOR SECTION

1. C
Difficulty: Easy

Category: Heart of Algebra / Linear Equations

Strategic Advice: To determine what the *y*-intercept could mean in the context of a word problem, examine the labels on the graph and note what each axis represents.

Getting to the Answer: According to the labels, the *y*-axis represents cost, and the *x*-axis represents the number of games played. The *y*-intercept, (0, 5), has an *x*-value of 0, which means 0 games were played, yet there is still a cost of $5. The cost must represent a flat fee that is charged before any games are played, such as an entrance fee to enter the arcade.

2. A
Difficulty: Easy

Category: Passport to Advanced Math / Exponents

Strategic Advice: To divide one rational expression by another, multiply the first expression by the reciprocal (the flip) of the second expression.

Getting to the Answer: Rewrite the division as multiplication, factor any factorable expressions, and then simplify if possible.

$$\begin{aligned}\frac{3x}{x+5} \div \frac{6}{4x+20} &= \frac{3x}{x+5} \cdot \frac{4x+20}{6} \\ &= \frac{3x}{\cancel{x+5}} \cdot \frac{4\cancel{(x+5)}}{6} \\ &= \frac{12x}{6} \\ &= 2x\end{aligned}$$

Note that the question also states that $x \neq -5$. This doesn't affect your answer—it is simply stated because the denominators of rational expressions cannot equal 0.

3. D
Difficulty: Easy

Category: Additional Topics in Math / Geometry

Strategic Advice: When the equation of a circle is written in the form $(x - h)^2 + (y - k)^2 = r^2$, the point (h, k) represents the center of the circle on a coordinate plane, and r represents the length of the radius.

Getting to the Answer: To find the area of a circle, use the formula $A = \pi r^2$. In the equation given in the question, r^2 is the constant on the right-hand side (25)—you don't even need to solve for r because the area formula involves r^2, not r. So, the area is $\pi(25)$ or 25π.

4. C
Difficulty: Easy

Category: Passport to Advanced Math / Functions

Strategic Advice: When using function notation, $f(x)$ is simply another way of saying y, so this question is asking you to find the values of x for which $y = 0$, or in other words, where the graph crosses the *x*-axis.

Getting to the Answer: The graph crosses the *x*-axis at the points (–2, 0) and (3, 0), so the values of x for which $f(x) = 0$ are –2 and 3.

5. B
Difficulty: Medium

Category: Heart of Algebra / Linear Equations

Strategic Advice: Cross-multiplying is a good strategy for a question like this, but simplifying the numerators first will make the calculations easier.

Getting to the Answer:

$$\frac{4(d+3)-9}{8}=\frac{10-(2-d)}{6}$$
$$\frac{4d+12-9}{8}=\frac{10-2+d}{6}$$
$$\frac{4d+3}{8}=\frac{8+d}{6}$$
$$6(4d+3)=8(8+d)$$
$$24d+18=64+8d$$
$$16d=46$$
$$d=\frac{46}{16}=\frac{23}{8}$$

6. D
Difficulty: Medium

Category: Passport to Advanced Math / Functions

Strategic Advice: This is a crossover question, so quickly skim the first couple of sentences. Then look for the relevant information in the last couple of sentences. It may also help to circle the portions of the graph that meet the given requirement.

Getting to the Answer: Because *greater* means *higher* on a graph, the statement $f(t) > g(t)$ translates to "Where is $f(t)$ above $g(t)$?" The solid curve represents f and the dashed curve represents g, so $f > g$ between the years 1960 and 1980 and again between the years 2000 and 2010. Look for these time intervals in the answer choices: $1960 < t < 1980$ and $2000 < t < 2010$.

7. D
Difficulty: Medium

Category: Passport to Advanced Math / Scatterplots

Strategic Advice: Use the shape of the data to predict the type of equation that might be used as a model. Then, use specific values from the graph to choose the correct equation.

Getting to the Answer: According to the graph, the population of the whales grew slowly at first and then more quickly. This means that an exponential model is probably the best fit, so you can eliminate A (linear) and B (quadratic). The remaining equations are both exponential, so choose a data point and see which equation is the closest fit. Be careful—the vertical axis represents *hundreds* of whales, and the question states that t represents the number of years since the study began, so $t = 0$ for 1995, $t = 3$ for 1998, and so on. If you use the data for 1995, which is the point (0, 100), the results are the same for both equations, so choose a different point. Using the data for 2007, $t = 2007 - 1995 = 12$, and the number of whales was 800. Substitute these values into C and D to see which one is true. Choice C is not true because $800 \neq 100 \times 2^{12}$. Choice (D) is correct because $800 = 100 \times 2^{\frac{12}{4}} = 100 \times 2^3 = 100 \times 8$ is true.

8. B
Difficulty: Medium

Category: Heart of Algebra / Linear Equations

Strategic Advice: Average rate of change is the same as slope, so use the slope formula.

Getting to the Answer: To find the average rate of change over the 5-year period, find the slope between the starting point (0, 1,200) and the ending point (5, 100).

$$m=\frac{y_2-y_1}{x_2-x_1}=\frac{100-1{,}200}{5-0}=\frac{-1{,}100}{5}=-220$$

The average rate of change is negative because the laptop decreases in value over time.

Note: Because the question involves *straight-line* depreciation, you could have used any two points on the graph to find the slope. As a general rule, however, you should use the endpoints of the given time interval.

9. B

Difficulty: Medium

Category: Passport to Advanced Math / Exponents

Strategic Advice: When multiplying polynomials, carefully multiply each term in the first factor by each term in the second factor. This question doesn't ask for the entire product, so check to make sure you answered the right question (the coefficient of x^2).

Getting to the Answer:

$$\left(6x^2 - \frac{2}{5}x + 1\right)\left(10x + \frac{1}{3}\right)$$
$$= 6x^2\left(10x + \frac{1}{3}\right) - \frac{2}{5}x\left(10x + \frac{1}{3}\right) + 1\left(10x + \frac{1}{3}\right)$$
$$= 60x^3 + \underline{2x^2 - 4x^2} - \frac{2}{15}x + 10x + \frac{1}{3}$$

The coefficient of x^2 is $2 + (-4) = -2$.

10. A

Difficulty: Medium

Category: Heart of Algebra / Linear Equations

Strategic Advice: Notice that there are no grid-lines and no numbers on the axes. This is a great clue that the numbers in the equations don't actually matter.

Getting to the Answer: The line is decreasing, so the slope (m) is negative. The line crosses the y-axis below 0, so the y-intercept (b) is also negative. Put each answer choice in slope-intercept form, one at a time, and examine the signs of m and b. Begin with (A):

$$-6x - 4y = 5$$
$$-4y = 6x + 5$$
$$y = \frac{6x}{-4} + \frac{5}{-4}$$
$$y = -\frac{3}{2}x - \frac{5}{4}$$

You don't need to check any of the other equations. Choice (A) has a negative slope and a negative y-intercept, so it is the correct equation.

11. C

Difficulty: Hard

Category: Heart of Algebra / Systems of Linear Equations

Strategic Advice: Graphically, a system of linear equations that has no solution indicates two parallel lines or, in other words, two lines that have the same slope. So, write each of the equations in slope-intercept form ($y = mx + b$) and set their slopes (m) equal to each other to solve for k. Before finding the slopes, multiply the top equation by 4 to make it easier to manipulate.

Getting to the Answer:

$$4\left(\frac{3}{4}x - \frac{1}{2}y = 12\right) \rightarrow 3x - 2y = 48 \rightarrow y = \frac{3}{2}x - 24$$
$$kx - 2y = 22 \rightarrow -2y = -kx + 22 \rightarrow y = \frac{k}{2}x - 11$$

The slope of the first line is $\frac{3}{2}$, and the slope of the second line is $\frac{k}{2}$. Set them equal and solve for k:

$$\frac{3}{2} = \frac{k}{2}$$
$$2(3) = 2(k)$$
$$6 = 2k$$
$$3 = k$$

12. B

Difficulty: Hard

Category: Heart of Algebra / Inequalities

Strategic Advice: Pay careful attention to units, particularly when a question involves rates. The \$4.00 for the first $\frac{1}{4}$ mile is a flat fee. Before you write the inequality, you need to find the per-mile rate for the remaining miles.

Getting to the Answer: The driver charges \$4.00 for the first $\frac{1}{4}$ mile, which is a flat fee, so write 4. The additional charge is \$1.50 per $\frac{1}{2}$ mile, or 1.50

times 2 = \$3.00 per mile. The number of miles after the first $\frac{1}{4}$ mile is $m-\frac{1}{4}$, so the cost of the trip, not including the first $\frac{1}{4}$ mile, is $3\left(m-\frac{1}{4}\right)$. This means the cost of the whole trip is $4+3\left(m-\frac{1}{4}\right)$. The clue "no more than \$10" means that much or less, so use the symbol ≤. The inequality is $4+3\left(m-\frac{1}{4}\right)\leq 10$, which simplifies to $3.25 + 3m \leq 10$.

13. A

Difficulty: Hard

Category: Passport to Advanced Math / Functions

Strategic Advice: Think about how the transformations affect the graph of $g(x)$ and draw a sketch of $h(x)$ on the same grid. Compare the new graph to each of the answer choices until you find one that is true.

Getting to the Answer: The graph of $h(x) = -g(x) + 1$ is a vertical reflection of $g(x)$, over the x-axis, that is then shifted up 1 unit. The graph looks like the dashed line in the following graph:

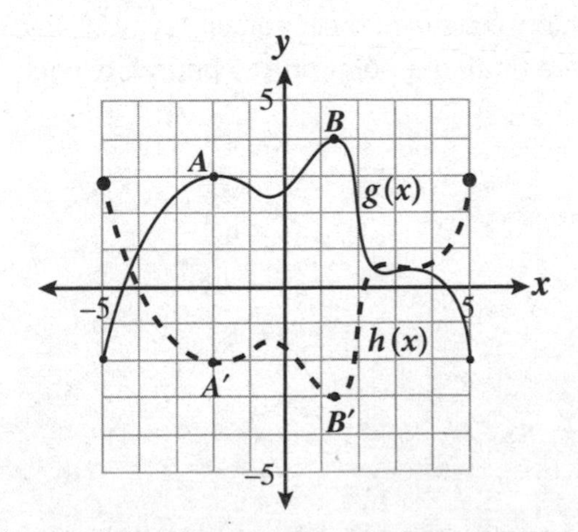

Now, compare the dashed line to each of the answer choices: the range of $h(x)$ is the set of y-values from lowest to highest (based on the dashed line). The lowest point occurs at point B' and has a y-value of -3; the highest value occurs at both ends of the graph and is 3, so the range is $-3 \leq y \leq 3$. This means (A) is correct and you can move on to the next question. Don't waste valuable time checking the other answer choices unless you are not sure about the range. (Choice B: The minimum value of $h(x)$ is -3, not -4. Choice C: The coordinates of point A on $h(x)$ are $(-2, -2)$, not $(2, 4)$. Choice D: The graph of $h(x)$ is decreasing, not increasing, between $x = -5$ and $x = -2$.)

14. D

Difficulty: Medium

Category: Additional Topics in Math / Imaginary Numbers

Strategic Advice: Multiply the two complex numbers just as you would two binomials (using FOIL). Then, combine like terms and use the definition $i^2 = -1$ to simplify the result.

Getting to the Answer:

$$\begin{aligned}(3+2i)(5-i) &= 3(5-i)+2i(5-i)\\ &= 15-3i+10i-2i^2\\ &= 15+7i-2(-1)\\ &= 15+7i+2\\ &= 17+7i\end{aligned}$$

The question asks for a in $a + bi$, so the correct answer is 17.

15. A

Difficulty: Hard

Category: Passport to Advanced Math / Exponents

Strategic Advice: Think of the rate given in the question in terms of the constant term you see on the right-hand side of the equation. Working together, the two treatment plants can filter the water in 72 hours. This is equivalent to saying that they can filter $\frac{1}{72}$ of the water in 1 hour.

Getting to the Answer: If $\frac{1}{72}$ is the portion of the water the two treatment plants can filter *together*, then each term on the left side of the equation represents the portion that each plant can filter *individually* in 1 hour. Because the new facility is 4 times as fast as the older facility, $\frac{4}{x}$ represents the portion of the water the new plant can filter in 1 hour, and $\frac{1}{x}$ represents the portion of the water the older plant can filter in 1 hour.

16. 20
Difficulty: Medium

Category: Heart of Algebra / Linear Equations

Strategic Advice: Only one equation is given, and it has two variables. This means that you don't have enough information to solve for either variable. Instead, look for the relationship between the variable terms in the equation and those in the expression that you are trying to find, $x + 2y$.

Getting to the Answer: First, move the y-term to the left side of the equation to make it look more like the expression you are trying to find. The expression doesn't have fractions, so clear the fractions in the equation by multiplying both sides by 4. This yields the expression that you are looking for, $x + 2y$, so no further work is required—just read the value on the right-hand side of the equation. The answer is 20.

$$\frac{1}{4}x = 5 - \frac{1}{2}y$$
$$\frac{1}{4}x + \frac{1}{2}y = 5$$
$$4\left(\frac{1}{4}x + \frac{1}{2}y\right) = 4(5)$$
$$x + 2y = 20$$

17. 1
Difficulty: Medium

Category: Heart of Algebra / Inequalities

Strategic Advice: This question is extremely difficult to answer unless you draw a sketch. It doesn't have to be perfect—you just need to get an idea of where the solution region is. Don't forget to flip the inequality symbol when you graph the second equation.

Getting to the Answer: Sketch the system.

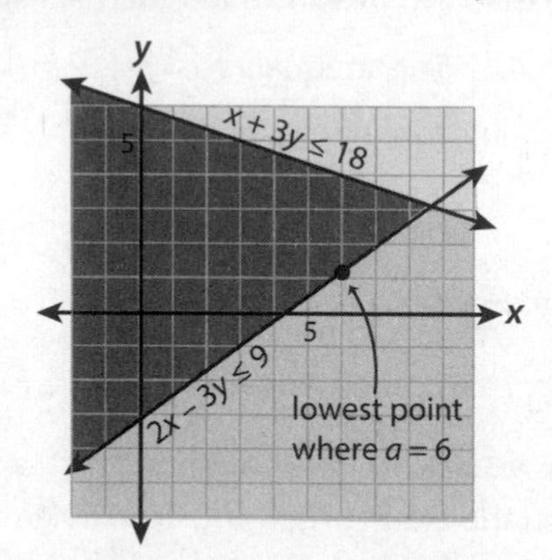

If (a, b) is a solution to the system, then a is the x-coordinate of any point in the darkest shaded region and b is the corresponding y-coordinate. When $a = 6$, the minimum possible value for b lies on the lower boundary line, $2x - 3y \leq 9$. It looks like the y-coordinate is 1, but to be sure, substitute $x = 6$ into the equation and solve for y. You can use = in the equation, instead of the inequality symbol, because you are finding a point on the boundary line.

$$2x - 3y = 9$$
$$2(6) - 3y = 9$$
$$12 - 3y = 9$$
$$-3y = -3$$
$$y = 1$$

18. 2
Difficulty: Hard

Category: Passport to Advanced Math / Exponents

Strategic Advice: Rewrite the radicals as fraction exponents: $\sqrt{x} = x^{\frac{1}{2}}$ and $\sqrt[3]{x} = x^{\frac{1}{3}}$.

Getting to the Answer: Write each factor in the expression in exponential form. Then use the rules of exponents to simplify the expression. Add the exponents of the factors that are being multiplied, and subtract the exponent of the factor that is being divided:

$$\frac{\sqrt{x} \cdot x^{\frac{5}{6}} \cdot x}{\sqrt[3]{x}} = \frac{x^{\frac{1}{2}} \cdot x^{\frac{5}{6}} \cdot x^{1}}{x^{\frac{1}{3}}}$$

$$= x^{\frac{1}{2}+\frac{5}{6}+\frac{1}{1}-\frac{1}{3}}$$

$$= x^{\frac{3}{6}+\frac{5}{6}+\frac{6}{6}-\frac{2}{6}}$$

$$= x^{\frac{12}{6}} = x^2$$

Because n is the power of x, the value of n is 2.

19. 14
Difficulty: Hard

Category: Additional Topics in Math / Geometry

Strategic Advice: The shaded region is the area of the larger triangle minus the area of the smaller triangle. Set up and solve an equation using the information from the figure. Before you grid in your answer, check that you answered the right question (height of larger triangle).

Getting to the Answer: You don't know the height of the smaller triangle, so call it h. You do know the area of the shaded region—it's 52 square units.

Larger triangle: base = 12; height = $h + 3 + 3$

Smaller triangle: base = 8; height = h

Shaded area = large area – small area

$$52 = \left[\left(\frac{1}{2}\right)(12)(h+6)\right] - \left[\left(\frac{1}{2}\right)(8)(h)\right]$$

$$52 = 6(h+6) - 4h$$

$$52 = 6h + 36 - 4h$$

$$52 = 2h + 36$$

$$16 = 2h$$

$$8 = h$$

The question asks for the height of the *larger* triangle, so the correct answer is 8 + 3 + 3 = 14.

20. 6
Difficulty: Hard

Category: Passport to Advanced Math / Quadratics

Strategic Advice: The highest power of x in the equation is 2, so the function is quadratic. Writing quadratic equations can be tricky and time-consuming. If you know the roots, you can use factors to write the equation. If you don't know the roots, you need to create a system of equations to find the coefficients of the variable terms.

Getting to the Answer: You don't know the roots of this equation, so start with the point that has the easiest values to work with, (0, 1), and substitute them into the equation $y = ax^2 + bx + c$:

$$1 = a(0)^2 + b(0) + c$$

$$1 = c$$

Now your equation looks like $y = ax^2 + bx + 1$. Next, use the other two points to create a system of two equations in two variables:

$$(-3, 10) \rightarrow 10 = a(-3)^2 + b(-3) + 1 \rightarrow 9 = 9a - 3b$$

$$(2, 15) \rightarrow 15 = a(2)^2 + b(2) + 1 \rightarrow 14 = 4a + 2b$$

You now have a system of equations to solve. None of the variables has a coefficient of 1, so use elimination to solve the system. If you multiply the top equation by 2 and the bottom equation by 3, the b-terms will eliminate each other:

$$2[9a - 3b = 9] \rightarrow 18a - 6b = 18$$

$$3[4a + 2b = 14] \rightarrow \underline{12a + 6b = 42}$$

$$30a = 60$$

$$a = 2$$

Now, find b by substituting $a = 2$ into either of the original equations. Using the top equation, you get:

$$9(2) - 3b = 9$$

$$18 - 3b = 9$$

$$-3b = -9$$

$$b = 3$$

The value of $a + b + c$ is 2 + 3 + 1 = 6.

MATH TEST: CALCULATOR SECTION

1. C

Difficulty: Easy

Category: Problem Solving and Data Analysis / Rates, Ratios, Proportions, and Percentages

Strategic Advice: You can use the formula $\text{Percent} = \frac{\text{part}}{\text{whole}} \times 100\%$ whenever you know two out of the three quantities.

Getting to the Answer: The clue "all" tells you that the "whole" is what you don't know. The percent is 96.5, and the part is 321,000,000.

$$96.5 = \frac{321{,}000{,}000}{w} \times 100\%$$
$$96.5w = 32{,}100{,}000{,}000$$
$$w = \frac{32{,}100{,}000{,}000}{96.5}$$
$$w = 332{,}642{,}487$$

The answer choices are rounded to the nearest thousand, so the answer is 332,642,000.

2. C

Difficulty: Easy

Category: Heart of Algebra / Linear Equations

Strategic Advice: A *one-time* fee does not depend on the variable and is therefore a constant. A unit rate, however, is always multiplied by the independent variable.

Getting to the Answer: The total cost consists of the site visit fee (a constant), an hourly cost (which depends on the number of hours), and the cost of the materials (which are taxed). The constant in the equation is 75 and is therefore the site visit fee; 45 is being multiplied by h (the number of hours), so \$45 must be the hourly rate. That leaves the remaining term, 1.06(82.5), which must be the cost of the materials (\$82.50) plus a 6% tax.

3. D

Difficulty: Easy

Category: Heart of Algebra / Inequalities

Strategic Advice: The intersection (overlap) of the two shaded regions is the solution to the system of inequalities. Check each point to see whether it lies in the region with the darkest shading. Don't forget to check that you answered the right question—you are looking for the point that is *not* a solution to the system.

Getting to the Answer: Each of the first three points clearly lies in the overlap. The point (3, 3) looks like it lies on the dashed line, which means it is *not* included in the solution. To check this, plug (3, 3) into the easier inequality: $3 \not> 3$ (3 is equal to itself, not greater than itself), so (D) is correct.

4. A

Difficulty: Easy

Category: Passport to Advanced Math / Quadratics

Strategic Advice: Quadratic equations can be written in several forms, each of which reveals something special about the graph. For example, the vertex form of a quadratic equation gives the minimum or maximum value of the function, while the standard form reveals the y-intercept.

Getting to the Answer: The factored form of a quadratic equation reveals the solutions to the equation, which graphically represent the x-intercepts. Choice (A) is the only equation written in this form and therefore must be correct. You can set each factor equal to 0 and solve to find that the x-intercepts of the graph are $x = \frac{5}{2}$ and $x = -1$.

5. B

Difficulty: Easy

Category: Problem Solving and Data Analysis / Rates, Ratios, Proportions, and Percentages

Strategic Advice: Break the question into steps. Before you can use the ratio, you need to find the percent of the students who answered either "Foreign Policy" or "Environment."

Getting to the Answer: The ratio given in the question is 5:3, so write this as 5 parts "Foreign Policy" and 3 parts "Environment." You don't know how big a *part* is, so call it x. This means that $5x + 3x =$ the percent of the students who answered either "Foreign Policy" or "Environment," which is 100% – all the other answers:

$$100 - (16 + 14 + 9 + 5) = 100 - 44 = 56$$
$$5x + 3x = 56$$
$$8x = 56$$
$$x = 7$$

Each part has a value of 7, and 3 parts answered "Environment," so the correct percentage is $3(7) = 21\%$.

6. C

Difficulty: Easy

Category: Problem Solving and Data Analysis / Scatterplots

Strategic Advice: Examine both the shape and the direction of the data to pick the best description of the association.

Getting to the Answer: A data set that has a linear association follows the path of a straight line; a data set that is exponential follows a path that is similar to linear data, but with a curve to it because the rate of increase (or decrease) changes over time. This data set has a curve to it, so "exponential" describes the association better than "linear." This means you can eliminate A and B. A positive association between two variables is one in which higher values of one variable correspond to higher values of the other variable, and vice versa. In other words, as the x-values of the data points go up, so do the y-values. This is indeed the case for this data set, so (C) is correct.

7. D

Difficulty: Easy

Category: Problem Solving and Data Analysis / Statistics and Probability

Strategic Advice: Your only choice for this question is to compare each statement to the figure. Don't waste time trying to figure out the exact value for each bar—an estimate is good enough to determine whether each statement is true.

Getting to the Answer: Choice A is incorrect because the price in 2008 was slightly less (not more) than \$3.50, while the price in 2013 was right around \$3.50. Choice B is incorrect because the price in 2003 was more than \$2.00, and the price in 2013 was not more than twice that (\$4.00). Choice C is incorrect because the price in 2008 was about \$3.25 and the price in 2009 was about \$2.75—this is not a difference of more than \$1.00. This means (D) must be correct. You don't have to check it—just move on. (Between 2003 and 2008, the change in price was about \$3.40 – \$2.30 = \$1.10; between 2008 and 2013, the change in price was only about \$3.50 – \$3.40 = \$0.10; the change in price was greater between 2003 and 2008.)

8. B

Difficulty: Medium

Category: Heart of Algebra / Systems of Linear Equations

Strategic Advice: Because none of the variable terms has a coefficient of 1, solve the system of equations using elimination by addition (combining the equations). Before you choose an answer, check that you answered the right question (the sum of x and y).

Getting to the Answer: Multiply the top equation by 2 to eliminate the terms that have y's in them:

$$\begin{aligned} 2[-2x+5y=1] &\rightarrow -4x+10y=2 \\ 7x-10y=-11 &\rightarrow \underline{7x-10y=-11} \\ 3x &= -9 \\ x &= -3 \end{aligned}$$

Now, substitute the result into either of the original equations and simplify to find *y*:

$$\begin{aligned} -2x+5y&=1 \\ -2(-3)+5y&=1 \\ 6+5y&=1 \\ 5y&=-5 \\ y&=-1 \end{aligned}$$

The question asks for the *sum*, so add *x* and *y* to get $-3 + (-1) = -4$.

9. A
Difficulty: Medium

Category: Heart of Algebra / Systems of Linear Equations

Strategic Advice: Take a quick peek at the answers just to see what variables are being used, but don't study the equations. Instead, write your own system using the same variables as given in the answer choices.

Getting to the Answer: One of the equations in the system should represent the sum of the two resistors ($R_1 + R_2$), which is equal to 294. This means you can eliminate C and D. The second equation needs to satisfy the condition that R_2 is 6 less than twice R_1, or $R_2 = 2R_1 - 6$. This means (A) is correct.

10. C
Difficulty: Medium

Category: Heart of Algebra / Linear Equations

Strategic Advice: Use the distributive property to simplify each of the terms that contains parentheses. Then use inverse operations to solve for *x*.

Getting to the Answer:

$$\begin{aligned} \frac{2}{\not{5}}(\not{5}x)+2(x-1)&=4(x+1)-2 \\ 2x+2x-2&=4x+4-2 \\ 4x-2&=4x+2 \\ -2&\neq 2 \end{aligned}$$

All of the variable terms cancel out, and the resulting numerical statement is false (because negative 2 does not equal positive 2), so there is no solution to the equation. Put another way, there is no value of *x* for which the equation is true.

11. B
Difficulty: Medium

Category: Additional Topics in Math / Geometry

Strategic Advice: Think about this question logically before you start writing things down—after it's transferred, the volume of the oil in the cylindrical container will be the same volume as the rectangular container, so you need to set the two volumes equal and solve for *h*.

Getting to the Answer: The volume of the rectangular container is $4 \times 9 \times 10$, or 360 cubic meters. The volume of a cylinder equals the area of its base times its height, or $\pi r^2 h$. Because the diameter is 6 meters, the radius, *r*, is half that, or 3 meters. Now we're ready to set up an equation and solve for *h* (which is the height of the cylinder or, in this case, the length of the transportation container):

$$\begin{aligned} \text{Volume of oil} &= \text{Volume of rectangular container} \\ \pi(3)^2 h &= 360 \\ 9\pi h &= 360 \\ h &= \frac{360}{9\pi} = \frac{40}{\pi} \end{aligned}$$

12. D
Difficulty: Medium

Category: Problem Solving and Data Analysis / Rates, Ratios, Proportions, and Percentages

Strategic Advice: Even though this question uses the word *percent*, you are never asked to find the actual percent itself. Set this question up as a proportion to get the answer more quickly. Remember, percent change equals amount of change divided by the original amount.

Getting to the Answer:

$$\frac{12-5}{5}=\frac{x-12}{12}$$
$$\frac{7}{5}=\frac{x-12}{12}$$
$$12(7)=5(x-12)$$
$$84=5x-60$$
$$144=5x$$
$$28.8=x$$

13. A
Difficulty: Medium

Category: Passport to Advanced Math / Exponents

Strategic Advice: Don't spend too much time reading the scientific explanation of the equation. Focus on the question at the very end—it's just asking you to solve the equation for d.

Getting to the Answer: First, cross-multiply to get rid of the denominator. Then, divide both sides of the equation by $4\pi b$ to isolate d^2. Finally, take the square root of both sides to find d:

$$b(4\pi d^2)=L$$
$$\frac{\cancel{b}(\cancel{4\pi} d^2)}{\cancel{4\pi b}}=\frac{L}{4\pi b}$$
$$d^2=\frac{L}{4\pi b}$$
$$\sqrt{d^2}=\sqrt{\frac{L}{4\pi b}}$$
$$d=\sqrt{\frac{L}{4\pi b}}$$

Unfortunately, this is not one of the answer choices, so you'll need to simplify further. You can take the square root of 4 (it's 2), but be careful—it's in the denominator of the fraction, so it comes out of the square root as $\frac{1}{2}$.

The simplified equation is $d=\frac{1}{2}\sqrt{\frac{L}{\pi b}}$.

14. D
Difficulty: Easy

Category: Problem Solving and Data Analysis / Statistics and Probability

Strategic Advice: You do not need to use all of the information presented in the table to find the answer. Read the question carefully to make sure you use only what you need.

Getting to the Answer: To calculate the percentage of men in each age group who reported being unemployed in January 2014, divide the number in *that* age group who were unemployed by the total number in *that* age group. There are six age groups but only four answer choices, so don't waste time on the age groups that aren't represented. Choice (D) is correct because $7 \div 152 \approx 0.046 = 4.6\%$, which is a lower percentage than that for any other age group (20 to 24 = 12.5%; 35 to 44 = 4.9%; 45 to 54 = 6.1%).

15. B
Difficulty: Medium

Category: Problem Solving and Data Analysis / Statistics and Probability

Strategic Advice: The follow-up survey targets only those respondents who said they were unemployed, so focus on that column in the table.

Getting to the Answer: There were 6 respondents out of 44 unemployed males who were between the ages of 45 and 54, so the probability is $\frac{6}{44}=0.1\overline{36}$, or about 13.6%.

16. B
Difficulty: Medium

Category: Passport to Advanced Math / Quadratics

Strategic Advice: Taking the square root is the inverse operation of squaring, and both sides of the equation are already perfect squares, so take their square roots. Then solve the resulting equations. Remember, there will be two equations to solve.

Getting to the Answer:

$$(x-1)^2 = \frac{4}{9}$$

$$\sqrt{(x-1)^2} = \sqrt{\frac{4}{9}}$$

$$x-1 = \pm\frac{\sqrt{4}}{\sqrt{9}}$$

$$x = 1 \pm \frac{2}{3}$$

Now, simplify each equation: $x = 1 + \frac{2}{3} = \frac{3}{3} + \frac{2}{3} = \frac{5}{3}$ and $x = 1 - \frac{2}{3} = \frac{3}{3} - \frac{2}{3} = \frac{1}{3}$.

17. D

Difficulty: Medium

Category: Heart of Algebra / Linear Equations

Strategic Advice: The key to answering this question is to determine how many darts land in each color ring. If there are 6 darts total and x land in a blue ring, the rest, or $6 - x$, must land in a red ring.

Getting to the Answer: Write the expression in words first: points per blue ring (5) times number of darts in blue ring (x), plus points per red ring (10) times number of darts in red ring ($6 - x$). Now, translate the words into numbers, variables, and operations: $5x + 10(6 - x)$. This is not one of the answer choices, so simplify the expression by distributing the 10 and then combining like terms: $5x + 10(6 - x) = 5x + 60 - 10x = 60 - 5x$.

18. A

Difficulty: Medium

Category: Problem Solving and Data Analysis / Statistics and Probability

Strategic Advice: This is a science crossover question. Read the first two sentences quickly—they are simply describing the context of the question. The last two sentences pose the question, so read those more carefully.

Getting to the Answer: In the sample, 184 out of 200 square feet were free of red tide after applying the spray. This is $\frac{184}{200} = 0.92 = 92\%$ of the area. For the whole beach, 0.92(10,000) = 9,200 square feet should be free of the red tide. Be careful—this is *not* the answer. The question asks how much of the beach would still be covered by red tide, so subtract to get 10,000 – 9,200 = 800 square feet.

19. A

Difficulty: Medium

Category: Passport to Advanced Math / Quadratics

Strategic Advice: The solution to a system of equations is the point(s) where their graphs intersect. You can solve the system algebraically by setting the equations equal to each other, or you can solve it graphically using your calculator. Use whichever method gets you to the answer more quickly.

Getting to the Answer: Both equations are given in calculator-friendly format ($y = \ldots$), so graphing them is probably the more efficient approach. The graph looks like:

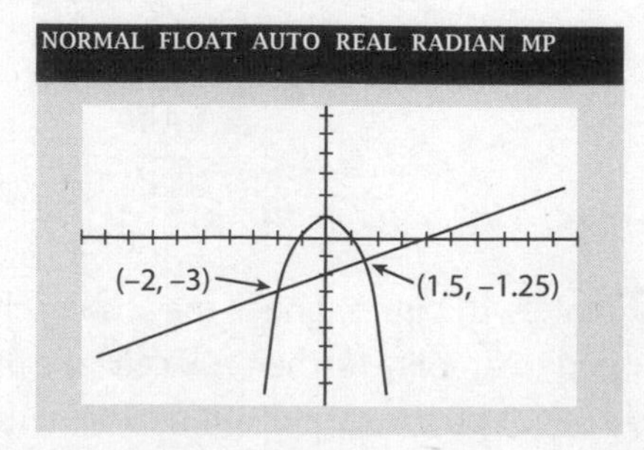

The solution point in the question is given as (a, b), so b represents the y-coordinate of the solution.

The y-coordinates of the points of intersection are −3 and −1.25, so choice (A) is correct.

20. A
Difficulty: Medium

Category: Passport to Advanced Math / Functions

Strategic Advice: Don't answer this question too quickly—you may be tempted to substitute 3 for x, but 3 is the output (range), not the input (domain).

Getting to the Answer: The given range value is an output value, so substitute 3 for $g(x)$ and use inverse operations to solve for x, which is the corresponding domain value.

$$g(x)=\frac{2}{3}x+7$$
$$3=\frac{2}{3}x+7$$
$$-4=\frac{2}{3}x$$
$$-12=2x$$
$$-6=x$$

You could also graph the function and find the value of x (the domain value) for which the value of y (the range value) is 3. The point on the graph is (−6, 3).

21. C
Difficulty: Medium

Category: Heart of Algebra / Linear Equations

Strategic Advice: Don't peek at the answers. Write your own equation using the initial cost and the rate of change in the value of the lawn mower. Remember—when something changes at a constant rate, it can be represented by a linear equation.

Getting to the Answer: When a linear equation in the form $y = mx + b$ is used to model a real-world scenario, m represents the constant rate of change, and b represents the starting amount. Here, the starting amount is easy—it's the purchase price, \$2,800. To find the rate of change, think of the initial cost as the value at 0 years, or the point (0, 2,800), and the salvage amount as the value at 8 years, or the point (8, 240). Substitute these points into the slope formula to find that $m=\frac{y_2-y_1}{x_2-x_1}=\frac{240-2,800}{8-0}=\frac{-2,560}{8}=-320$, so the equation is $y = -320x + 2,800$.

22. D
Difficulty: Medium

Category: Problem Solving and Data Analysis / Functions

Strategic Advice: Determine whether the change in the number of bacteria is a common difference (linear function) or a common ratio (exponential function) or if the number of bacteria changes direction (quadratic or polynomial function).

Getting to the Answer: The question tells you that the number of bacteria is reduced by half every hour after the antibiotic is applied. The microbiologist started with 20,000, so after one hour, there are 10,000 left, or $20,000\times\frac{1}{2}$. After 2 hours, there are 5,000 left, or $20,000\times\frac{1}{2}\times\frac{1}{2}$, and so on. The change in the number of bacteria is a common ratio $\left(\frac{1}{2}\right)$, so the best model is an exponential function of the form $y=a\left(\frac{1}{2}\right)^x$. In this scenario, a is 20,000.

23. B
Difficulty: Medium

Category: Problem Solving and Data Analysis / Rates, Ratios, Proportions, and Percentages

Strategic Advice: Let the units in this question guide you to the solution. The speeds of the airplanes are given in miles per hour, but the question asks about the number of miles each airplane can travel in 12 seconds, so convert miles per hour to miles per second.

Getting to the Answer:

Slower airplane:

$$\frac{600 \text{ mi}}{\cancel{\text{hr}}} \times \frac{1 \cancel{\text{hr}}}{60 \cancel{\text{min}}} \times \frac{1 \cancel{\text{min}}}{60 \cancel{\text{sec}}} \times 12 \cancel{\text{sec}} = 2 \text{ mi}$$

Faster airplane:

$$\frac{720 \text{ mi}}{\cancel{\text{hr}}} \times \frac{1 \cancel{\text{hr}}}{60 \cancel{\text{min}}} \times \frac{1 \cancel{\text{min}}}{60 \cancel{\text{sec}}} \times 12 \cancel{\text{sec}} = 2.4 \text{ mi}$$

The faster plane can travel $2.4 - 2 = 0.4$ miles farther, which is the same as $\frac{2}{5}$ miles.

24. C
Difficulty: Medium

Category: Heart of Algebra / Inequalities

Strategic Advice: The best way to answer this question is to pretend you are the worker. How much more would you earn for one hour in Oregon than in Idaho? If you worked 35 hours per week, how much more would this be? If you worked 40 hours per week, how much more would this be?

Getting to the Answer: Based on the data in the table, a worker would earn $9.10 – $7.25 = $1.85 more for one hour of work in Oregon than in Idaho. If he worked 35 hours per week, he would earn 35(1.85) = $64.75 more. If he worked 40 hours per week, he would earn 40(1.85) = $74 more. So, the worker would earn somewhere between $64.75 and $74 more per week, which can be expressed as the compound inequality $64.75 \le x \le 74$.

25. D
Difficulty: Medium

Category: Problem Solving and Data Analysis / Rates, Ratios, Proportions, and Percentages

Strategic Advice: This is another question where the units can help you find the answer. Use the number of vehicles owned to find the total number of miles driven to find the total number of gallons of gas used to find the total tax paid. Phew!

Getting to the Answer:

$$1.75 \cancel{\text{ vehicles}} \times \frac{11{,}340 \text{ miles}}{\cancel{\text{vehicle}}} = 19{,}845 \text{ miles}$$

$$19{,}845 \cancel{\text{ miles}} \times \frac{1 \text{ gallon of gas}}{21.4 \cancel{\text{ miles}}} = 927.336 \text{ gallons}$$

$$927.336 \cancel{\text{ gallons}} \times \frac{\$0.184}{\cancel{\text{gallon}}} = \$170.63$$

26. C
Difficulty: Medium

Category: Problem Solving and Data Analysis / Scatterplots

Strategic Advice: The average rate of change of a function over a given interval, from *a* to *b*, compares the change in the outputs, $f(b) - f(a)$, to the change in the inputs, $b - a$. In other words, it is the slope of the line that connects the endpoints of the interval, so you can use the slope formula.

Getting to the Answer: Look at the quadratic model, not the data points, to find that the endpoints of the given interval, week 2 to week 8, are (2, 280) and (8, 400). The average rate of change is $\frac{400-280}{8-2} = \frac{120}{6} = 20$.

On average, the dolphin's weight increased by 20 pounds per week.

27. A
Difficulty: Hard

Category: Additional Topics in Math / Geometry

Strategic Advice: In this question, information is given in both the diagram and the text. You need to relate the text to the diagram, one piece of information at a time, to calculate how long the lifeguard ran along the beach and how long he swam. Before you find the swim time, you need to know how *far* he swam.

Getting to the Answer: Whenever you see a right triangle symbol in a diagram, you should think Pythagorean theorem or, in this question, special right triangles. All multiples of 3-4-5 triangles are right triangles, so the length of the lifeguard's swim is the hypotenuse of a 30-40-50 triangle, or 50 feet. Add this number to the diagram. Now calculate the times using the distances and the speeds given. Don't forget the 1 second that the lifeguard paused.

Run time $= 60\ \not{ft} \times \frac{1\text{ sec}}{12\ \not{ft}} = \frac{60}{12} = 5\text{ sec}$

Pause time = 1 sec

Swim time $= 50\ \not{ft} \times \frac{1\text{ sec}}{5\ \not{ft}} = \frac{50}{5} = 10\text{ sec}$

Total time = 5 + 1 + 10 = 16 seconds

28. B

Difficulty: Hard

Category: Heart of Algebra / Linear Equations

Strategic Advice: Write an equation in words first and then translate from English to math. Finally, rearrange your equation to find what you're interested in, which is the initial amount of gasoline.

Getting to the Answer: Call the initial amount A. After you've written your equation, solve for A.

Amount now (x) = Initial amount (A) minus y, plus 50

$$x = A - y + 50$$
$$x + y - 50 = A$$

The initial amount was $x + y - 50$ gallons. Note that you could also use Picking Numbers to answer this question.

29. B

Difficulty: Hard

Category: Problem Solving and Data Analysis / Statistics and Probability

Strategic Advice: When a question involves reading data from a graph, it is sometimes better to skip an answer choice if it involves long calculations. Skim the answer choices for this question—A involves finding two averages, each of which is composed of 7 data values. Skip this choice for now.

Getting to the Answer: Start with (B). Be careful—you are not looking for places where the line segments are increasing. The y-axis already represents the change in prices, so you are simply counting the number of positive values for the imports (5) and for the exports (4). There are more for the imports, so (B) is correct and you don't need to check any of the other statements. Move on to the next question.

30. D

Difficulty: Hard

Category: Passport to Advanced Math / Exponents

Strategic Advice: The key to answering this question is deciding what you're trying to find. The question tells you that x represents the athlete's swim rate, and you are looking for the number of kilometers he swam in one hour—these are the same thing. If you find x (in kilometers per hour), you will know how many kilometers he swam in one hour.

Getting to the Answer: Set the equation equal to the total time, 16.2, and solve for x. To do this, write the variable terms over a common denominator, $10x$, and combine them into a single term. Then cross-multiply and go from there.

$$16.2 = \frac{10}{10}\left(\frac{3.86}{x}\right) + \frac{180.2}{10x} + \frac{2}{2}\left(\frac{42.2}{5x}\right)$$
$$16.2 = \frac{38.6}{10x} + \frac{180.2}{10x} + \frac{84.4}{10x}$$
$$16.2 = \frac{303.2}{10x}$$
$$10x(16.2) = 303.2$$
$$162x = 303.2$$
$$x = \frac{303.2}{162} \approx 1.87$$

31. 1

Difficulty: Easy

Category: Heart of Algebra / Linear Equations

Strategic Advice: Choose the best strategy to answer the question. If you distribute the $\frac{2}{3}$, it creates messy calculations. Instead, clear the fraction by multiplying both sides of the equation by 3. Then use the distributive property and inverse operations to solve for *x*.

Getting to the Answer:

$$\begin{aligned}\frac{2}{3}(5x+7)&=8x\\ \cancel{3}\cdot\frac{2}{\cancel{3}}(5x+7)&=3\cdot 8x\\ 2(5x+7)&=24x\\ 10x+14&=24x\\ 14&=14x\\ 1&=x\end{aligned}$$

32. 192

Difficulty: Medium

Category: Passport to Advanced Math / Exponents

Strategic Advice: This looks like a word problem, but don't let it intimidate you. Once you read it, you'll see that it boils down to substituting a few given values for the variables and solving the equation.

Getting to the Answer: Before you start substituting values, quickly check that the units given match the units required to use the equation—they do, so proceed. The patient's weight (*w*) is 150 and the patient's BSA is $2\sqrt{2}$, so the equation becomes $2\sqrt{2}=\sqrt{\frac{150h}{3,600}}$. The only variable left in the equation is *h*, and you are trying to find the patient's height, so you're ready to solve the equation. To do this, square both sides of the equation and then continue using inverse operations. Be careful when you square the left side—you must square both the 2 and the root 2.

$$\begin{aligned}2\sqrt{2}&=\sqrt{\frac{150h}{3,600}}\\ \left(2\sqrt{2}\right)^2&=\left(\sqrt{\frac{150h}{3,600}}\right)^2\\ 2^2\left(\sqrt{2}\right)^2&=\frac{150h}{3,600}\\ 4(2)&=\frac{150h}{3,600}\\ 28,800&=150h\\ 192&=h\end{aligned}$$

33. 3.4

Difficulty: Medium

Category: Problem Solving and Data Analysis / Statistics and Probability

Strategic Advice: The test average is the same as the mean of the data. The *mean* is the sum of all the values divided by the number of values. Break the question into short steps to keep your calculations organized. Before gridding in your answer, make sure you answered the right question (how much the final test average changes).

Getting to the Answer:

Step 1: Find the original test average:

$$\frac{86+92+81+64+83}{5}=\frac{406}{5}=81.2$$

Step 2: Find the average of the tests after replacing the lowest score (64) with the next to lowest score (81):

$$\frac{86+92+81+81+83}{5}=\frac{423}{5}=84.6$$

Step 3: Subtract the original average from the new average: 84.6 – 81.2 = 3.4.

34. 40

Difficulty: Hard

Category: Additional Topics in Math / Geometry

Strategic Advice: Because $\overline{AB}$, $\overline{CD}$, and $\overline{EF}$ are diameters, the sum of x, y, and the interior angle of the shaded region is 180 degrees. The question tells you that the shaded region is $\frac{1}{5}$ of the circle, so the interior angle must equal $\frac{1}{5}$ of the degrees in the whole circle, or $\frac{1}{5}$ of 360.

Getting to the Answer: Use what you know about y (that it is equal to $2x - 12$) and what you know about the shaded region (that it is $\frac{1}{5}$ of 360 degrees) to write and solve an equation:

$$x+y+\frac{1}{5}(360)=180$$
$$x+(2x-12)+72=180$$
$$3x+60=180$$
$$3x=120$$
$$x=40$$

35. 14
Difficulty: Hard

Category: Heart of Algebra / Linear Equations

Strategic Advice: When you know the slope and one point on a line, you can use $y = mx + b$ to write the equation. Substitute the slope for m and the coordinates of the point for x and y and then solve for b, the y-intercept of the line.

Getting to the Answer: The slope is given as $-\frac{7}{4}$, so substitute this for m. The point is given as (4, 7), so $x = 4$ and $y = 7$. Now, find b:

$$y=mx+b$$
$$7=-\frac{7}{\cancel{4}}(\cancel{4})+b$$
$$7=-7+b$$
$$14=b$$

The y-intercept of the line is 14.

You could also very carefully graph the line using the given point and the slope. Start at (4, 7) and move toward the y-axis by rising 7 and running *to the left* 4 (because the slope is negative). You should land at the point (0, 14).

36. 45
Difficulty: Hard

Category: Problem Solving and Data Analysis / Rates, Ratios, Proportions, and Percentages

Strategic Advice: Make a chart that represents rate, time, and distance and fill in what you know. Then use your table to solve for distance. If it took Rory t hours to get to the airport, and the total trip took 2 hours and 30 minutes (or 2.5 hours), how long (in terms of t) did the return trip take?

Getting to the Answer:

	Rate	**Time**	**Distance**
To airport	45 mph	t	d
Back to home	30 mph	$2.5 - t$	d

Now use the formula $d = r \times t$ for both parts of the trip: $d = 45t$ and $d = 30(2.5 - t)$. Because both are equal to d, you can set them equal to each other and solve for t:

$$45t = 30(2.5 - t)$$
$$45t = 75 - 30t$$
$$75t = 75$$
$$t = 1$$

Now plug back in to solve for d:

$$d = 45t$$
$$d = 45(1)$$
$$d = 45$$

37. 10
Difficulty: Medium

Category: Problem Solving and Data Analysis / Rates, Ratios, Proportions, and Percentages

Strategic Advice: You don't need to know chemistry to answer this question. All the information you need is in the table. Use the formula $\text{Percent} = \frac{\text{part}}{\text{whole}} \times 100\%$.

Getting to the Answer: To use the formula, find the part of the mass represented by the carbon; there is 1 mole of carbon, and it has a mass of 12.011 grams. Next, find the whole mass of the mole of chloroform; 1 mole carbon (12.011 g) + 1 mole hydrogen (1.008 g) + 3 moles chlorine (3 × 35.453 = 106.359 g) = 12.011 + 1.008 + 106.359 = 119.378. Now use the formula:

$$\begin{aligned}\text{Percent} &= \frac{12.011}{119.378} \times 100\% \\ &= 0.10053 \times 100\% \\ &= 10.053\%\end{aligned}$$

Before you grid in your answer, make sure you follow the directions—round to the nearest whole percent, which is 10.

38. 12
Difficulty: Hard

Category: Problem Solving and Data Analysis / Rates, Ratios, Proportions, and Percentages

Strategic Advice: This part of the question contains several steps. Think about the units given in the question and how you can use what you know to find what you need.

Getting to the Answer: Start with grams of chloroform; the chemist starts with 1,000 and uses 522.5, so there are 1,000 – 522.5 = 477.5 grams left. From the previous question, you know that 1 mole of chloroform has a mass of 119.378 grams, so there are 477.5 ÷ 119.378 = 3.999, or about 4 moles of chloroform left. Be careful—you're not finished yet. The question asks for the number of moles of *chlorine*, not chloroform. According to the table, each mole of chloroform contains 3 moles of chlorine, so there are 4 × 3 = 12 moles of chlorine left.